Lecture Notes in Computer Science 11438

Commenced Publication in 1973
Founding and Former Series Editors:
Gerhard Goos, Juris Hartmanis, and Jan van Leeuwen

More information about this series at http://www.springer.com/series/7409

Leif Azzopardi · Benno Stein ·
Norbert Fuhr · Philipp Mayr ·
Claudia Hauff · Djoerd Hiemstra (Eds.)

Advances in Information Retrieval

41st European Conference on IR Research, ECIR 2019
Cologne, Germany, April 14–18, 2019
Proceedings, Part II

 Springer

Editors
Leif Azzopardi (iD)
University of Strathclyde
Glasgow, UK

Norbert Fuhr (iD)
Universität Duisburg-Essen
Duisburg, Germany

Claudia Hauff (iD)
Delft University of Technology
Delft, The Netherlands

Benno Stein (iD)
Bauhaus Universität Weimar
Weimar, Germany

Philipp Mayr (iD)
GESIS - Leibniz Institute
for the Social Sciences
Cologne, Germany

Djoerd Hiemstra (iD)
University of Twente
Enschede, The Netherlands

ISSN 0302-9743   ISSN 1611-3349   (electronic)
Lecture Notes in Computer Science
ISBN 978-3-030-15718-0   ISBN 978-3-030-15719-7   (eBook)
https://doi.org/10.1007/978-3-030-15719-7

Library of Congress Control Number: 2019934339

LNCS Sublibrary: SL3 – Information Systems and Applications, incl. Internet/Web, and HCI

This Springer imprint is published by the registered company Springer Nature Switzerland AG
The registered company address is: Gewerbestrasse 11, 6330 Cham, Switzerland

Preface

The 41st European Conference on Information Retrieval (ECIR) was held in Cologne, Germany, during April 14–18, 2019, and brought together hundreds of researchers from Europe and abroad. The conference was organized by GESIS – Leibniz Institute for the Social Sciences and the University of Duisburg-Essen—in cooperation with the British Computer Society's Information Retrieval Specialist Group (BCS-IRSG).

These proceedings contain the papers, presentations, workshops, and tutorials given during the conference. This year the ECIR 2019 program boasted a variety of novel work from contributors from all around the world and provided new platforms for promoting information retrieval-related (IR) activities from the CLEF Initiative. In total, 365 submissions were fielded across the tracks from 50 different countries.

The final program included 39 full papers (23% acceptance rate), 44 short papers (29% acceptance rate), eight demonstration papers (67% acceptance rate), nine reproducibility full papers (75% acceptance rate), and eight invited CLEF papers. All submissions were peer reviewed by at least three international Program Committee members to ensure that only submissions of the highest quality were included in the final program. As part of the reviewing process we also provided more detailed review forms and guidelines to help reviewers identify common errors in IR experimentation as a way to help ensure consistency and quality across the reviews.

The accepted papers cover the state of the art in IR: evaluation, deep learning, dialogue and conversational approaches, diversity, knowledge graphs, recommender systems, retrieval methods, user behavior, topic modelling, etc., and also included novel application areas beyond traditional text and Web documents such as the processing and retrieval of narrative histories, images, jobs, biodiversity, medical text, and math. The program boasted a high proportion of papers with students as first authors, as well as papers from a variety of universities, research institutes, and commercial organizations.

In addition to the papers, the program also included two keynotes, four tutorials, four workshops, a doctoral consortium, and an industry day. The first keynote was presented by this year's BCS IRSG Karen Sparck Jones Award winner, Prof. Krisztian Balog, On Entities and Evaluation, and the second keynote was presented by Prof. Markus Strohmaier, On Ranking People. The tutorials covered a range of topics from conducting lab-based experiments and statistical analysis to categorization and deep learning, while the workshops brought together participants to discuss algorithm selection (AMIR), narrative extraction (Text2Story), Bibliometrics (BIR), as well as social media personalization and search (SoMePeAS). As part of this year's ECIR we also introduced a new CLEF session to enable CLEF organizers to report on and promote their upcoming tracks. In sum, this added to the success and diversity of ECIR and helped build bridges between communities.

The success of ECIR 2019 would not have been possible without all the help from the team of volunteers and reviewers. We wish to thank all our track chairs for coordinating the different tracks along with the teams of meta-reviewers and reviewers who helped ensure the high quality of the program. We also wish to thank the demo chairs: Christina Lioma and Dagmar Kern; student mentorship chairs: Ahmet Aker and Laura Dietz; doctoral consortium chairs: Ahmet Aker, Dimitar Dimitrov and Zeljko Carevic; workshop chairs: Diane Kelly and Andreas Rauber; tutorial chairs: Guillaume Cabanac and Suzan Verberne; industry chair: Udo Kruschwitz; publicity chair: Ingo Frommholz; and sponsorship chairs: Jochen L. Leidner and Karam Abdulahhad. We would like to thank our webmaster, Sascha Schüller and our local chair, Nina Dietzel along with all the student volunteers who helped to create an excellent online and offline experience for participants and attendees.

ECIR 2019 was sponsored by: DFG (Deutsche Forschungsgemeinschaft), BCS (British Computer Society), SIGIR (Special Interest Group on Information Retrieval), City of Cologne, Signal Media Ltd, Bloomberg, Knowledge Spaces, Polygon Analytics Ltd., Google, Textkernel, MDPI Open Access Journals, and Springer. We thank them all for their support and contributions to the conference.

Finally, we wish to thank all the authors, reviewers, and contributors to the conference.

April 2019

Leif Azzopardi
Benno Stein
Norbert Fuhr
Philipp Mayr
Claudia Hauff
Djoerd Hiemstra

Organization

General Chairs

Norbert Fuhr   Universität Duisburg-Essen, Germany
Philipp Mayr   GESIS – Leibniz Institute for the Social Sciences, Germany

Program Chairs

Leif Azzopardi   University of Glasgow, UK
Benno Stein   Bauhaus-Universität Weimar, Germany

Short Papers and Poster Chairs

Claudia Hauff   Delft University of Technology, The Netherlands
Djoerd Hiemstra   University of Twente, The Netherlands

Workshop Chairs

Diane Kelly   University of Tennessee, USA
Andreas Rauber   Vienna University of Technology, Austria

Tutorial Chairs

Guillaume Cabanac   University of Toulouse, France
Suzan Verberne   Leiden University, The Netherlands

Demo Chairs

Christina Lioma   University of Copenhagen, Denmark
Dagmar Kern   GESIS – Leibniz Institute for the Social Sciences, Germany

Industry Day Chair

Udo Kruschwitz   University of Essex, UK

Proceedings Chair

Philipp Mayr   GESIS – Leibniz Institute for the Social Sciences, Germany

Publicity Chair

Ingo Frommholz   University of Bedfordshire, UK

Sponsor Chairs

Jochen L. Leidner   Thomson Reuters/University of Sheffield, UK
Karam Abdulahhad   GESIS – Leibniz Institute for the Social Sciences,
                              Germany

Student Mentoring Chairs

Ahmet Aker   Universität Duisburg-Essen, Germany
Laura Dietz   University of New Hampshire, USA

Website Infrastructure

Sascha Schüller   GESIS – Leibniz Institute for the Social Sciences,
                              Germany

Local Chair

Nina Dietzel   GESIS – Leibniz Institute for the Social Sciences,
                              Germany

Program Committee

Mohamed Abdel Maksoud   Codoma.tech Advanced Technologies, Egypt
Ahmed Abdelali   Research Administration
Karam Abdulahhad   GESIS - Leibniz institute for the Social Sciences,
                              Germany
Dirk Ahlers   Norwegian University of Science and Technology,
                              Norway
Qingyao Ai   University of Massachusetts Amherst, USA
Ahmet Aker   University of Duisburg Essen, Germany
Elif Aktolga   Apple Inc., USA
Dyaa Albakour   Signal Media, UK
Giambattista Amati   Fondazione Ugo Bordoni, Italy
Linda Andersson   Vienna University of Technology, Austria
Avi Arampatzis   Democritus University of Thrace, Greece
Ioannis Arapakis   Telefonica Research, Spain
Jaime Arguello   The University of North Carolina at Chapel Hill, USA
Leif Azzopardi   University of Strathclyde, UK
Ebrahim Bagheri   Ryerson University, USA
Krisztian Balog   University of Stavanger, Norway
Alvaro Barreiro   University of A Coruña, Spain

Alberto Barrón-Cedeño	Qatar Computing Research Institute, Qatar
Srikanta Bedathur	IIT Delhi, India
Alejandro Bellogin	Universidad Autonoma de Madrid, Spain
Patrice Bellot	Aix-Marseille Université - CNRS (LSIS), France
Pablo Bermejo	Universidad de Castilla-La Mancha, Spain
Catherine Berrut	LIG, Université Joseph Fourier Grenoble I, France
Prakhar Biyani	Yahoo
Pierre Bonnet	CIRAD, France
Gloria Bordogna	National Research Council of Italy – CNR, Italy
Dimitrios Bountouridis	Delft University of Technology, The Netherlands
Pavel Braslavski	Ural Federal University, Russia
Paul Buitelaar	Insight Centre for Data Analytics, National University of Ireland Galway, Ireland
Guillaume Cabanac	IRIT - Université Paul Sabatier Toulouse 3, France
Fidel Cacheda	Universidade da Coruña, Spain
Sylvie Calabretto	LIRIS, France
Pável Calado	Universidade de Lisboa, Portugal
Arthur Camara	Delft University of Technology, The Netherlands
Ricardo Campos	Polytechnic Institute of Tomar, Portugal
Fazli Can	Bilkent University, Turkey
Iván Cantador	Universidad Autónoma de Madrid, Spain
Cornelia Caragea	University of Illinois at Chicago, USA
Zeljko Carevic	GESIS, Germany
Claudio Carpineto	Fondazione Ugo Bordoni, Italy
Pablo Castells	Universidad Autónoma de Madrid, Spain
Long Chen	University of Glasgow, UK
Max Chevalier	IRIT, France
Manoj Chinnakotla	Microsoft, India
Nurendra Choudhary	International Institute of Information Technology, Hyderabad, India
Vincent Claveau	IRISA - CNRS, France
Fabio Crestani	University of Lugano (USI), Switzerland
Bruce Croft	University of Massachusetts Amherst, USA
Zhuyun Dai	Carnegie Mellon University, USA
Jeffery Dalton	University of Glasgow, UK
Martine De Cock	University of Washington, USA
Pablo de La Fuente	Universidad de Valladolid, Spain
Maarten de Rijke	University of Amsterdam, The Netherlands
Arjen de Vries	Radboud University, The Netherlands
Yashar Deldjoo	Polytechnic University of Milan, Italy
Kuntal Dey	IBM India Research Lab, India
Emanuele Di Buccio	University of Padua, Italy
Giorgio Maria Di Nunzio	University of Padua, Italy
Laura Dietz	University of New Hampshire, USA
Dimitar Dimitrov	GESIS, Germany
Mateusz Dubiel	University of Strathclyde, UK

Jiaul Paik	IIT Kharagpur, India
Joao Palotti	Vienna University of Technology/Qatar Computing Research Institute, Austria/Qatar
Girish Palshikar	Tata Research Development and Design Centre, India
Javier Parapar	University of A Coruña, Spain
Gabriella Pasi	Università degli Studi di Milano Bicocca, Italy
Arian Pasquali	University of Porto, Portugal
Bidyut Kr. Patra	VTT Technical Research Centre, Finland
Pavel Pecina	Charles University in Prague, Czech Republic
Gustavo Penha	UFMG
Avar Pentel	University of Tallinn, Estonia
Raffaele Perego	ISTI-CNR, Italy
Vivien Petras	Humboldt-Universität zu Berlin, Germany
Jeremy Pickens	Catalyst Repository Systems, USA
Karen Pinel-Sauvagnat	IRIT, France
Florina Piroi	Vienna University of Technology, Austria
Benjamin Piwowarski	CNRS, Pierre et Marie Curie University, France
Vassilis Plachouras	Facebook, UK
Bob Planque	Vrije Universiteit Amsterdam, The Netherlands
Senja Pollak	University of Ljubljana, Slovenia
Martin Potthast	Leipzig University, Germany
Georges Quénot	Laboratoire d'Informatique de Grenoble, CNRS, France
Razieh Rahimi	University of Tehran, Iran
Nitin Ramrakhiyani	TCS Research, Tata Consultancy Services Ltd., India
Jinfeng Rao	University of Maryland, USA
Andreas Rauber	Vienna University of Technology, Austria
Traian Rebedea	University Politehnica of Bucharest, Romania
Navid Rekabsaz	Idiap Research Institute, Switzerland
Steffen Remus	University of Hamburg, Germany
Paolo Rosso	Universitat Politècnica de València, Spain
Dmitri Roussinov	University of Strathclyde, UK
Stefan Rueger	Knowledge Media Institute, UK
Tony Russell-Rose	UXLabs, UK
Alan Said	University of Skövde, Sweden
Mark Sanderson	RMIT University, Australia
Eric Sanjuan	Laboratoire Informatique d'Avignon- Université d'Avignon, France
Rodrygo Santos	Universidade Federal de Minas Gerais, Brazil
Kamal Sarkar	Jadavpur University, Kolkata, India
Fabrizio Sebastiani	Italian National Council of Research, Italy
Florence Sedes	IRIT P. Sabatier University, France
Giovanni Semeraro	University of Bari, Italy
Procheta Sen	Indian Statistical Institute, India
Armin Seyeditabari	UNC Charlotte, USA
Mahsa Shahshahani	University of Amsterdam, The Netherlands
Azadeh Shakery	University of Tehran, Iran

Manish Shrivastava	International Institute of Information Technology, Hyderabad, India
Ritvik Shrivastava	Columbia University, USA
Rajat Singh	International Institute of Information Technology, Hyderabad, India
Eero Sormunen	University of Tampere, Finland
Laure Soulier	Sorbonne Universités UPMC-LIP6, France
Rene Spijker	Cochran, The Netherlands
Efstathios Stamatatos	University of the Aegean, Greece
Benno Stein	Bauhaus-Universität Weimar, Germany
L. Venkata Subramaniam	IBM Research, India
Hanna Suominen	The ANU, Australia
Pascale Sébillot	IRISA, France
Lynda Tamine	IRIT, France
Thibaut Thonet	University of Grenoble Alpes, France
Marko Tkalcic	Free University of Bozen-Bolzano, Italy
Nicola Tonellotto	ISTI-CNR, Italy
Michael Tschuggnall	Institute for computer science, DBIS, Innsbruck, Austria
Theodora Tsikrika	Information Technologies Institute, CERTH, Greece
Denis Turdakov	Institute for System Programming RAS
Ferhan Ture	Comcast Labs, USA
Yannis Tzitzikas	University of Crete and FORTH-ICS, Greece
Sumithra Velupillai	KTH Royal Institute of Technology, Sweden
Suzan Verberne	Leiden University, The Netherlands
Vishwa Vinay	Adobe Research Bangalore, India
Marco Viviani	Università degli Studi di Milano-Bicocca - DISCo, Italy
Stefanos Vrochidis	Information Technologies Institute, Greece
Shuohang Wang	Singapore Management University, Singapore
Christa Womser-Hacker	Universität Hildesheim, Germany
Chenyan Xiong	Carnegie Mellon University; Microsoft, USA
Grace Hui Yang	Georgetown University, USA
Peilin Yang	Twitter Inc., USA
Tao Yang	University of California at Santa Barbara, USA
Andrew Yates	Max Planck Institute for Informatics, Germany
Hai-Tao Yu	University of Tsukuba, Japan
Hamed Zamani	University of Massachusetts Amherst, USA
Eva Zangerle	University of Innsbruck, Austria
Fattane Zarrinkalam	Ferdowsi University, Iran
Dan Zhang	Facebook, USA
Duo Zhang	Kunlun Inc.
Shuo Zhang	University of Stavanger, Norway
Sicong Zhang	Georgetown University, USA
Guoqing Zheng	Carnegie Mellon University, USA
Leonardo Zilio	Université catholique de Louvain, Belgium
Guido Zuccon	The University of Queensland, Australia

Sponsors

 Information Retrieval
Specialist Group

City of Cologne

KÖLNER
WISSENSCHAFTSRUNDE

SIGNAL

Bloomberg
Engineering

 Springer

KnowledgeSpaces™   POLYGON ANALYTICS™

textkernel

Machine Intelligence for People and Jobs

informatics

information

Open Access Journals by MDPI

Google

Contents – Part II

Demonstration Papers

CLEF Organizers Lab Track

Doctoral Consortium Papers

Workshops

Tutorials

Contents – Part I

Reproducibility (Systems)

Reproducibility (Application)

Neural IR

Cross Lingual IR

QA and Conversational Search

Topic Modeling

Metrics

Image IR

Short Papers

Short Papers (Continued)

Open-Set Web Genre Identification Using Distributional Features and Nearest Neighbors Distance Ratio

Dimitrios Pritsos[1], Anderson Rocha[2], and Efstathios Stamatatos[1(⊠)]

[1] University of the Aegean, 83200 Karlovassi, Samos, Greece
{dpritsos,stamatatos}@aegean.gr
[2] Institute of Computing, University of Campinas (Unicamp), Campinas, SP, Brazil

Abstract. Web genre identification can boost information retrieval systems by providing rich descriptions of documents and enabling more specialized queries. The open-set scenario is more realistic for this task as web genres evolve over time and it is not feasible to define a universally agreed genre palette. In this work, we bring to bear a novel approach to web genre identification underpinned by distributional features acquired by doc2vec and a recently-proposed open-set classification algorithm—the nearest neighbors distance ratio classifier. We present experimental results using a benchmark corpus and a strong baseline and demonstrate that the proposed approach is highly competitive, especially when emphasis is given on precision.

Keywords: Web genre identification · Open-set classification ·
Distributional features

1 Introduction

Web Genre Identification (WGI) is a multi-class text classification task aiming at the association of web pages to labels (e.g., blog, e-shop, personal home page, etc.) corresponding to their form, communicative purpose, and style rather than their content. WGI can enhance the potential of information retrieval (IR) systems by allowing more complex and informative queries, whereby topic-related keywords and genre labels are combined to better express the information need of users and grouping search results by genre [16,31]. Moreover, WGI is specially useful to enhance performance of Natural Language Processing (NLP) methods, such as part-of-speech tagging (POS) [22] and text summarization [36] by empowering genre-specific model development.

In spite of WGI's immediate applications, there are certain fundamental difficulties hardening its deployment in practice. First, there is a lack of both a consensus on the exact definition of genre [5] and a genre palette that comprises all available genres and sub-genres [17,18,33,34] to aim for. New web genres appear on-the-fly and existing genres evolve over time [4]. Furthermore, it is

© Springer Nature Switzerland AG 2019
L. Azzopardi et al. (Eds.): ECIR 2019, LNCS 11438, pp. 3–11, 2019.
https://doi.org/10.1007/978-3-030-15719-7_1

not clear whether a whole web page should belong to a single genre or sections of the same web page can belong to different genres [8,15]. Finally, style of documents is affected by both genre-related choices and author-related choices [24,34].

Instead of aiming to anticipate all possible web-genres possible to appear in a practical scenario, it would be wiser to consider a proper handling of genres of interest while properly handling "unseen" genres. In this vein, WGI can be viewed as an open-set classification task to better deal with incomplete genre palettes [2,26–28,37]. This scheme requires strong generalization in comparison to the traditional closed-set setup—the one in which all genres of interest are known or defined a priori. One caveat, though, is that open-set classification methods tend to perform better while operating in not-so-high dimensional manifolds. However, to date, most common and effective stylometric features in prior art, e.g., word and character n-grams, yield high-dimensional spaces [10,34].

Aiming at properly bringing to bear the powerful algorithm modeling of open-set classification to the WGI setup, in this paper, we apply a recently-proposed open-set classification algorithm, the *Nearest Neighbors Distance Ratio* (NNDR) [19], to WGI. To produce a compact representation of web pages—more amenable to the open-set modelling—we rely upon *Distributional Features* (DF) [39] in this paper. Finally, we are also using an evaluation methodology that is more appropriate for the open-set classification framework with unstructured noise [27].

We organize the remaining of this paper into four more sections. Sect. 2 presents previous work on WGI while Sect. 3 describes the proposed approach. Sect. 4 discusses the experimental setup and obtained results. Finally, Sect. 6 draws the main conclusions of this study and presents some future work directions.

2   Related Work

Most previous studies in WGI consider the case where all web pages should belong to a predefined taxonomy of genres [7,10,14,32]. Putting this setup under the vantage point of machine learning, it is the same as assuming what is known as a closed-set problem definition. However, this naïve assumption is not appropriate for most applications related to WGI as it is not possible to construct a universal genre palette a priori nor force web pages to always fall into any of the predefined genre labels. Such web pages are considered *noise* and include web documents where multiple genres co-exist [13,33].

Santini [33] defines *structured noise* as the collection of web pages belonging to several genres, unknown during training. Such structured noise can be used as a negative class for training a binary classifier [38]. However, it is highly unlikely that such a collection represents the real distribution of pages of the web at large. On the other hand, *unstructured noise* is a random collection of pages [33] for which no genre labels are available. The effect of noise in WGI was first studied in [6,11,13,35].

Open-set classification models for WGI were first described in [28,37]. However, these models were only tested in noise-free corpora [26]. Asheghi [2] showed that it is much more challenging to perform WGI in the noisy web setup in comparison to noise-free corpora. Recently, *Ensemble Methods* were shown to achieve high effectiveness in open-set WGI setups [27].

Great attention historically on WGI has been given to the appropriate definition of features that are capable of capturing genre characteristics—which includes but are not limited to character n-grams or word n-grams, part-of-speech histograms, the frequency of the most discriminative words, etc. [10,12–14,17,23,24,34]. Additionally, some additional useful features might come from exploiting HTML structure and/or the hyperlink functionality of web pages [1,3,7,29,40]. Recently deep learning methods have also been tested in genre detection setups with promising results [39].

3 Proposed Approach—Open-Set Web Genre Identification

3.1 Distributional Features Learning

In this study, we rely upon a Doc2Vec text representation to provide distributional features for the WGI problem [20,21,30]. In particular, we have implemented a special module inside our package, named *Html2Vec*[1] where a whole corpus can be used as input and one *Bag-of-Words Paragraph Vector* (PV-BOW) is returned per web-page of the corpus. PV-BOW consists of a *Neural Network* (NNet) comprising a *softmax* multi-class classifier approximating $\max \frac{1}{T} \sum_{T-k}^{a=k} \log p(t_a | t_{a-k}, ..., t_{a+k})$. PV-BOW is trained using *stochastic gradient-descent* where the gradient is obtained via *back-propagation*. Given a sequence of training n-grams (word or character) $t_1, t_2, t_3, ..., t_T$, the objective function of the NNet is the maximized *average log-probability* $p(t_a | t_{a-k}, ..., t_{a+k}) = \frac{e^{y_{t_a}}}{\sum_i e^{y_i}}$.

For training the PV-BOW in this study, for each iteration, of the stochastic gradient descent, a *text window* is sampled with size w_{size}. Then a random term (n-gram) is sampled from the text window and form a classification task given the paragraph vector. Thus $y = b + s(t_1, t_2, t_3, ..., t_{w_{size}})$, where $s()$ is the sequence of word-n-grams or character-n-grams of the sampled window. Each type of n-grams is used separately as suggested in [25]. This model provides us with a representation of web pages of pre-defined dimensionality DF_{dim}.

3.2 Nearest Neighbors Distance Ratio Classifier

The Nearest Neighbors Distance Ratio (NNRD) classifier is an open-set classification algorithm introduced in [19], which in turn, is an extension upon the *Nearest Neighbors* (NN) algorithm. NNRD calculates the distance of a new sample s to its nearest neighbor t and to the closest training sample u belonging to

---

[1] https://github.com/dpritsos/html2vec.

a different class with respect to t. Then, if the ratio $d(s,t)/d(s,u)$ is higher than a threshold, the new sample is classified to the class of s. Otherwise, it is left unclassified.

It is remarkable that, in contrast to other open-set classifiers, training of NNDR requires both known samples (belonging to classes known during training) and unknown examples (belonging to other/unknown classes) of interest. In more details, the *Distance Ratio Threshold* (DRT) used to classify new samples is adjusted by maximizing the *Normalized Accuracy* (NA) $NA = \lambda A_{KS} + (1-\lambda)A_{US}$, where A_{KS} is the accuracy on known samples and A_{US} is the accuracy on unknown samples. The parameter λ regulates the mistakes trade-off on the known and unknown samples prediction. Since usually in training phase only known classes are available, Mendes et al. [19] propose an approach to repeatedly split available training classes into two sets (known and "simulated" unknown). In our implementation of NNDR, we use cosine distance rather than the Euclidean distance because previous work found this type of distance more suitable for WGI [27].[2]

4   Experiments

4.1   Corpus

Our experiments are based on *SANTINIS*, a benchmark corpus already used in previous work in WGI [18,27,32]. This dataset comprises 1,400 English web-pages evenly distributed into seven genres (blog, eshop, FAQ, frontpage, listing, personal home page, search page) as well as 80 BBC web-pages evenly categorized into four additional genres (DIY mini-guide, editorial, features, short-bio). In addition, the dataset comprises a random selection of 1,000 English web-pages taken from the SPIRIT corpus [9]. The latter can be viewed as *unstructured noise* since genre labels are missing.

4.2   Experimental Setup

To represent web-pages, we use features exclusively related to textual information, excluding any structural information, URLs, etc. The following representation schemes are examined: Character 4-grams (C4G), Word unigrams (W1G), and Word 3-grams (W3G). For each of these schemes, we use either Term-Frequency (TF) weights or DF features. The feature space for TF is defined by a vocabulary V_{TF}, which is extracted based on the most frequent terms of the training set—we consider $V_{TF} = \{5k, 10k, 50k, 100k\}$. The DF space is predefined in the PV-BOW model—we consider $DF_{dim} = \{50, 100, 250, 500, 1000\}$.

In PV-BOW, the terms with very low-frequency in the training set are discarded. In this study, we examine $TF_{min} = \{3, 10\}$ as cutoff frequency threshold. The text window size is selected from $W_{size} = \{3, 8, 20\}$. The remaining

---

[2] https://github.com/dpritsos/OpenNNDR.

parameters of PV-BOW are set as follows: $\alpha = 0.025$, $epochs = \{1, 3, 10\}$ and $decay = \{0.002, 0.02\}$.

Regarding the NNRD open-set classifier, there are two parameters, *lambda* and DRT, and their considered values are: $\lambda = \{0.2, 0.5, 0.7\}$, $DRT = \{0.4, 0.6, 0.8, 0.9\}$. All aforementioned parameters are adjusted based on grid-search using only the training part of the corpus.

For a proper comparison with prior art, we use two open-set WGI approaches with good previously reported results as baselines: Random Feature Subset Ensemble (RFSE) and one-class SVM (OCSVM) [27,28]. All parameters of these methods have been been adjusted as suggested in [27] (based on the same corpus).

We follow the open-set evaluation framework with unstructured noise introduced in [27]. In particular, the open-set F1 score [19] is calculated over the known classes (the noisy class is excluded). The reported evaluation results are obtained by performing 10-fold cross-validation and, in each fold, we include the full set of 1,000 pages of noise. This setup is comparable to previous studies [27].

Table 1. Performance of baselines and NNDR on the SANTINIS coprus. All evaluation scores are macro-averaged.

Model	Features	Dim.	Precision	Recall	AUC	F1
RFSE	TF-C4G	50k	0.739	**0.780**	0.652	0.759
RFSE	TF-W1G	50k	0.776	0.758	**0.657**	**0.767**
RFSE	TF-W3G	50k	0.797	0.722	0.615	0.758
OCSVM	TF-C4G	5k	0.662	0.367	0.210	0.472
OCSVM	TF-W1G	5k	0.332	0.344	0.150	0.338
OCSVM	TF-W3G	10k	0.631	0.654	0.536	0.643
NNDR	TF-C4G	5k	0.664	0.403	0.291	0.502
NNDR	TF-W1G	5k	0.691	0.439	0.348	0.537
NNDR	TF-W3G	10k	0.720	0.664	0.486	0.691
NNDR	DF-C4G	50	**0.829**	0.600	0.455	0.696
NNDR	DF-W1G	50	0.733	0.670	0.541	0.700
NNDR	DF-W3G	100	0.827	0.615	0.564	0.706

5 Results

We apply the baselines and NNDR in the SANTINIS corpus. In the training phase, we use only the 11 known genre classes while in test phase, we also consider an additional class (unstructured noise). Table 1 shows the performance of tested methods when either TF or DF representation schemes, based on C4G, W1G, or W4G features, are used.

First, we compare NNDR using TF features with baselines, also using this kind of features. In this case, NNDR outperforms OCSVM. On the other hand, RFSE performed better than NNDR for Macro-F1 and Macro-AUC. This is consistent for any kind of features (C4G, W1G, or W3G). There is notable difference in the dimensionality of representation used by the examined approaches though. RFSE relies upon a 50k-D manifold while NNDR and OCSVM are based on much lower dimensional spaces. It has to be noted that RFSE builds an ensemble by iteratively selecting a subset of the available features (randomly). That way, it internally reduces the dimensionality for each constituent base classifier. On the other hand, NNDR seems to be confused when thousands of features are considered as it is based on distance calculations.

Next, we compare NNDR models using either TF or DF features. There is a notable improvement when DFs are used in associated with the open-set NNDR classifier. The dimensionality of DF is much lower than TF and this seems to be crucial to improve the performance of NNDR. This is consistent for all three feature types (C4G, W1G, and W3G). NNDR with TF scheme is competitive only when W3G features are used. It has also to be noted that in all cases the selected value of parameter DRT is 0.8. This indicates that NNDR is a very robust algorithm.

Finally, the proposed approach using NNDR and DF outperforms OCSVM but it is outperformed by the strong baseline RFSE in both macro-AUC and macro F1. However, when precision is concerned, NNDR is much better. A closer look at the comparison of the two methods is provided in Fig. 1, where precision curves in 11-standard recall levels are depicted. The precision value at r_j level is interpolated as follows: $P(r_j) = max_{r_j \leq r \leq r+j+1}(P(r))$.

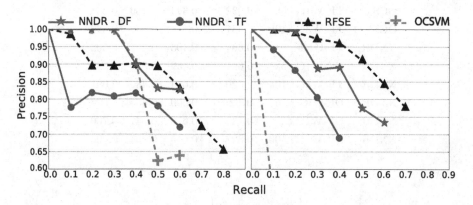

Fig. 1. Precision curves in 11-standard recall levels of the examined open-set classifiers using either W3G features (left) or W1G features (right).

The NNDR-DF model maintains very high precision scores for low levels of recall. The difference between NNDR-DF and RFSE at that point is clearer when W3G features are used. NNDR-TF is clearly worse than both NNDR-DF and

RFSE. In addition, OCSVM is competitive in terms of precision only when W3G features are used but its performance drops abruptly in comparison to that of NNDR-DF. Note that the point where the curves end indicates the percentage of corpus that is left unclassified (assigned to unknown class). RFSE manages to recognize correctly larger part of the corpus, more than 70%, with respect to NNDR-DF that reaches 60%.

6 Conclusions

It seems that distributional features provide a significant enhancement to the NNDR open-set method. The low-dimensionality of DF is crucial to boost the performance of NNDR. Yet, RFSE proves to be a hard-to-beat baseline at the expense of relying upon a much higher representation space (usually in the thousands of features). However, with respect to precision, the proposed approach is much more conservative and it prefers to leave web-pages unclassified rather than guessing an inaccurate genre label. Depending on the application of WGI, precision can be considered much more important than recall and this is where our proposed algorithm shines.

Further research could focus on more appropriate distance measures within NNDR specially with recent data-driven features obtained with powerful NLP convolutional and recurrent deep networks. Moreover, alternative types of distributional features could be used (e.g., topic modeling). Finally, a combination of NNDR with RFSE models could be studied as they seem to exploit complementary views of the same problem.

Acknowledgement. Prof. Rocha thanks the financial support of FAPESP DéjàVu (Grant #2017/12646-3) and CAPES DeepEyes Grant.

References

1. Abramson, M., Aha, D.W.: What's in a URL? Genre classification from URLs. Intelligent techniques for web personalization and recommender systems. AAAI Technical report. Association for the Advancement of Artificial Intelligence (2012)
2. Asheghi, N.R.: Human Annotation and Automatic Detection of Web Genres. Ph.D. thesis, University of Leeds (2015)
3. Asheghi, N.R., Markert, K., Sharoff, S.: Semi-supervised graph-based genre classification for web pages. In: TextGraphs-9, p. 39 (2014)
4. Boese, E.S., Howe, A.E.: Effects of web document evolution on genre classification. In: Proceedings of the 14th ACM International Conference on Information and Knowledge Management, pp. 632–639. ACM (2005)
5. Crowston, K., Kwaśnik, B., Rubleske, J.: Problems in the use-centered development of a taxonomy of web genres. In: Mehler, A., Sharoff, S., Santini, M. (eds.) Genres on the Web. Text, Speech and Language Technology, vol. 42, pp. 69–84. Springer, Dordrecht (2011). https://doi.org/10.1007/978-90-481-9178-9_4
6. Dong, L., Watters, C., Duffy, J., Shepherd, M.: Binary cybergenre classification using theoretic feature measures. In: 2006 IEEE/WIC/ACM International Conference on Web Intelligence (WI 2006), pp. 313–316 (2006)

7. Jebari, C.: A pure URL-based genre classification of web pages. In: 2014 25th International Workshop on Database and Expert Systems Applications (DEXA), pp. 233–237. IEEE (2014)

8. Jebari, C.: A combination based on OWA operators for multi-label genre classification of web pages. Procesamiento del Lenguaje Nat. **54**, 13–20 (2015)

9. Joho, H., Sanderson, M.: The spirit collection: an overview of a large web collection. SIGIR Forum **38**(2), 57–61 (2004)

10. Kanaris, I., Stamatatos, E.: Learning to recognize webpage genres. Inf. Process. Manage. **45**(5), 499–512 (2009)

11. Kennedy, A., Shepherd, M.: Automatic identification of home pages on the web. In: Proceedings of the 38th Annual Hawaii International Conference on System Sciences, HICSS 2005, p. 99c. IEEE (2005)

12. Kumari, K.P., Reddy, A.V., Fatima, S.S.: Web page genre classification: impact of n-gram lengths. Int. J. Comput. Appl. **88**(13), 13–17 (2014)

13. Levering, R., Cutler, M., Yu, L.: Using visual features for fine-grained genre classification of web pages. In: Proceedings of the 41st Annual Hawaii International Conference on System Sciences, pp. 131–131. IEEE (2008)

14. Lim, C.S., Lee, K.J., Kim, G.C.: Multiple sets of features for automatic genre classification of web documents. Inf. Process. Manage. **41**(5), 1263–1276 (2005)

15. Madjarov, G., Vidulin, V., Dimitrovski, I., Kocev, D.: Web genre classification via hierarchical multi-label classification. In: Jackowski, K., Burduk, R., Walkowiak, K., Woźniak, M., Yin, H. (eds.) IDEAL 2015. LNCS, vol. 9375, pp. 9–17. Springer, Cham (2015). https://doi.org/10.1007/978-3-319-24834-9_2

16. Malhotra, R., Sharma, A.: Quantitative evaluation of web metrics for automatic genre classification of web pages. Int. J. Syst. Assur. Eng. Manage. **8**(2), 1567–1579 (2017)

17. Mason, J., Shepherd, M., Duffy, J.: An n-gram based approach to automatically identifying web page genre. In: HICSS, pp. 1–10. IEEE Computer Society (2009)

18. Mehler, A., Sharoff, S., Santini, M.: Genres on the Web: Computational Models and Empirical Studies. Text, Speech and Language Technology. Springer, Heidelberg (2010). https://doi.org/10.1007/978-90-481-9178-9

19. Mendes Júnior, P.R., et al.: Nearest neighbors distance ratio open-set classifier. Mach. Learn. **106**, 1–28 (2016)

20. Mikolov, T., Chen, K., Corrado, G., Dean, J.: Efficient estimation of word representations in vector space. arXiv preprint arXiv:1301.3781 (2013)

21. Mikolov, T., Sutskever, I., Chen, K., Corrado, G.S., Dean, J.: Distributed representations of words and phrases and their compositionality. In: Advances in Neural Information Processing Systems, pp. 3111–3119 (2013)

22. Nooralahzadeh, F., Brun, C., Roux, C.: Part of speech tagging for French social media data. In: COLING 2014, 25th International Conference on Computational Linguistics, Proceedings of the Conference: Technical Papers, 23–29 August 2014, Dublin, Ireland, pp. 1764–1772 (2014)

23. Onan, A.: An ensemble scheme based on language function analysis and feature engineering for text genre classification. J. Inf. Sci. **44**(1), 28–47 (2018)

24. Petrenz, P., Webber, B.: Stable classification of text genres. Comput. Linguist. **37**(2), 385–393 (2011)

25. Posadas-Durán, J.P., Gómez-Adorno, H., Sidorov, G., Batyrshin, I., Pinto, D., Chanona-Hernández, L.: Application of the distributed document representation in the authorship attribution task for small corpora. Soft Comput. **21**(3), 627–639 (2017)

26. Pritsos, D., Stamatatos, E.: The impact of noise in web genre identification. In: Mothe, J., et al. (eds.) CLEF 2015. LNCS, vol. 9283, pp. 268–273. Springer, Cham (2015). https://doi.org/10.1007/978-3-319-24027-5_27
27. Pritsos, D., Stamatatos, E.: Open set evaluation of web genre identification. Lang. Resour. Eval. **52**, 1–20 (2018)
28. Pritsos, D.A., Stamatatos, E.: Open-set classification for automated genre identification. In: Serdyukov, P., et al. (eds.) ECIR 2013. LNCS, vol. 7814, pp. 207–217. Springer, Heidelberg (2013). https://doi.org/10.1007/978-3-642-36973-5_18
29. Priyatam, P.N., Iyengar, S., Perumal, K., Varma, V.: Don't use a lot when little will do: genre identification using URLs. Res. Comput. Sci. **70**, 207–218 (2013)
30. Řehůřek, R., Sojka, P.: Software framework for topic modelling with large corpora. In: Proceedings of the LREC 2010 Workshop on New Challenges for NLP Frameworks, ELRA, Valletta, Malta, pp. 45–50, May 2010. http://is.muni.cz/publication/884893/en
31. Rosso, M.A.: User-based identification of web genres. J. Am. Soc. Inf. Sci. Technol. **59**(7), 1053–1072 (2008). https://doi.org/10.1002/asi.20798
32. Santini, M.: Automatic identification of genre in web pages. Ph.D. thesis, University of Brighton (2007)
33. Santini, M.: Cross-testing a genre classification model for the web. In: Mehler, A., Sharoff, S., Santini, M. (eds.) Genres on the Web. Text, Speech and Language Technology, vol. 42, pp. 87–128. Springer, Dordrecht (2011). https://doi.org/10.1007/978-90-481-9178-9_5
34. Sharoff, S., Wu, Z., Markert, K.: The Web Library of Babel: evaluating genre collections. In: Proceedings of the Seventh Conference on International Language Resources and Evaluation, pp. 3063–3070 (2010)
35. Shepherd, M.A., Watters, C.R., Kennedy, A.: Cybergenre: automatic identification of home pages on the web. J. Web Eng. **3**(3–4), 236–251 (2004)
36. Stewart, J.G.: Genre oriented summarization. Ph.D. thesis, Carnegie Mellon University (2009)
37. Stubbe, A., Ringlstetter, C., Schulz, K.U.: Genre as noise: noise in genre. Int. J. Doc. Anal. Recogn. (IJDAR) **10**(3–4), 199–209 (2007)
38. Vidulin, V., Luštrek, M., Gams, M.: Using genres to improve search engines. In: Proceedings of the International Workshop Towards Genre-Enabled Search Engines, pp. 45–51 (2007)
39. Worsham, J., Kalita, J.: Genre identification and the compositional effect of genre in literature. In: Proceedings of the 27th International Conference on Computational Linguistics, pp. 1963–1973 (2018)
40. Zhu, J., Zhou, X., Fung, G.: Enhance web pages genre identification using neighboring pages. In: Bouguettaya, A., Hauswirth, M., Liu, L. (eds.) WISE 2011. LNCS, vol. 6997, pp. 282–289. Springer, Heidelberg (2011). https://doi.org/10.1007/978-3-642-24434-6_23

Exploiting Global Impact Ordering
for Higher Throughput
in Selective Search

Michał Siedlaczek[✉][ID], Juan Rodriguez[ID], and Torsten Suel[ID]

Computer Science and Engineering, New York University, New York, USA
{michal.siedlaczek,jcr365,torsten.suel}@nyu.edu

Abstract. We investigate potential benefits of exploiting a global impact ordering in a selective search architecture. We propose a generalized, ordering-aware version of the learning-to-rank-resources framework [9] along with a modified selection strategy. By allowing partial shard processing we are able to achieve a better initial trade-off between query cost and precision than the current state of the art. Thus, our solution is suitable for increasing query throughput during periods of peak load or in low-resource systems.

Keywords: Selective search · Global ordering · Shard selection

1 Introduction

Substantial advances have recently been made in selective search—a type of federated search where a collection is clustered into topical shards, and a *selection algorithm* selects for each query a small set of relevant shards for processing. The latest state of the art proposed by Dai et al. [9] uses a learning-to-rank technique from complex document ranking [18] to select shards during query processing.

We propose a generalization of this approach that exploits a global impact ordering of the documents, in addition to topic-based clustering. Query-independent global impact scores model the overall quality of documents in a collection. They have previously been used in unsafe early termination techniques, such as tiering, which limit query cost by disregarding documents deemed unimportant, unless a query is found to require more exhaustive processing. However, to our knowledge, this is the first attempt at combining selective search and global ordering-based early termination.

Contributions. We produce global orderings for GOV2 and Clueweb09-B collections; describe a modified selective search architecture that can exploit the orderings, and expand the state-of-the-art solution to allow for partial shard selection; compare our results with the baseline, and show that we achieve a better quality-efficiency trade-off at low query costs; discuss a number of research opportunities that are motivated by our work.

L. Azzopardi et al. (Eds.): ECIR 2019, LNCS 11438, pp. 12–19, 2019.
https://doi.org/10.1007/978-3-030-15719-7_2

2   Related Work

Selective Search. A number of researchers have recently explored selective search architectures, where a collection is clustered into topical shards [10,16]. Then, when processing a query, we select a limited number of shards that are predicted to be most relevant to the query topic. The literature identifies three classes of shard selection algorithms: Term-based methods [1,7] model the language distribution of each shard. Sample-based methods [16,24,25] use results from a centralized sample index, which contains a small random sample (up to 2%) of the entire index. Supervised methods [9,13] use labeled data to learn models that predict shard relevance. The current state of the art, proposed by Dai et al. [9], belongs to the last group. We derive our solution from their model, and also use it as a baseline for comparison.

Global Impact Ordering and Index Tiering. A global impact ordering is a query-independent ordering of the documents in a collection in terms of quality or importance. There are a number of ways to compute such orderings, including Pagerank [20], spam scores [8], performance of a document on past queries [2, 12], or machine-learned orderings [22]. Global orderings can be exploited for faster query processing by organizing index structures such that higher-quality documents appear first during index traversal.

One well known approach that exploits a global ordering is called *index tiering* [4,17,19,21,23]. Here, the collection is partitioned into two or more subsets called tiers based on document quality, and an inverted index is built for each of them. Each query is then first evaluated on the first tier with the highest-quality documents, and only evaluated on additional tiers if results on earlier tiers are considered insufficient. An alternative approach simply assigns IDs to documents in descending order of quality, and then stops index traversal once enough high quality documents have been evaluated for a query [3,12]. We refer to this method as *global rank cutoff* (GRC).

Our Approach. We apply GRC to a selective search environment by ordering documents inside each cluster by global impact ordering. Our ordering is determined by the performance of documents on past queries as in [2,12], which appears to provide a stronger ordering than Pagerank or spam scores, though limited improvements may be possible with an ML-based approach that combines a number of features [22]. We note that index tiering and GRC are orthogonal and complementary to safe early termination techniques such as WAND [6], BMW [11], or Max-Score [26]. In fact, previous work [15] has shown that these techniques provide benefits in selective search architectures, and we would expect them to also be profitable in our proposed architecture.

3   Ordering-Aware Selective Search

Clustering. Our index relies on the collection of documents being clustered into a number of topical shards. Since this step is orthogonal to our work, we adopt the clusters used by Dai et al. [9,10].

Reordering. Documents in each shard are reordered according to a global ordering obtained by counting document hits on a query log: Given a set of queries Q and a document d, we define a hit count $h(d) = \sum_{q \in Q} I_q(d)$, where $I_q(d)$ is equal to 1 if d is among the top k results retrieved for query q, or 0 otherwise (we used $k = 1000$). For Clueweb09-B, we used 40,000 queries from the TREC 2009 Million Query Track. For GOV2, we used a set of 2 million queries randomly generated by a language model learned from a sample of several TREC query tracks. In each shard, we then assign document IDs to documents based on decreasing hit count. Evaluation is performed on a set of queries that is disjoint from those used to compute the ordering or train the language model.

Ordered Partitioning. The idea behind reordering is to facilitate global rank cutoff within topical clusters. In this paper, we approximate it by partitioning consecutive document IDs in each shard into b ranges of equal size, effectively setting up b cutoff points where processing can stop. We call these partitions *buckets* and number them with consecutive integers, where 1 denotes the first (highest quality) bucket and b denotes the last one. This approach allows us to extend selective search with global ordering while trading off model accuracy and complexity. Large values of b model GRC well but make both learning and feature extraction more expensive. Structures with small b are fast and resemble discrete tiers. When $b = 1$, we get the standard selective search approach, as done in previous work. From this point forward, we refer to a solution with b buckets as B_b. In particular, B_1 denotes the baseline.

Learning Model. We modify the learning-to-rank-resources model proposed by Dai et al. [9]; see that paper for more details. We discard CSI-based features, as they were shown to provide only negligible improvement but are expensive to compute [9]. To the shard-level features (shard popularity and term-based statistics), we add only one bucket-level feature, namely the bucket number, from 1 to b. Thus, we train and predict a ranking of buckets instead of shards. This ranking is further converted into the final selection as described below.

Bucket Cost. Given a query, every bucket i in a shard s has an associated processing cost $c(s, i)$. We consider two *cost models*: (1) uniform cost: $c(s, i) = 1/b$, used during shard selection, and (2) posting cost: the number of postings for the query terms within bucket (s, i), used during evaluation. We also present one set of experiments where we add an additional per-shard access cost to the posting cost, in order to model the overhead of forwarding a query to a shard.

Shard Selection. In constrast to previous shard selection algorithms, our approach requires us to select shards as well as a cut-off point for early termination inside each selected shard (Fig. 1). Our model predicts a bucket-level ranking, but to model GRC, we need to make sure to select a prefix of the bucket list in each shard. Thus, given a budget T per query, we iterate over all buckets from highest to lowest ranked, until the remaining query budget t (initially $t = T$) drops to 0. For a bucket at a position i in a shard s, we consider all *unselected* buckets (s, j) for $j \leq i$, and denote the sum of their costs as C. If $C \leq t$, then we update the cost $t \leftarrow t - C$ and mark all buckets (s, j), $j \leq i$ as selected.

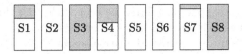

Fig. 1. An illustration of order-aware shard selection. Based on the predicted bucket ranking, portions of shards are selected (shaded area). While strongly relevant shards (S3, S8) may be fully processed, others could terminate early using GRC (S1, S4, S7), or be ignored entirely (S2, S5, S6).

4   Methodology

Data Sets. We conduct our experiments on GOV2 and Clueweb09-B collections, consisting of approximately 25 million and 50 million documents, respectively. Following previous work [9], we remove the documents with Waterloo spam score [8] below 50 from Clueweb09-B. The resulting collection consists of nearly 38 million documents. We use the default HTML parser from the BUbiNG library, and stem terms using the Porter2 algorithm. No stopwords are removed. We use 150 queries from the TREC 04-05 Terabyte Track (GOV2) and 200 queries from the TREC 09-12 Web Track (Clueweb09-B) topics for training our model. We evaluate our method using 10-fold cross-validation.

Training Model. Following [9], we use the SVM^{rank} library [14] with linear kernel to train our ranking prediction model. It implements the pair-wise approach to learning to rank [18]. We compute the *relevance-based* ground truth as described in [9].

Evaluation Metrics. Following [9], we report search accuracy in terms of P@10 and the more recall-oriented MAP@1000. (Due to space restrictions, we exclude NDCG@30, as it exhibited similar behavior to MAP@1000.) Query cost is reported in terms of the number of postings as described in Sect. 3.

Search Architecture. The GOV2 collection is clustered into 199 shards, and Clueweb09-B into 123 shards, as done in previous work [9,10]. We process our queries using the MG4J search engine [5] with default BM25 scoring ($k_1 = 1.2$, $b = 0.5$). When scoring documents in shards, we use global values for collection size, posting count, term frequencies, and occurrence counts, in order to obtain scores that are comparable between shards.

5   Experiments

We experimented with $b = 1, 10, 20$, where $b = 1$ is equivalent to the current state of the art without using a global ordering in [9], which serves as the baseline. Since the baseline disregards differences in query cost due to inverted list lengths during shard selection, we also do so, using the uniform cost model for this step. Following previous work, we report the resulting posting costs in the evaluation.

We express budgets in terms of the numbers of selected shards, where one shard is translated into b buckets for $b > 1$. We report results for budgets up to

20 shards. However, our goal is to limit the cost with little decrease in quality; therefore, we are mostly interested in very low costs, below what has already been shown to work on a par with exhaustive search. Figure 2 shows the trade-offs between quality measures (P@10 and MAP@1000) and query cost in terms of processed postings. Reported values are averaged over all queries.

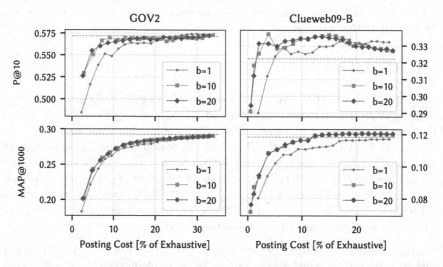

Fig. 2. Quality-cost trade-off of bucket selection for different bucket counts and data sets. Shards were selected under the uniform cost model and with budgets from 1 to 20 shards. For Clueweb09-B, additional budgets of 0.25 and 0.5 were tested for $b > 1$. The quality of exhaustive search is indicated by a dashed line.

Both measures improve much faster for B_{10} and B_{20} than for B_1. For instance, P@10 in Clueweb09-B improves upon exhaustive search for a budget as small as 1 shard. At low budgets, we need to process only about a third of what is needed in B_1 to achieve the same quality under P@10. Similarly, achieving the MAP@1000 quality level of exhaustive search requires much fewer postings to be traversed, with smaller improvements observed for GOV2. These improvements are achieved by enabling partial shard access; instead of traversing unimportant documents in highly relevant shards, we allocate budget towards traversing highly important documents in slightly less topically relevant shards. This is achieved by learning of a model that automatically adjusts the cutoff levels.

Shard Access Overhead. While our results show that we can decrease resource requirements, there is an important caveat. One expected side effect of the proposed solution is that for the average query, we contact more shards than for $b = 1$. However, dispatching a query to more shards and collecting the results is likely to result in some amount of overhead. We now address this concern.

As shown in Fig. 3, B_{10} and B_{20} contact about 2 to 3 times as many shards as B_1. We now estimate how this would impact the performance of our measures.

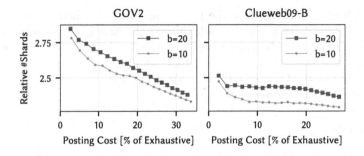

Fig. 3. Average number of shards used in query processing, relative to the baseline.

To do this, we add an overhead cost for contacting a shard equal to 10% of the average cost of an exhaustive query on a single shard. We believe this is a conservative upper bound to the overhead of contacting a shard. Figure 4 compares the performance of the no-overhead case (in red) to the 10%-overhead case (in blue), and shows that the improvement over the baseline is still significant.

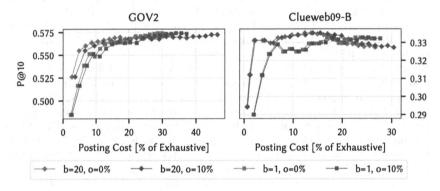

Fig. 4. Quality-cost trade-off of bucket selection with per-shard access overheads of 0% and 10% of the average cost of an exhaustive query on a single shard. (Color figure online)

6 Conclusions

In this paper, we have described how to combine selective search with global ordering-based early termination. Our approach can significantly improve throughput in scenarios where resources are scarce, or where a system experiences peak loads. Although our solution results in additional overhead for contacting more shards, we showed that overall costs still decrease significantly. Our results motivates a number of research questions that we are currently pursuing, including the use of better global orderings based on machine learning, use of additional bucket-level features, the performance of the schemes when used to generate candidates for reranking under a complex ranker, and possible schemes for adapting to high loads under realistic service level agreements (SLAs).

Acknowledgement. This research was partially supported by NSF Grant IIS-1718680 and a grant from Amazon.

References

1. Aly, R., Hiemstra, D., Demeester, T.: Taily: shard selection using the tail of score distributions. In: Proceedings of the 36th Annual International ACM SIGIR Conference on Research and Development in Information Retrieval, pp. 673–682 (2013)
2. Anagnostopoulos, A., Becchetti, L., Leonardi, S., Mele, I., Sankowski, P.: Stochastic query covering. In: Proceedings of the 4th ACM International Conference on Web Search and Data Mining, pp. 725–734 (2011)
3. Asadi, N., Lin, J.: Effectiveness/efficiency tradeoffs for candidate generation in multi-stage retrieval architectures. In: Proceedings of the 36th Annual International ACM SIGIR Conference on Research and Development in Information Retrieval, pp. 997–1000 (2013)
4. Baeza-Yates, R., Murdock, V., Hauff, C.: Efficiency trade-offs in two-tier web search systems. In: Proceedings of the 32nd Annual International ACM SIGIR Conference on Research and Development in Information Retrieval, pp. 163–170. ACM (2009)
5. Boldi, P., Vigna, S.: MG4J at TREC 2005. In: The Fourteenth Text REtrieval Conference (TREC 2005) Proceedings (2005)
6. Broder, A.Z., Carmel, D., Herscovici, M., Soffer, A., Zien, J.: Efficient query evaluation using a two-level retrieval process. In: Proceedings of the 12th International Conference on Information and Knowledge Management, pp. 426–434 (2003)
7. Callan, J.P., Lu, Z., Croft, W.B.: Searching distributed collections with inference networks. In: Proceedings of the 18th Annual International ACM SIGIR Conference on Research and Development in Information Retrieval, pp. 21–28 (1995)
8. Cormack, G.V., Smucker, M.D., Clarke, C.L.A.: Efficient and effective spam filtering and re-ranking for large web datasets. Inf. Retrieval **14**(5), 441–465 (2011)
9. Dai, Z., Kim, Y., Callan, J.: Learning to rank resources. In: Proceedings of the 40th International ACM SIGIR Conference on Research and Development in Information Retrieval, pp. 837–840 (2017)
10. Dai, Z., Xiong, C., Callan, J.: Query-biased partitioning for selective search. In: Proceedings of the 25th ACM International Conference on Information and Knowledge Management, pp. 1119–1128 (2016)
11. Ding, S., Suel, T.: Faster top-k document retrieval using block-max indexes. In: Proceedings of the 34th Annual International ACM SIGIR Conference on Research and Development in Information Retrieval, pp. 993–1002 (2011)
12. Garcia, S., Williams, H.E., Cannane, A.: Access-ordered indexes. In: Proceedings of the 27th Australasian Conference on Computer Science, pp. 7–14 (2004)
13. Hong, D., Si, L., Bracke, P., Witt, M., Juchcinski, T.: A joint probabilistic classification model for resource selection. In: Proceedings of the 33rd International ACM SIGIR Conference on Research and Development in Information Retrieval, pp. 98–105 (2010)
14. Joachims, T.: Training linear SVMs in linear time. In: Proceedings of the 12th ACM SIGKDD International Conference on Knowledge Discovery and Data Mining, pp. 217–226 (2006)

15. Kim, Y., Callan, J., Culpepper, J.S., Moffat, A.: Does selective search benefit from WAND optimization? In: Ferro, N., Crestani, F., Moens, M.-F., Mothe, J., Silvestri, F., Di Nunzio, G.M., Hauff, C., Silvello, G. (eds.) ECIR 2016. LNCS, vol. 9626, pp. 145–158. Springer, Cham (2016). https://doi.org/10.1007/978-3-319-30671-1_11

16. Kulkarni, A., Callan, J.: Selective search: efficient and effective search of large textual collections. ACM Trans. Inf. Syst. (TOIS) **33**(4), 17 (2015)

17. Leung, G., Quadrianto, N., Tsioutsiouliklis, K., Smola, A.J.: Optimal web-scale tiering as a flow problem. In: Advances in Neural Information Processing Systems, pp. 1333–1341 (2010)

18. Liu, T.Y.: Learning to rank for information retrieval. Found. Trends Inf. Retrieval **3**(3), 225–331 (2009)

19. Ntoulas, A., Cho, J.: Pruning policies for two-tiered inverted index with correctness guarantee. In: Proceedings of the 30th Annual International ACM SIGIR Conference on Research and Development in Information Retrieval, pp. 191–198 (2007)

20. Page, L., Brin, S., Motwani, R., Winograd, T.: The PageRank citation ranking: Bringing order to the web. Technical report 1999–66 (1999)

21. Panigrahi, D., Gollapudi, S.: Document selection for tiered indexing in commerce search. In: Proceedings of the 6th ACM International Conference on Web Search and Data Mining, pp. 73–82. ACM (2013)

22. Richardson, M., Prakash, A., Brill, E.: Beyond PageRank: machine learning for static ranking. In: Proceedings of the 15th International Conference on World Wide Web, pp. 707–715 (2006)

23. Risvik, K.M., Aasheim, Y., Lidal, M.: Multi-tier architecture for web search engines. In: Proceedings of the First Conference on Latin American Web Congress, pp. 132–143 (2003)

24. Si, L., Callan, J.: Relevant document distribution estimation method for resource selection. In: Proceedings of the 26th Annual International ACM SIGIR Conference on Research and Development in Information Retrieval, pp. 298–305 (2003)

25. Thomas, P., Shokouhi, M.: SUSHI: scoring scaled samples for server selection. In: Proceedings of the 32nd Annual International ACM SIGIR Conference on Research and Development in Information Retrieval, pp. 419–426 (2009)

26. Turtle, H., Flood, J.: Query evaluation: strategies and optimizations. Inf. Process. Manage. **31**(6), 831–850 (1995)

Cross-Domain Recommendation via Deep Domain Adaptation

Heishiro Kanagawa[1]([✉]), Hayato Kobayashi[2,4], Nobuyuki Shimizu[2],
Yukihiro Tagami[2], and Taiji Suzuki[3,4]

[1] Gatsby Unit, UCL, London, UK
heishiro.kanagawa@gmail.com
[2] Yahoo Japan Corporation, Tokyo, Japan
{hakobaya,nobushim,yutagami}@yahoo-corp.jp
[3] The University of Tokyo, Tokyo, Japan
[4] RIKEN Center for Advanced Intelligence Project, Tokyo, Japan
taiji@mist.i.u-tokyo.ac.jp

Abstract. The behavior of users in certain services indicates their preferences, which may be used to make recommendations for other services they have never used. However, the cross-domain relation between items and user preferences is not simple, especially when there are few or no common users and items across domains. We propose a content-based cross-domain recommendation method for cold-start users that does not require user- or item-overlap. We formulate recommendations as an extreme classification task, and the problem is treated as an instance of unsupervised domain adaptation. We assess the performance of the approach in experiments on large datasets collected from Yahoo! JAPAN video and news services and find that it outperforms several baseline methods including a cross-domain collaborative filtering method.

Keywords: Cross-domain recommendation · Deep domain adaptation

1 Introduction

Conventional recommender systems are known to be ineffective in cold-start scenarios, e.g., when the user is new to the service, or the goal is to recommend items from a service that the user has not used, because they require knowledge of the user's past interactions [32]. On the other hand, as the variety of Web services has increased, information about cold-start users can often be obtained from their activities in other services. Therefore, cross-domain recommender systems (CDRSs), which utilize such information from other domains, have gained research attention as a promising solution to the cold-start problem [4].

This paper addresses a problem of cross-domain recommendation in which we cannot expect users or items to be shared across domains. Our task is to

This work was conducted during the first author's internship at Yahoo! JAPAN.
TS was partially supported by MEXT Kakenhi (18H03201) and JST-CREST.

recommend items in a domain (service) to users who have not used it before but have a history in another domain (see Sect. 5 for details). This situation is common in practice. For instance, when one service is not known (e.g., when it is newly built) to users of another more popular service, there are few overlapping users, which was actually observed in the dataset of this study. Another probable instance is when user identities are anonymized to preserve privacy [17]; there are no shared users in this case because we cannot know if two users are identical across domains.

The case in which domains have a user- or item-overlap has been extensively studied because an overlap helps us learn relations between users and items of different domains [24,25,27,36]. The problem, however, is harder when there is little or no overlap as learning of user-item relations becomes more challenging. Previous studies dealt with this challenge by using specific forms of auxiliary information that are not necessarily available including user search queries on Bing for recommendations in other Microsoft services [8], tags on items given by users such as in a photo sharing service, e.g., Flickr [1,2], and external knowledge repositories such as Wikipedia [9,19,20,23]. Methods that use the content information of items like this are called content-based (CB) methods. Collaborative filtering (CF) methods, which do not require auxiliary information, have also been proposed [12,17,43]. Despite their broader applicability, a drawback of CF methods compared to CB ones is that they suffer from data sparsity and require a substantial amount of user interactions.

Our main contribution is to fill in the gap between CB and CF approaches. We propose a CB method that (a) only uses content information generally available to content service providers and (b) does not require shared users or items. Our approach is the application of unsupervised domain adaptation (DA). Although unsupervised DA has shown fruitful results in machine learning [3,11,33], its connection with CDRSs has been largely unexplored. Our formulation by extreme multi-class classification [6,30], where labels (items) corresponding to a user are predicted, allows us the use of this technique. We use a neural network (NN) architecture for unsupervised DA, the domain separation network (DSN) [3], because NNs can learn efficient features, as shown in the successful applications to recommender systems [6,8,14,28,38–41,44]. We provide a practical case study through experiments on large datasets collected from existing commercial services of Yahoo! JAPAN. We compare our method with several baselines including the state-of-the-art CF method [17]. We show that (a) the CF method does not actually perform well in this real application, and (b) our approach shows the highest performance in DCG.

2   Related Work

Approaches to CDRS can be categorized into two types [4]. The first type aggregates knowledge by combining datasets from multiple domains in a common format, e.g., a common rating matrix [24]. As a result, user-item overlaps or a specific data format are typically assumed [1,2,8,25–27,35]. In particular, the deep-learning approach [15] uses, as in our method, unstructured texts to learn user and item representations but assumes a relatively large number of common users. The second type links domains by transferring learned knowledge. This line of research has been limited to matrix-factorization based CF methods because sharing latent factors across domains allows knowledge transfer [12,17,43]. Our method is a CB method and solves the aforementioned data sparsity problem of CF methods. Also, DA provides an alternative method of knowledge transfer to shared latent factors.

There are works labeled as cross-domain recommendation that solve a different task. Studies including [7,22,29,42] have dealt with sparsity reduction. The task is to improve recommendation quality within a target domain and is not cross-selling from a different domain. This is performed, for example, by conducting CF on a single domain having sparse data with the help of other domains, such as using a shared latent factor learned in dense domains [7,22]. Although the work [42] uses DA, it cannot solve the same problem as ours.

Finally, online reinforcement learning can be used to obtain feedback from cold-start users [31]. However, it is difficult to deploy it in commercial applications because the performance is harder to evaluate compared to offline methods such as ours since the model constantly changes over time.

3 Problem Setting

Let X be the input feature space, such as $\mathbb{R}^d$ with a positive integer d. Let $Y = \{1, \ldots, L\}$ be a collection of items that we wish to recommend. We are given two datasets from two different domains. The first is a labeled dataset $\mathcal{D}_S = \{(\mathbf{x}_i^S, y_i^S)\}_{i=1}^{N_S}$. Each element denotes a user-item pair with $y_i^S \in Y$ being a label representing the item, and $\mathbf{x}_i^S \in X$ being a feature vector that represents the previous history of a user and is formed by the content information of the items in the history. This domain, from which the labeled dataset comes, is called the *source domain*. The second dataset[1] is $\mathcal{D}_T^X = \{\mathbf{x}_i^T\}_{i=1}^{N_T}$, where $\mathbf{x}_i^T \in X$ denotes the feature vector of a user's history consisting of items in the other domain. We call this data domain the *target domain*. The goal is to recommend items in the source domain to users in the target domain. Our approach is to construct a classifier $\eta : X \to Y$ such that $\eta(\mathbf{x}^T)$ gives the most probable item in Y for a new user history $\mathbf{x}^T$ from the target domain.

4 Unsupervised Domain Adaptation

We define the problem of unsupervised DA [3,11]. We are given two datasets $\mathcal{D}_S \overset{\text{i.i.d.}}{\sim} P_S$ and $\mathcal{D}_T^X \overset{\text{i.i.d.}}{\sim} P_T^X$ as in the previous section. Here, P_S and P_T are probability distributions over $X \times Y$, which are assumed to be similar but different. P_T^X is the marginal distribution over X in domain T. The goal of

---

[1] Superscript X denotes that the data is missing labels associated to their input vectors.

unsupervised DA is to obtain a classifier η with a low risk defined by $R_T(\eta) = \Pr_{(\mathbf{x},y)\sim P_T}(\eta(\mathbf{x}) \neq y)$ only using the datasets $\mathcal{D}_S$ and $\mathcal{D}_T^X$.

Domain Separation Networks: We introduce the neural network architecture, domain separation networks (DSNs) [3]. DSN, separately from domain-specific features, learns predictive domain-invariant features, which gives a shared representation of inputs from different domains. Figure 1 shows the architecture of a DSN model, which consists of the following components: a shared encoder $E_c(\mathbf{x}; \theta_c)$, private encoders $E_p^S(\mathbf{x}; \theta_p^S)$, $E_p^T(\mathbf{x}; \theta_p^T)$, a shared decoder $D(\mathbf{h}; \theta_d)$, and a classifier $G(\mathbf{h}; \theta_g)$, where θ_c, θ_p^S, θ_p^T, θ_d, and θ_g denote parameters. Given an input vector $\mathbf{x}$, which comes from the source domain or the target domain, a shared encoder function E_c maps it to a hidden representation $\mathbf{h}_c$, which represents features shared across domains. For the source (target) domain, a DSN has a private encoder E_p^S (E_p^T) that maps an input vector to a hidden representation $\mathbf{h}_p^S$ $(\mathbf{h}_p^T)$, which serves as a feature vector specific to the domain.

A common decoder D reconstructs an input vector $\mathbf{x}$ of the source (or target) domain from the sum of shared and private hidden representations: $\mathbf{h}_c$ and $\mathbf{h}_p^S$ $(\mathbf{h}_p^T)$. The classifier G takes a shared hidden representation $\mathbf{h}_c$ as input and predicts the corresponding label.

Training is performed by minimizing the following objective function L_{DSN} with respect to parameters θ_c, θ_p^S, θ_p^T, θ_d, and θ_g: $L_{\mathrm{DSN}} = L_{\mathrm{task}} + \alpha L_{\mathrm{recon}} + \beta L_{\mathrm{diff}} + \gamma L_{\mathrm{sim}}$, where α, β, and γ are parameters that control the effect of the asso-

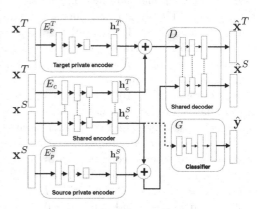

Fig. 1. Architecture of DSN

ciated terms. L_{task} is a classification loss (cross entropy loss). L_{recon} is a reconstruction loss (squared Euclidean distance). The loss L_{diff} encourages the shared and private encoders to extract different types of features. As in [3], we impose a soft subspace orthogonality loss. The similarity loss L_{sim} encourages the shared encoder to produce representations that are hardly distinguishable. We use the domain adversarial similarity loss [10,11].

For the output layer of the classifier, we use the softmax activation. We suggest a list of items in descending order of the output probabilities.

5   Experiments

We conduct performance evaluation on a pair of two real-world datasets in collaboration with Yahoo! JAPAN. The datasets consist of the browsing logs of a video on demand service (VIDEO) and a news aggregator (NEWS). The task is to recommend videos to NEWS users who have not used the VIDEO service.

Dataset Description: For VIDEO, we create a labeled dataset where a label is a video watched by a particular user, and the input is the list of videos previously watched by the user. Similarly, for NEWS, we build an unlabeled dataset that consists of histories of news articles representing users. The number of instances for each dataset is roughly 11 million. For evaluation, we form a test dataset of size $38,250$ from users who used both services, where each instance is a pair of a news history (input) and a video (label).

Items in both services have text attributes. For VIDEO, we use the following attributes: title, category, short description, and cast information. For NEWS, article title and category are used. We treat a history as a document comprised of attribute words and represent it with the TF-IDF scheme. We form a vocabulary set for each dataset according to the TF-IDF value. Combining the two sets gives a common vocabulary set of 50,000 words.

Experimental Settings: We construct a DSN consisting of fully-connected layers [2]. The exponential linear unit (ELU) [5] is used as the activation function for all layers. The weight and bias parameters for the fully-connected layers are initialized following [13]. We apply dropout [34] to all fully-connected layers with the rates 0.25 for all the encoders and 0.5 for the decoder and the classifier. We also impose an L2 penalty on the weight parameters with the regularization parameter chosen from $\{10^{-1}, 10^{-2}, 10^{-3}, 10^{-4}\}$. The parameters α, β, and γ in L_{DSN} are set to 10^{-3}, 10^{-2}, and 100, respectively. The parameter γ is chosen by the similarity of the marginals [3] and agrees with the setting in [3]. The ADAM optimizer [21] is used, and the initial learning rate is set to 10^{-3}.

We use discounted cumulative gain (DCG) [18] for evaluation. DCG gives a higher score when a correct item appears earlier in a suggested item list and thus measures ranking quality. This is defined by $\text{DCG@}M = \sum_{m=1}^{M} I(\hat{y}_m = y)/\log(m+1)$, where the function $I(\hat{y}_m = y)$ returns 1 when the m-th suggested item $\hat{y}_m$ is y and otherwise 0. The use of metrics such as precision is not appropriate since negative feedback is not well-defined in implicit feedback; unobserved items may express users' distaste or just have been unseen, whereas we can treat viewed items as true positive.

We evaluate the method for five different training and test dataset pairs, which are randomly sampled from the datasets. We sample 80% of both the whole training and test datasets. For each pair, we compute DCG averaged over the test set. We use a validation set that consists of logs of common users on the same dates as the training data for model selection to investigate attainable performance of domain adaptation. Note that this setting still keeps our target situation where there are not enough overlapping users for training.

---

[2] The unit size of each hidden layer is as follows: $E_c = E_p^S = E_p^T = (256 - 128 - 128 - 64)$, $D = (128 - 128 - 256)$, and $G = (256 - 256 - 256 - 64)$. Left is the input.

[3] We train a classifier that detects the domain of the input represented by the shared encoder and test the classification accuracy. This is due to the adversarial training.

Table 1. Performance comparison of DSN, NN, CdMF, POP. (DCG) of DSN and NN denotes models that are chosen according to the value of DCG@100. Similarly, (CEL) denotes the cross entropy loss. Each entry is Mean ± Std.

Method	DCG@1	DCG@10	DCG@50	DCG@100
DSN (DCG)	**0.0618 ± 0.0212**	**0.2133 ± 0.0154**	**0.2873 ± 0.0151**	**0.2945 ± 0.0153**
DSN (CEL)	0.0406 ± 0.0212	0.1668 ± 0.0384	0.2583 ± 0.0230	0.2655 ± 0.0229
NN (DCG)	0.0415 ± 0.0211	0.1938 ± 0.0131	0.2735 ± 0.0102	0.2797 ± 0.0107
NN (CEL)	0.0282 ± 0.0301	0.1616 ± 0.0279	0.2473 ± 0.0247	0.2556 ± 0.0238
CdMF	0.0005 ± 0.0000	0.0040 ± 0.0000	0.0135 ± 0.0000	0.0644 ± 0.0004
POP	0.0398 ± 0.0007	0.2099 ± 0.0012	0.2790 ± 0.0016	0.2871 ± 0.0010

Baseline Models: The following baseline methods are compared.

- **Most Popular Items (POP).** This suggests the most popular items in the training data. The comparison with this shows how well the proposed approach achieves personalization.
- **Cross-domain Matrix Factorization (CdMF)** [17]. This is the state-of-the-art collaborative filtering method for CDRSs requiring no user- or item-over lap. To alleviate sparsity, we eliminate users with history logs fewer than 5 for the video data and 20 for the news data from the training data. We construct a user-item matrix for each domain, with the value of observed entries 1 and of unobserved entries 0. The unobserved entries are randomly subsampled as the size of the whole unseen entries is too large to be processed. As with our method, we use 80% of the instances for training and the remaining 20% for validation. The hyper-parameters α, β_0, and ν_0 are set at the same values as in [17]. The method requires inference of latent variables by Gibbs sampling; each sampling step involves computing matrix inverses. Due to the computational complexity, we were unable to optimize these hyper-parameters, and therefore they are fixed.
- **Neural Network (NN).** To investigate the effect of DA, the same network without DA is evaluated. This is obtained by minimizing the same loss as DSN except that β, γ in L_{DSN} are set to 0. It can be thought of as a strong content-based single-domain method as more robust features are learned with the reconstruction loss L_{recon} than only with the task loss L_{task}(as in [39], with the roles of item and user swapped).

Experimental Results: We report the results in Table 1. CdMF underperformed other methods. As mentioned in [16], this is likely because CdMF cannot process the implicit feedback or accurately capture the popularity structure in the dataset. The DSN model chosen by the cross-entropy loss (DSN (CEL)) also had a worse performance than POP. We also chose the model based on the values of DCG@100 on the validation set (denoted by DSN (DCG)), which showed the best performance in all cut-off number settings. This result can be interpreted as follows. As predicting the most probable item is hard, the cross-entropy loss does not give a useful signal for model selection. On the other hand,

DCG provides the quality of ranking, and therefore a classifier which captures the joint distribution of items and users well is more likely to be chosen. The improved performance of DSN (DCG) over NN (DCG) supports this claim since DA improves the learning of the joint distribution. This implies that replacing the loss with a ranking loss such as [37] could improve ranking quality.

Conclusion and Future Work: Our evaluation demonstrated the potential of the proposed DA-based approach. However, extreme classification is still challenging and should be addressed as a future work. One possible direction would be incorporation of item information as in [39], as this would make items more distinguishable and enable predictions of unobserved items.

References

1. Abel, F., Araújo, S., Gao, Q., Houben, G.-J.: Analyzing cross-system user modeling on the social web. In: Auer, S., Díaz, O., Papadopoulos, G.A. (eds.) ICWE 2011. LNCS, vol. 6757, pp. 28–43. Springer, Heidelberg (2011). https://doi.org/10.1007/978-3-642-22233-7_3
2. Abel, F., Herder, E., Houben, G.J., Henze, N., Krause, D.: Cross-system user modeling and personalization on the social web. User Model. User-Adap. Inter. **23**(2), 169–209 (2013). https://doi.org/10.1007/s11257-012-9131-2
3. Bousmalis, K., Trigeorgis, G., Silberman, N., Krishnan, D., Erhan, D.: Domain separation networks. In: Lee, D.D., Sugiyama, M., Luxburg, U.V., Guyon, I., Garnett, R. (eds.) Advances in Neural Information Processing Systems 29, pp. 343–351. Curran Associates, Inc. (2016). http://papers.nips.cc/paper/6254-domain-separation-networks.pdf
4. Cantador, I., Fernández-Tobías, I., Berkovsky, S., Cremonesi, P.: Cross-domain recommender systems. In: Ricci, F., Rokach, L., Shapira, B. (eds.) Recommender Systems Handbook, pp. 919–959. Springer, Boston (2015). https://doi.org/10.1007/978-1-4899-7637-6_27
5. Clevert, D., Unterthiner, T., Hochreiter, S.: Fast and accurate deep network learning by exponential linear units (ELUs). CoRR abs/1511.07289 (2015). http://arxiv.org/abs/1511.07289
6. Covington, P., Adams, J., Sargin, E.: Deep neural networks for YouTube recommendations. In: Proceedings of the 10th ACM Conference on Recommender Systems, RecSys 2016, pp. 191–198. ACM, New York (2016). https://doi.org/10.1145/2959100.2959190
7. Cremonesi, P., Quadrana, M.: Cross-domain recommendations without overlapping data: myth or reality? In: Proceedings of the 8th ACM Conference on Recommender Systems, RecSys 2014, pp. 297–300. ACM, New York (2014). https://doi.org/10.1145/2645710.2645769
8. Elkahky, A.M., Song, Y., He, X.: A multi-view deep learning approach for cross domain user modeling in recommendation systems. In: Proceedings of the 24th International Conference on World Wide Web, WWW 2015, pp. 278–288. International World Wide Web Conferences Steering Committee, Republic and Canton of Geneva, Switzerland (2015). https://doi.org/10.1145/2736277.2741667

9. Fernández-Tobías, I., Cantador, I., Kaminskas, M., Ricci, F.: A generic semantic-based framework for cross-domain recommendation. In: Proceedings of the 2nd International Workshop on Information Heterogeneity and Fusion in Recommender Systems, HetRec 2011, pp. 25–32. ACM, New York (2011). https://doi.org/10.1145/2039320.2039324

10. Ganin, Y., Lempitsky, V.S.: Unsupervised domain adaptation by backpropagation. In: Proceedings of the 32nd International Conference on Machine Learning, ICML 2015, Lille, France, 6–11 July 2015, pp. 1180–1189 (2015). http://jmlr.org/proceedings/papers/v37/ganin15.html

11. Ganin, Y., et al.: Domain-adversarial training of neural networks. J. Mach. Learn. Res. **17**(59), 1–35 (2016). http://jmlr.org/papers/v17/15-239.html

12. Gao, S., Luo, H., Chen, D., Li, S., Gallinari, P., Guo, J.: Cross-domain recommendation via cluster-level latent factor model. In: Blockeel, H., Kersting, K., Nijssen, S., Železný, F. (eds.) ECML PKDD 2013. LNCS (LNAI), vol. 8189, pp. 161–176. Springer, Heidelberg (2013). https://doi.org/10.1007/978-3-642-40991-2_11

13. He, K., Zhang, X., Ren, S., Sun, J.: Delving deep into rectifiers: surpassing human-level performance on ImageNet classification. CoRR abs/1502.01852 (2015). http://arxiv.org/abs/1502.01852

14. He, X., Liao, L., Zhang, H., Nie, L., Hu, X., Chua, T.S.: Neural collaborative filtering. In: Proceedings of the 26th International Conference on World Wide Web, WWW 2017, pp. 173–182. International World Wide Web Conferences Steering Committee, Republic and Canton of Geneva, Switzerland (2017). https://doi.org/10.1145/3038912.3052569

15. Hu, G., Zhang, Y., Yang, Q.: MTNet: a neural approach for cross-domain recommendation with unstructured text. In: KDD 2018 Deep Learning Day (2018). https://www.kdd.org/kdd2018/files/deep-learning-day/DLDay18_paper_5.pdf

16. Hu, Y., Koren, Y., Volinsky, C.: Collaborative filtering for implicit feedback datasets. In: Proceedings of the 2008 Eighth IEEE International Conference on Data Mining, ICDM 2008, pp. 263–272. IEEE Computer Society, Washington (2008). https://doi.org/10.1109/ICDM.2008.22

17. Iwata, T., Takeuchi, K.: Cross-domain recommendation without shared users or items by sharing latent vector distributions. In: Proceedings of the 18th International Conference on Artificial Intelligence and Statistics (AISTATS), AISTATS 2015, USA, pp. 379–387 (2015)

18. Järvelin, K., Kekäläinen, J.: Cumulated gain-based evaluation of IR techniques. ACM Trans. Inf. Syst. **20**(4), 422–446 (2002). https://doi.org/10.1145/582415.582418

19. Kaminskas, M., Fernández-Tobías, I., Cantador, I., Ricci, F.: Ontology-based identification of music for places. In: Cantoni, L., Xiang, Z.P. (eds.) Information and Communication Technologies in Tourism 2013, pp. 436–447. Springer, Heidelberg (2013). https://doi.org/10.1007/978-3-642-36309-2_37

20. Kaminskas, M., Fernández-Tobías, I., Ricci, F., Cantador, I.: Knowledge-based identification of music suited for places of interest. Inf. Technol. Tourism **14**(1), 73–95 (2014). https://doi.org/10.1007/s40558-014-0004-x

21. Kingma, D.P., Ba, J.: Adam: a method for stochastic optimization. CoRR abs/1412.6980 (2014). http://arxiv.org/abs/1412.6980

22. Li, B., Yang, Q., Xue, X.: Can movies and books collaborate? Cross-domain collaborative filtering for sparsity reduction. In: Proceedings of the 21st International Joint Conference on Artificial Intelligence, IJCAI 2009, pp. 2052–2057. Morgan Kaufmann Publishers Inc., San Francisco (2009)

23. Loizou, A.: How to recommend music to film buffs: enabling the provision of recommendations from multiple domains, May 2009. https://eprints.soton.ac.uk/66281/
24. Loni, B., Shi, Y., Larson, M., Hanjalic, A.: Cross-domain collaborative filtering with factorization machines. In: de Rijke, M., et al. (eds.) ECIR 2014. LNCS, vol. 8416, pp. 656–661. Springer, Cham (2014). https://doi.org/10.1007/978-3-319-06028-6_72
25. Low, Y., Agarwal, D., Smola, A.J.: Multiple domain user personalization. In: Proceedings of the 17th ACM SIGKDD International Conference on Knowledge Discovery and Data Mining, KDD 2011, pp. 123–131. ACM, New York (2011). https://doi.org/10.1145/2020408.2020434
26. Man, T., Shen, H., Jin, X., Cheng, X.: Cross-domain recommendation: an embedding and mapping approach. In: Proceedings of the Twenty-Sixth International Joint Conference on Artificial Intelligence, IJCAI 2017, pp. 2464–2470 (2017). https://doi.org/10.24963/ijcai.2017/343
27. Nakatsuji, M., Fujiwara, Y., Tanaka, A., Uchiyama, T., Ishida, T.: Recommendations over domain specific user graphs. In: Proceedings of the 2010 Conference on ECAI 2010: 19th European Conference on Artificial Intelligence, pp. 607–612. IOS Press, Amsterdam (2010)
28. Oord, A.V.D., Dieleman, S., Schrauwen, B.: Deep content-based music recommendation. In: Proceedings of the 26th International Conference on Neural Information Processing Systems, NIPS 2013, pp. 2643–2651. Curran Associates Inc., USA (2013)
29. Pan, W., Xiang, E.W., Liu, N.N., Yang, Q.: Transfer learning in collaborative filtering for sparsity reduction. In: Proceedings of the Twenty-Fourth AAAI Conference on Artificial Intelligence, AAAI 2010, pp. 230–235. AAAI Press (2010)
30. Partalas, I., et al.: LSHTC: a benchmark for large-scale text classification. CoRR abs/1503.08581 (2015). http://arxiv.org/abs/1503.08581
31. Rubens, N., Elahi, M., Sugiyama, M., Kaplan, D.: Active learning in recommender systems. In: Ricci, F., Rokach, L., Shapira, B. (eds.) Recommender Systems Handbook, pp. 809–846. Springer, Boston (2015). https://doi.org/10.1007/978-1-4899-7637-6_24
32. Sahebi, S., Brusilovsky, P.: Cross-domain collaborative recommendation in a cold-start context: the impact of user profile size on the quality of recommendation. In: Carberry, S., Weibelzahl, S., Micarelli, A., Semeraro, G. (eds.) UMAP 2013. LNCS, vol. 7899, pp. 289–295. Springer, Heidelberg (2013). https://doi.org/10.1007/978-3-642-38844-6_25
33. Sener, O., Song, H.O., Saxena, A., Savarese, S.: Learning transferrable representations for unsupervised domain adaptation. In: Lee, D.D., Sugiyama, M., Luxburg, U.V., Guyon, I., Garnett, R. (eds.) Advances in Neural Information Processing Systems 29, pp. 2110–2118. Curran Associates, Inc. (2016). http://papers.nips.cc/paper/6360-learning-transferrable-representations-for-unsupervised-domain-adaptation.pdf
34. Srivastava, N., Hinton, G., Krizhevsky, A., Sutskever, I., Salakhutdinov, R.: Dropout: a simple way to prevent neural networks from overfitting. J. Mach. Learn. Res. 15, 1929–1958 (2014). http://jmlr.org/papers/v15/srivastava14a.html
35. Sun, M., Li, F., Zhang, J.: A multi-modality deep network for cold-start recommendation. Big Data Cogn. Comput. 2(1), 7 (2018). https://doi.org/10.3390/bdcc2010007. http://www.mdpi.com/2504-2289/2/1/7

36. Tang, J., Wu, S., Sun, J., Su, H.: Cross-domain collaboration recommendation. In: Proceedings of the 18th ACM SIGKDD International Conference on Knowledge Discovery and Data Mining, KDD 2012, pp. 1285–1293. ACM, New York (2012). https://doi.org/10.1145/2339530.2339730

37. Valizadegan, H., Jin, R., Zhang, R., Mao, J.: Learning to rank by optimizing NDCG measure. In: Bengio, Y., Schuurmans, D., Lafferty, J.D., Williams, C.K.I., Culotta, A. (eds.) Advances in Neural Information Processing Systems 22, pp. 1883–1891. Curran Associates, Inc. (2009). http://papers.nips.cc/paper/3758-learning-to-rank-by-optimizing-ndcg-measure.pdf

38. Wang, H., SHI, X., Yeung, D.Y.: Collaborative recurrent autoencoder: recommend while learning to fill in the blanks. In: Lee, D.D., Sugiyama, M., Luxburg, U.V., Guyon, I., Garnett, R. (eds.) Advances in Neural Information Processing Systems 29, pp. 415–423. Curran Associates, Inc. (2016)

39. Wang, H., Wang, N., Yeung, D.Y.: Collaborative deep learning for recommender systems. In: Proceedings of the 21th ACM SIGKDD International Conference on Knowledge Discovery and Data Mining, KDD 2015, pp. 1235–1244. ACM, New York (2015). https://doi.org/10.1145/2783258.2783273

40. Wu, C.Y., Ahmed, A., Beutel, A., Smola, A.J., Jing, H.: Recurrent recommender networks. In: Proceedings of the Tenth ACM International Conference on Web Search and Data Mining, WSDM 2017, pp. 495–503. ACM, New York (2017). https://doi.org/10.1145/3018661.3018689

41. Wu, Y., DuBois, C., Zheng, A.X., Ester, M.: Collaborative denoising auto-encoders for top-n recommender systems. In: Proceedings of the Ninth ACM International Conference on Web Search and Data Mining, WSDM 2016, pp. 153–162. ACM, New York (2016). https://doi.org/10.1145/2835776.2835837

42. Zhang, Q., Wu, D., Lu, J., Liu, F., Zhang, G.: A cross-domain recommender system with consistent information transfer. Decis. Support Syst. **104**, 49–63 (2017). https://doi.org/10.1016/j.dss.2017.10.002. http://www.sciencedirect.com/science/article/pii/S016792361730180X

43. Zhang, Y., Cao, B., Yeung, D.Y.: Multi-domain collaborative filtering. In: Proceedings of the Twenty-Sixth Conference on Uncertainty in Artificial Intelligence, UAI 2010, pp. 725–732. AUAI Press, Arlington (2010)

44. Zheng, Y., Tang, B., Ding, W., Zhou, H.: A neural autoregressive approach to collaborative filtering. In: Proceedings of the 33rd International Conference on International Conference on Machine Learning, ICML 2016, vol. 48, pp. 764–773. JMLR.org (2016)

It's only Words and Words Are All I Have

Manash Pratim Barman[1], Kavish Dahekar[2], Abhinav Anshuman[3],
and Amit Awekar[4(✉)]

[1] Indian Institute of Information Technology Guwahati, Guwahati, India
[2] SAP Labs, Bengaluru, India
[3] Dell India R&D Center, Bengaluru, India
[4] Indian Institute of Technology Guwahati, Guwahati, India
awekar@iitg.ac.in

Abstract. The central idea of this paper is to demonstrate the strength
of lyrics for music mining and natural language processing (NLP) tasks
using the distributed representation paradigm. For music mining, we
address two prediction tasks for songs: genre and popularity. Existing
works for both these problems have two major bottlenecks. First, they
represent lyrics using handcrafted features that require intricate knowledge of language and music. Second, they consider lyrics as a weak indicator of genre and popularity. We overcome both the bottlenecks by
representing lyrics using distributed representation. In our work, genre
identification is a multi-class classification task whereas popularity prediction is a binary classification task. We achieve an F1 score of around
0.6 for both the tasks using only lyrics. Distributed representation of
words is now heavily used for various NLP algorithms. We show that
lyrics can be used to improve the quality of this representation.

Keywords: Distributed representation · Music mining

1 Introduction

The dramatic growth in streaming music consumption in the past few years has
fueled the research in music mining [2]. More than 85% of online music sub-
scribers search for lyrics [1]. It indicates that lyrics are an important part of the
musical experience. This work is motivated by the observation that lyrics are
not yet used to their true potential for understanding music and language computationally. There are three main components to experiencing a song: visual
through video, auditory though music, and linguistic through lyrics. As compared to video and audio components, lyrics have two main advantages when
it comes to analyzing songs. First, the purpose of the song is mainly conveyed
through the lyrics. Second, lyrics as a text data require far fewer resources to
analyze computationally. In this paper, we focus on lyrics to demonstrate their
value for two broad domains: music mining and NLP.

A line from song Words by Bee Gees.

© Springer Nature Switzerland AG 2019
L. Azzopardi et al. (Eds.): ECIR 2019, LNCS 11438, pp. 30–36, 2019.
https://doi.org/10.1007/978-3-030-15719-7_4

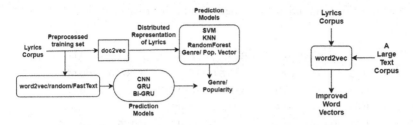

Fig. 1. Genre & popularity prediction   **Fig. 2.** Improving word vectors

A recent trend in NLP is to move away from handcrafted features in favor of distributed representation. Methods such as word2vec and doc2vec have achieved tremendous success for various NLP tasks in conjunction with Deep Learning [14]. Given a song, we focus on two prediction tasks: genre and popularity. We apply distributed representation learning methods to jointly learn the representation of lyrics as well as genre & popularity labels. Using these learned vectors, we experiment with various traditional supervised machine learning and Deep Learning models. We apply the same methodology for popularity prediction. Please refer to Figs. 1 and 2 for overview of our approach.

Our work has three research contributions. First, this is the first work that demonstrates the strength of distributed representation of lyrics for music mining and NLP tasks. Second, contrary to existing work, we show that lyrics alone can be good indicators of genre and popularity. Third, the quality of words vectors can be improved by capitalizing on knowledge encoded in lyrics.

2   Dataset

Lyrics are protected by copyright and cannot be shared directly. Most researchers in the past have used either small datasets that are manually curated or large datasets that represent lyrics as a bag of words [3–5,7,11–13]. Small datasets are not enough for training distributed representation. Bag of words representation lacks information about the order of words in lyrics. Such datasets cannot be used for training distributed representation. To get around this problem, we harvested lyrics from user-generated content on the Web. Our dataset contains around 400,000 songs in English. We had to do extensive preprocessing to remove text that is not part of lyrics. We also had to detect and remove duplicate lyrics. Metadata about lyrics that is genre and popularity was obtained from Fell and Sporleder [4]. However, for genre and popularity prediction, we were constrained to use only a subset of dataset due to class imbalance problem.

3   Genre Prediction

Our dataset contains songs from eight genres: Metal, Country, Religious, Rap, R&B, Reggae, Folk, and Blues. Our dataset had a severe problem of class imbalance with genres such as Rap dominating. Using complete dataset was resulting in prediction models that were highly biased towards the dominant classes.

Table 1. F1-scores for genre prediction. Highest value for each genre is in bold.

Model	Genre								
	Metal	Country	Religious	Rap	R& B	Reggae	Folk	Blues	Average
SVM	0.575	0.493	0.634	**0.815**	0.534	0.608	0.437	0.532	0.579
KNN	0.463	0.457	0.557	0.729	0.457	0.547	0.428	0.515	0.519
Random Forest	0.552	0.536	0.644	0.791	0.525	0.599	0.474	0.559	0.585
Genre Vector	**0.605**	**0.551**	0.641	0.738	**0.541**	**0.716**	**0.475**	**0.59**	**0.607**
CNN	0.543	0.466	**0.668**	0.801	0.504	0.628	0.471	0.563	0.580
GRU	0.479	0.467	0.558	0.745	0.462	0.601	0.355	0.531	0.525
Bi-GRU	0.494	0.471	0.567	0.752	0.492	0.609	0.372	0.488	0.531

Hence, we use undersampling technique to generate balanced training and test datasets. We repeated this method to generate ten different versions of training and test datasets. Each version of dataset had about 8000 songs with about 1000 songs for each genre. Lyrics of each genre were randomly split into two partitions: 80% for training and 20% for testing. Experimental results reported here are average across these ten datasets. We did not observe any significant variance in results across different instances of training and test datasets, indicating the robustness of the results.

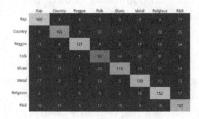

Fig. 3. Confusion matrix for genre. Rows: true label, Columns: predicted label.

Distributed representation of lyrics and genres were jointly learned using doc2vec model [6]. This model gave eight genre vectors (a vector representation for each genre) and vector representation for each song in the training and test dataset. We experimented with vector dimensionality and found 300 as the optimal dimensionality for our task. Using this vector representation, we experimented with both traditional machine learning models (SVM, KNN, Random Forest and Genre Vectors) and deep learning models (CNN, GRU, and Bidirectional GRU) for genre prediction task. Please refer to Table 1. For the KNN model, the genre of a test instance was determined based on genres of K nearest neighbors in the training dataset. Nearest neighbors were determined using cosine similarity. We tried three parameter values for K: 10, 25, and 50. However, there was no significant difference in results. For Genre Vector model, the genre of a test instance was determined based on the cosine similarity of test instance with vectors obtained for each genre. We can observe that Rap is the easiest genre to predict as rap songs have a distinctive vocabulary. The Folk genre is the most difficult to identify. For each genre, the worst performing model is the KNN, indicating that the local neighborhood of a test instance is not the best indicator of the genre. On average, Genre Vector model performs the best.

Please refer to Fig. 3. This figure represents confusion matrix for one version of training dataset using Genre Vector model. Each row of the matrix sums to around 200 as the number of instances in test dataset per genre were around 200. We can notice that confusion relationships are asymmetric. We say that a genre X is confused with genre Y if the genre prediction model identifies many songs of genre X as having genre Y. For example, observe the row corresponding to the Folk genre. It is mainly confused with the Religious genre as about 16% of Folk songs are identified as Religious. However for the Religious genre, Folk does not appear as one of the top confused genres. Similarly, genre Reggae is most confused with R&B. However, R&B is least confused with Reggae.

4   Popularity Prediction

Only a subset of songs had user ratings data available with ratings ranging from 1 to 5 [4]. For two genres: Folk and Blues, we did not get popularity data for enough number of songs. For popularity prediction task, the number of genres was thus reduced to six. Number of songs per genre are: Metal (15254), Country (2640), Religious (3296), Rap (19774), R&B (6144), and Reggae (294). Songs of each genre were randomly partitioned into two disjoint sets: 80% for training and 20% for testing. To ensure robustness of results, we performed experiments on ten such versions of the dataset. Experimental results reported here are average across ten runs. We model popularity prediction as a binary classification problem. For each genre, we divided songs into two categories: low popularity (rated 1, 2, or 3) and high popularity(rated 4 or 5). The number of songs included in each class were balanced to avoid any over fitting of model.

Considering the distinctive nature of each genre, we built a separate model per genre for popularity prediction. For each genre using the doc2vec model, we generated two popularity vectors (one each for low and high popularity) and vector representation for each song in training and testing dataset. Similar to the genre prediction task, we experimented with seven prediction models. Please refer to Table 2. We can observe that Deep Learning based models perform better than other models. However, for every genre, the gap between the best and worst model has narrowed down as compared to the genre prediction task.

5   Improving Word Vectors with Lyrics

A large text corpus such as Wikipedia is necessary to train distributed representation of words. Lyrics are a poetic creation that requires significant creativity.

Table 2. F1-scores for popularity prediction. Highest value for each genre is in bold.

Model	Genre					
	Country	Metal	Rap	Reggae	Religious	R& B
SVM	0.6238	0.6756	0.7301	0.7539	0.5681	0.6342
KNN	0.5871	0.6351	0.7071	0.7713	0.5387	0.5814
Random Forest	0.6201	0.6683	0.7176	0.7647	0.5635	0.646
Popularity Vector	0.6180	**0.6776**	0.7663	0.7820	0.5886	0.6401
CNN	**0.632**	0.6717	0.7652	**0.8011**	**0.5933**	**0.6661**
GRU	0.5801	0.6479	0.7505	0.5187	0.5661	0.6434
Bi-GRU	0.6037	0.6581	**0.7684**	0.6613	0.5517	0.5886

Table 3. Results of word analogy tasks. Highest value for each task is in bold.

Tasks	Lyrics	Wikipedia	Sampled Wiki	Lyrics+Wiki
(1) capital-common-countries	10.95	87.75	**89.43**	87.94
(2) capital-world	07.26	**90.35**	79.25	90.00
(3) currency	02.94	**05.56**	01.85	**05.56**
(4) city-in-state	07.87	66.55	61.71	**66.73**
(5) family	81.05	**94.74**	82.82	94.15
(6) gram1-adjective-to-adverb	08.86	35.71	25.53	**36.51**
(7) gram2-opposite	19.88	**51.47**	33.46	51.10
(8) gram3-comparative	83.56	**91.18**	80.66	90.00
(9) gram4-superlative	53.33	75.72	59.49	**77.83**
(10) gram5-present-participle	74.32	73.33	60.82	**75.81**
(11) gram6-nationality-adjective	06.29	97.01	92.60	**97.08**
(12) gram7-past-tense	54.44	68.07	65.47	**69.00**
(13) gram8-plural	73.19	87.60	76.81	**89.52**
(14) gram9-plural-verbs	60.00	71.85	63.32	**72.62**
Overall across all tasks	50.33	75.71	66.6	**78.11**

Knowledge encoded in them can be utilized by training distributed representation of words. For this task, we used our entire dataset of 400K songs. Using the word2vec model, we generated four sets of word vectors. The four training datasets were: Lyrics only ($D1$, 470 MB), Complete Wikipedia ($D2$, 13 GB), Sampled Wikipedia ($D3$, 470 MB), and Lyrics combined with Wikipedia ($D4$, 13.47 GB). For dataset $D3$, we randomly sampled pages from Wikipedia till we collected dataset of a size comparable to our Lyrics dataset. For dataset $D3$, we created ten such sampled versions of Wikipedia. Results given here for $D3$ are average across ten such datasets.

To compare these four sets of word embeddings, we used 14 tasks of word analogy tests proposed by Mikolov [8]. Please refer to Table 3. Each cell in the table represents accuracy (in percentage) of a particular word vector set for a particular word analogy task. First five tasks in the table consist of finding a related pair of words. These can be grouped as semantic tests. Next nine tasks (6 to 14) check syntactic properties of word vectors using various grammar related tests. These can be grouped as syntactic tests.

By sheer size, we expect $D2$ to beat our dataset $D1$. However, we can observe that for tasks 5, 8, 12, 13, and 14 $D1$ gives results comparable to $D2$. For task 10, $D1$ is able to beat $D2$ despite the significant size difference. Datasets $D3$ and $D1$ are comparable in size. For task 10, $D1$ significantly outperforms $D3$. For all other tasks, the performance gap between $D3$ and $D1$ is reduced noticeably. We can observe that $D1$ performs better on syntactic tests than semantic tests. However, the main takeaway from this experiment is that dataset $D4$ performs the best for a majority of the tasks. Also, $D4$ is the best performing dataset overall. These results indicate that lyrics can be used in conjunction with large text corpus to further improve distributed representation of words.

6 Related Work

Existing works that have used lyrics for genre and popularity prediction can be partitioned into two categories. First, that use lyrics in augmentation with acoustic features of the song [5,7] and second, that do not use acoustic features [3,4,10,13]. However, all of them represent lyrics using either handcrafted features or bag-of-words models. Identifying features manually requires intricate knowledge of music, and such features vary with the underlying dataset. Mikolov and Le have shown that distributed representation of words and documents is superior to bag-of-words models [6,9]. To the best of our knowledge, this is the first work that capitalizes on such representation of lyrics for genre and popularity prediction. However, our results cannot be directly compared with existing works as datasets, set of genres, the definition of popularity, and distribution of target classes are not identical. Still, our results stand in contrast with existing works that have concluded that lyrics alone are a weak indicator of genre and popularity. These works report significantly low performance of lyrics for genre prediction task. For example, Rauber et al. report an accuracy of 34% [10], Doraisamy et al. report an accuracy of 40% [13], McKay et al. report an accuracy of 43% [7], and Hu et al. reported accuracy of abysmal 19% [5]. The accuracy of our method is around 63%.

7 Conclusion and Future Work

This work has demonstrated that using distributed representation; lyrics can serve as a good indicator of genre and popularity. Lyrics can also be useful to improve distributed representation of words. Deep Learning based models can deliver better results if larger training datasets are available. Our method can be easily integrated with recent music mining algorithms that use an ensemble of lyrical, audio, and social features.

References

1. Lyrics take centre stage in streaming music, a midia research white paper (2017). https://www.nielsen.com/us/en/insights/reports/2018/2017-music-us-year-end-report.html
2. Nielsen 2017 U.S. music year-end report. https://www.midiaresearch.com/app/uploads/2018/01/Lyrics-Take-Centre-Stage-In-Streaming-%E2%80%93-LyricFind-Report.pdf
3. Logan, P.M.B., Kositsky, A.: Semantic analysis of song lyrics. In: IEEE International Conference on Multimedia and Expo (ICME), pp. 159–168, June 2004
4. Fell, M., Sporleder, C.: Lyrics-based analysis and classification of music. In: COLING (2014)
5. Hu, Y., Ogihara, M.: Genre classification for million song dataset using confidence-based classifiers combination. In: ACM SIGIR International Conference on Research and Development in Information Retrieval, pp. 1083–1084 (2012)
6. Le, Q., Mikolov, T.: Distributed representations of sentences and documents. In: International Conference on Machine Learning, pp. 1188–1196 (2014)
7. McKay, C., Burgoyne, J.A., Hockman, J., Smith, J.B., Vigliensoni, G., Fujinaga, I.: Evaluating the genre classification performance of lyrical features relative to audio, symbolic and cultural features. In: ISMIR (2010)
8. Mikolov, T., Chen, K., Corrado, G., Dean, J.: Efficient estimation of word representations in vector space. CoRR, abs/1301.3781 (2013)
9. Mikolov, T., Sutskever, I., Chen, K., Corrado, G.S., Dean, J.: Distributed representations of words and phrases and their compositionality. In: Advances in Neural Information Processing Systems, pp. 3111–3119 (2013)
10. Rauber, A., Mayer, R., Neumayer, R.: Rhyme and style features for musical genre classification by song lyrics. In: International Conference on Music Information Retrieval (ISMIR), pp. 337–342, June 2008
11. Rauber, A., Mayer, R., Neumayer, R.: Combination of audio and lyrics features for genre classification in digital audio collections. In: Proceedings of the 16th ACM International Conference on Multimedia, pp. 159–168, October 2008
12. Doraisamy, S., Ying, T.C., Abdullah, L.N.: Genre and mood classification using lyric features. In: 2012 International Conference on Information Retrieval and Knowledge (CAMP). IEEE, March 2012
13. Ying, T.C., Doraisamy, S., Abdullah, L.N.: Genre and mood classification using lyric features. In: International Conference on Information Retrieval Knowledge Management, pp. 260–263 (2012)
14. Young, T., Hazarika, D., Poria, S., Cambria, E.: Recent trends in deep learning based natural language processing [review article]. IEEE Comput. Int. Mag. **13**(3), 55–75 (2018)

Modeling User Return Time Using Inhomogeneous Poisson Process

Mohammad Akbari[1,2(✉)], Alberto Cetoli[2], Stefano Bragaglia[2],
Andrew D. O'Harney[2], Marc Sloan[2], and Jun Wang[1]

[1] University College London, London, UK
{m.akbari,jun.wang}@ucl.ac.uk
[2] Context Scout, London, UK
{alberto,stefano,andy,marc}@contextscout.com

Abstract. For Intelligent Assistants (IA), user activity is often used as
a lag metric for user satisfaction or engagement. Conversely, predictive
leading metrics for engagement can be helpful with decision making and
evaluating changes in satisfaction caused by new features. In this paper,
we propose User Return Time (URT), a fine grain metric for gauging user
engagement. To compute URT, we model continuous inter-arrival times
between users' use of service via a log Gaussian Cox process (LGCP),
a form of inhomogeneous Poisson process which captures the irregular
variations in user usage rate and personal preferences typical of an IA.
We show the effectiveness of the proposed approaches on predicting the
return time of users on real-world data collected from an IA. Experimental results demonstrate that our model is able to predict user return
times reasonably well and considerably better than strong baselines that
make the prediction based on past utterance frequency.

Keywords: User Return Time Prediction · Intelligent Assistant

1 Introduction

Intelligent Assistants (IAs) are software agents that interact with users to complete a specific task. The success of an IA is directly linked to long-term user
engagement which can be measured by observing when users return to reuse the
IA. Indeed, the repeating usage of service is a typical characteristic for IAs, thus,
monitoring a user's usage pattern is a necessary measure of engagement.

Predicting when usage will next occur is therefore a useful indicator of
whether a user will continue to be engaged and can serve as a reference to compare against when introducing new features, as well as a method for managing
churn for business purposes. For example, predicted return times of customers
can be utilized for clustering customers according to their activity and narrow
their interest to investigate a specific groups of users with a short or long inter-
arrival time for target marketing [3,13]. Furthermore, it helps the service to

© Springer Nature Switzerland AG 2019
L. Azzopardi et al. (Eds.): ECIR 2019, LNCS 11438, pp. 37–44, 2019.
https://doi.org/10.1007/978-3-030-15719-7_5

prepare content for the customers in advance to better serve them in engagement utterance. For example, a target marketing and advertising program can be planned for the next engagement of the user.

Modeling the inter-arrival time of user engagement events is a challenging task due to the complex temporal patterns exhibited. Users typically engage with IA systems in an ad hoc fashion, starting tasks at different time points with irregular frequency. For example, Fig. 1 shows the inter-arrival times (denoted by black crosses) of two users in one week. Notice the existence of regions of both high and low density of inter-arrival times over a one week interval. Users do not arrive in evenly spaced intervals but instead they usually arrive in times that are clustered due to completing several tasks in bursts. This may be attributed to a user's personal preferences and their tasks' priorities. Retrospective studies in modeling inter-arrival times between events treat events as independent and exponentially distributed over time with constant rate [2]. They hence fail to perform accurate predictions when there exists time-varying patterns between events [15]. Further, they mostly focus on web search queries [1].

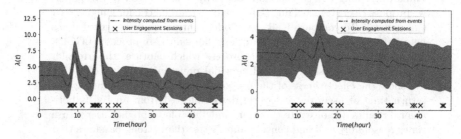

Fig. 1. Intensity functions (dotted lines) and corresponding predicted user inter-arrival times for different users (black crosses). Light regions depict the uncertainty of estimations.

To this purpose, we define User Return Time (URT) as the predicted inter-arrival time until next user activity, allowing us to predict user engagement to aid in creating an optimal IA system. To do so, we propose to model inter-arrival times between a user usage sessions with a doubly stochastic process. More specifically, we leverage the log-Gaussian Cox process (LGCP), an inhomogeneous Poisson process (IPP), to model the inter-arrival times between events. LGCP models return times that are generated by an intensity function which varies across time. It assumes a non-parametric form for the intensity function allowing the model complexity to depend on the data set. We evaluate the proposed model using a real-world dataset from an IA, and demonstrate that it provides good predictions for inter-arrival return times, improving upon the baselines. Even though the main application is return time estimation for an IA, one could apply the proposed approach to other events in e-commerce systems such as the arrival times of requests, return time of customers, etc.

The contribution of this paper can be summarized as,

- We propose User Return Time as a measure to predict user engagement with an IA.
- We leverage a doubly-stochastic process, i.e. Log-Gaussian Cox processes, to predict URT for an IA, and show that it effectively captures the time-varying patterns of usage between events.
- We verified the effectiveness of the proposed method on a real-world dataset and found that Gaussian and Periodic kernels can make accurate estimations.

2 Problem Statement

Let $\mathcal{U} = \{u_1, u_2, \ldots, u_n\}$ denote a set of n different users who used the IA to assist with different tasks. Let $\mathcal{H}_i = \{t_{i,j}\}_{j=1}^{n_i}$ denote the task history of the i-th user, where $t_{i,j}$ represents the start time of the j-th task performed by the i-th user and n_i is the number of tasks performed by user i. Our aim is to estimate a user's next start time from their past interactions with the IA, i.e., predict $t_{i,k+1}$ based on the start times of the previous sessions, i.e. $\{t_{i,j}\}_{j=1}^{k}$.

Based on the above discussion, we formally define the problem of **Predicting User Return Time** as: *Given a user history of session start times $\mathcal{H}_i$, predict the next times that the user will use the IA.*

3 Model

Poisson processes have been widely adopted for estimating the cross-interval times between different events such as failure of devices, social media events and purchase in e-commerce sites [12]. The Homogeneous Poisson process (HPP) is a class of point processes that assumes the events are generating with a constant intensity rate, i.e. λ, (with respect to the time and the product features). However, in an IA scenario, user engagement incidence often occurs with a varying-rate over time where there are several spikes of usage in a short period and a long-period of absence (when the user performs other tasks). Thus, we exploit the Inhomogeneous Poisson process (IPP) [7] that can model events happening at a variable rate by considering the intensity to be a function of time, i.e. $\lambda(t)$. For example, Fig. 1 shows intensity functions learned from two different IPP models. Notice how the generated inter-event times vary according to the intensity function values.

To model inter-arrival times, we employed a log-Gaussian Cox process which models the intensity function of point processes as a stochastic function [8]. LGCP learns the intensity function, $\lambda(t)$, non-parametrically via a latent function sampled from Gaussian processes [10]. Here, to impose non-negativity to the intensity function (as an interval cannot be negative), we assume an exponential form for the intensity function, i.e., $\lambda(t) = \exp(f(t))$. We adopted a non-parametric approach to model the intensity function which utilizes Bayesian inference to train a model, where the complexity of the model is learned from the training data available. In the next section, we explain the details of the proposed model and how to learn model parameters from training data.

3.1　Modeling Inter-arrival Time

An inhomogeneous Poisson process (unlike HPP) uses a time varying intensity function and hence, the distribution of inter-arrival times is not independent and identically distributed [11]. In IPP, the number of tasks y (signals of a returning user to the IA) occurring in an interval $[p, q]$ is Poisson distributed with rate $\int_p^q \lambda(t)dt$,

$$p(y|\lambda(t), [p, q]) = \text{Poisson}\left(y\Big| \int_p^q \lambda(t)dt\right) = \frac{\left(\int_p^q \lambda(t)dt\right)^y \exp\left(-\int_p^q \lambda(t)dt\right)}{y!} \quad (1)$$

Assume that the k-th event occurred at time $t_k = p$ and we are interested in the inter-arrival time $\delta_k = t_{k+1} - t_k$ of the next event. The arrival time of next event t_{k+1} can be obtained as $t_{k+1} = t_k + \delta_k$. The cumulative distribution for δ_k, which provides the probability that an event occurs by time $p + q$ can be obtained as,

$$p(\delta_k \leq q) = 1 - p(\delta_k > q|\lambda(t), t_k = p)$$
$$= 1 - \exp\left(-\int_p^{p+q} \lambda(t)dt\right) = 1 - \exp\left(-\int_0^q \lambda(p+t)dt\right). \quad (2)$$

The derivation is obtained by considering a Poisson probability for zero counts with rate parameter given by $\int_p^{p+q} \lambda(t)dt$ and applying integration by substitution to obtain Eq. (2). The probability density function of the random variable t_n is obtained by taking the derivative of Eq. (2) with respect to q,

$$p(\delta_k = q) = \lambda(p+q)\exp\left(-\int_0^q \lambda(p+t)dt\right). \quad (3)$$

We associate a distinct intensity function $\lambda_i(t) = \exp(f_i(t))$ to each user u_i as they have different temporal preferences. The latent function $f_i(t)$ is modeled to come from a zero mean Gaussian process (GP) [10] prior. The Squared Exponential (SE) kernel is a common choice for GPs and it is defined as,

$$k(t_i, t_j) = \sigma \exp\left(-\frac{(t_i - t_j)^2}{l}\right). \quad (4)$$

Equation (4) imposes smoothness over time on the intensity function. We also experiment with periodic kernels which allow the modelling of functions that repeat themselves exactly. Periodic kernels can model complex periodic structure relating to the working week by finding a proper periodicity hyperparameter in the kernel. The periodic kernel is defined as,

$$k_{periodic}(t_i, t_j) = \sigma^2 \exp\left(-\frac{2\sin^2 \pi|t_i - t_j|/r}{l^2}\right), \quad (5)$$

where σ and l are the output variance and length-scale, respectively, and r is the periodicity hyperparameter.

3.2   Inference

We learn the model parameters by maximizing the marginal likelihood over all users in the dataset. The likelihood of the return times over the dataset is then obtained by taking the product of return times over all users.

4   Experiments

4.1   Dataset

We conducted an empirical study in which we used the access log of a commercial web IA to construct our dataset. The IA has been developed as an extension for the Chrome Web browser that assists recruiters to automatically find and segregate job candidate information such as skills and contact details, based on the information found on popular social networking platforms such as LinkedIn.

For the purpose of this experiment the tool was altered to log the interactions of users with the tool when they performed their tasks. Various interactions were logged with a time stamp, based on which we constructed our dataset. Table 1 depicts a sample log record for a specific user from our dataset. Inspired by [4,5], we split action sequences into sessions based on a time gap of 15 min which is commonly used in information retrieval and web search to identify sessions. We collected all logs of a random set of users within a period of 3 months, from the beginning of April 2018 until the end of June 2018. The dataset consists of 2, 999, 593 interaction events committed by 133 distinct users.

Table 1. Example log of a user's interaction sequence. The first three interactions occurred within a single task. The final interaction indicated the start of a new task.

User	Action	Time stamp
484	New web page	2018-07-01 17:35:25
484	Opened IA	2018-07-01 17:35:27
484	Clicked Contact Button	2018-07-01 17:35:51
484	New web page	2018-07-01 18:05:25

4.2   Baselines and Evaluation Metrics

Here the proposed model is compared against several methods to evaluate their effectiveness in predicting URT. We discuss the advantages, assumptions and limitations of each and provide empirical results on a real-world dataset. We examine the following distinct models: (1) **Linear Regression**: a linear regression model which is trained on a historical window of URT. We used the last 20 (computed empirically) inter-arrival times as features. (2) **HPP**: we also used a homogeneous Poisson process (HPP) [6] which models an exponentially distributed inter-arrival times with a fix rate λ. The rate parameter was learned based on the maximum likelihood approach. (3) **HP**: we compare against the Hawkes Process (HP) [14], a self exciting point process where an occurrence of an event increases the probability of the event arriving soon afterwards. We consider a univariate Hawkes process where the intensity function is modeled as $\lambda(t) = \mu + \sum_{t_i < t} k(t_i, t)$. We apply Ogata's thinning algorithm for generating arrival times using Hawkes process [9]. (4) **GP**: we also exploited GP as a time-series baseline, where the inter-arrival time is modeled as a function of the time of occurrence of last action. We examine two commonly used kernels, SE and periodic, where kernel parameters are learned by maximizing the likelihood. (5) **RNN**: the final baseline was selected from deep learning approaches, where we used an LSTM model with two Bi-LSTM layers and input length of 20 for predicting URT (we used the same window size as with linear regression).

In order to evaluate the model, we use mean absolute error (MAE) and root mean square error (RMSE) between the actual and the predicted times for each user in hours. Since the data varies in size for each user, we take the micro average of the errors to obtain the final result.

4.3   Results

Table 2 compares the predictive performance of LGCP against various baselines. We find that the standard kernel used in GP models, the SE kernel, performs poorly as expected due to the complex temporal patterns exhibited by users in their session start times. The SE kernel typically models smoothly varying functions and is not suitable to model this situation. The periodic kernel could model the periodicity in the data (for instance, users tend to be more active on weekends) and are found to perform better than SE in both GP and LGCP. The LGCP models with periodic kernel outperforms the baseline approaches such as HPP, linear regression, GP regression. RNN model can outperform all baselines except LGCP which shows that deep models can capture complex user behavior in using IA.

Table 2. Comparison of different approaches in terms of MAE and RMSE.

Method	Kernel	MAE	RMSE
Linear regression		53.27($\pm$48.91)	86.10($\pm$79.13)
HPP		43.44($\pm$51.13)	63.70($\pm$53.60)
HP		28.12($\pm$28.47)	61.24($\pm$51.22)
RNN		19.11($\pm$28.01)	45.52($\pm$34.20)
GPR	SE	35.52($\pm$40.79)	42.51($\pm$54.22)
	Periodic	19.82($\pm$21.04)	38.30($\pm$48.71)
LGCP	SE	26.52($\pm$29.79)	54.12($\pm$58.37)
	Periodic	15.52($\pm$18.79)	32.54($\pm$38.15)

5 Conclusions

In this paper we proposed to use User Return Time as a predictive measure of engagement with an IA, for which the log-Gaussian Cox process was proposed as an appropriate prediction model. Through our experiments, we demonstrated that this model does indeed offer better predictive performance due to its ability to capture the complex temporal behaviour typical of IA users.

This approach can be generalized to model problems other than URT prediction for an IA, e.g. purchase time prediction, advertisement campaigns, and disaster management. The effectiveness of using an RNN in this research also shows that it is worth investigating the potential of deep sequence models for prediction in these scenarios. In future, we plan to examine user and context features to improve prediction performance.

References

1. Chakraborty, S., Radlinski, F., Shokouhi, M., Baecke, P.: On correlation of absence time and search effectiveness. In: Proceedings of the International ACM SIGIR Conference on Research & Development in Information Retrieval (2014)
2. Doerr, C., Blenn, N., Van Mieghem, P.: Lognormal infection times of online information spread. PloS One **8**, e64349 (2013)
3. Du, N., Wang, Y., He, N., Sun, J., Song, L.: Time-sensitive recommendation from recurrent user activities. In: Advances in Neural Information Processing Systems (2015)
4. Halfaker, A., et al.: User session identification based on strong regularities in inter-activity time. In: Proceedings of the International Conference on World Wide Web (2015)
5. He, D., Göker, A.: Detecting session boundaries from web user logs. In: Proceedings of the BCS-IRSG 22nd Annual Colloquium on Information Retrieval Research (2000)
6. Hosseini, S.A., et al.: Recurrent Poisson factorization for temporal recommendation. In: Proceedings of the 23rd ACM SIGKDD International Conference on Knowledge Discovery and Data Mining (2017)
7. Lee, S., Wilson, J.R., Crawford, M.M.: Modeling and simulation of a nonhomogeneous Poisson process having cyclic behavior. Commun. Stat.-Simul. Comput. **20**, 777–809 (1991)
8. Møller, J., Syversveen, A.R., Waagepetersen, R.P.: Log Gaussian cox processes. Scandinavian J. Stat. **25**, 451–482 (1998)
9. Ogata, Y.: On Lewis' simulation method for point processes. IEEE Trans. Inf. Theory **27**, 23–31 (1981)
10. Rasmussen, C.E.: Gaussian processes in machine learning. In: Bousquet, O., von Luxburg, U., Rätsch, G. (eds.) ML -2003. LNCS (LNAI), vol. 3176, pp. 63–71. Springer, Heidelberg (2004). https://doi.org/10.1007/978-3-540-28650-9_4
11. Ross, S.M.: Introduction to probability models (2014)
12. Takeshi Sakaki, M.O., Matsuo, Y.: Earthquake shakes Twitter users: real-time event detection by social sensors (2010)

13. Yang, C., Shi, X., Jie, L., Han, J.: I know you'll be back: interpretable new user clustering and churn prediction on a mobile social application. In: Proceedings of the 24th ACM SIGKDD International Conference on Knowledge Discovery & Data Mining (2018)
14. Yang, S.H., Zha, H.: Mixture of mutually exciting processes for viral diffusion. In: International Conference on Machine Learning (2013)
15. Zhang, R., Walder, C., Rizoiu, M.A., Xie, L.: Efficient non-parametric Bayesian Hawkes processes. In: Proceedings of the International Conference on World Wide Web (2018)

Inductive Transfer Learning for Detection of Well-Formed Natural Language Search Queries

Bakhtiyar Syed[1]([✉]), Vijayasaradhi Indurthi[1], Manish Gupta[1,2], Manish Shrivastava[1], and Vasudeva Varma[1]

[1] IIIT Hyderabad, Hyderabad, India
{syed.b,vijaya.saradhi}@research.iiit.ac.in,
{manish.gupta,m.shrivastava,vv}@iiit.ac.in
[2] Microsoft, Hyderabad, India
gmanish@microsoft.com

Abstract. Users have been trained to type keyword queries on search engines. However, recently there has been a significant rise in the number of verbose queries. Often times such queries are not well-formed. The lack of well-formedness in the query might adversely impact the downstream pipeline which processes these queries. A well-formed natural language question as a search query aids heavily in reducing errors in downstream tasks and further helps in improved query understanding. In this paper, we employ an inductive transfer learning technique by fine-tuning a pretrained language model to identify whether a search query is a well-formed natural language question or not. We show that our model trained on a recently released benchmark dataset spanning 25,100 queries gives an accuracy of 75.03% thereby improving by ~5 absolute percentage points over the state-of-the-art.

1 Introduction

Traditionally users have been trained to put up keyword queries on search engines mainly because search has been traditionally driven by "unigram match". However, recently, with the increasing popularity of voice based search, verbose queries have become quite popular [9]. Also, in-vogue deep learning algorithms have enabled search engines to process such verbose natural language (NL) queries effectively. But not all verbose queries from users exhibit proper structure. Such queries often lack a coherent structure and may sometimes violate grammar rules, thus mandating tailor-made processing [2,4,11,14]. This makes it challenging for NL Processing (NLP) tools trained on formal text to extract the relevant information required to understand the user's intention behind the query [1].

Identifying whether a search query is well-formed [8] is an important task which also aids in various downstream tasks like understanding the user's intent

---

B. Syed and V. Indurthi—Authors contributed equally.

© Springer Nature Switzerland AG 2019
L. Azzopardi et al. (Eds.): ECIR 2019, LNCS 11438, pp. 45–52, 2019.
https://doi.org/10.1007/978-3-030-15719-7_6

(in case of personal assistants and chatbots [15, 18]) and generating better related query suggestions in search engines.

A possible approach for improving the accuracy of downstream processing while still using malformed queries is to train models using labeled data with malformed queries. But this approach has two main drawbacks. First, getting annotations to generate such training data which captures all possible malformed variants can be quite expensive. Second, since there is frequent change in the nature and domain of these queries [3, 12, 17], any model which is trained on these queries will drift fairly quickly. Another possible approach is to use grammars for identification of query well-formedness. Ideally, grammars such as the grammar on English resource [5] should be able to identify whether a query is a well-formed question or not easily. But in practice, such grammars are very precise and are not able to accurately parse more than half of web search queries.

The idea of using inductive transfer learning in natural language processing, akin to allied areas in computer vision like object detection, image segmentation, etc. has been garnering attention. Most of the current research focused on training deep learning models from scratch require huge volumes of training data and are also computationally expensive. Recent advancements in training and fine-tuning language models (LMs) are being used for a variety of NLP applications and have shown significant promise, primarily in text classification tasks [10]. In this work, we show that inductive transfer learning is greatly beneficial in identifying well-formed natural language questions. We also perform ablation studies to show the effectiveness of each of the modules used in the inductive transfer learning technique. Our experiments show an accuracy of 75.03% on the benchmark dataset for the task improving by ~5 absolute percentage points over the state-of-the-art method [8].

2   Related Work

Faruqui and Das [8] introduce the task and provide a coherent understanding of why many of the current techniques are not suitable for detecting well-formed questions. They combine various NLP features like word, character and Part-of-Speech (POS) n-grams with a simple neural network for the task. We show improvements over their method in Sect. 5. Attempts to identify well-formed questions by parsing them using grammar like the English resource grammar [5] are not very effective as the grammar is highly precise and fails to parse significant fraction of the web queries. Two other related tasks are Grammatical Error Prediction (GEP) and Correction (GEC) [18, 19]. While GEP is simply the task of classifying whether a given sentence is grammatical, GEC is a more complex task which involves identifying parts of ungrammatical text and correcting the same to produce grammatically correct text. Although our work is similar to GEP, past research has explored GEP for fully formed sentences which may not have a search intent. Thus, unlike GEP/GEC, we focus on well-formedness check for web search queries which contain a specific user intent.

3   Proposed Approach: Inductive Transfer Learning (ITL)

In this section, we first define the problem formally and then discuss three important phases of our inductive transfer learning approach in detail.

The architecture diagram of the proposed approach is illustrated in Fig. 1.

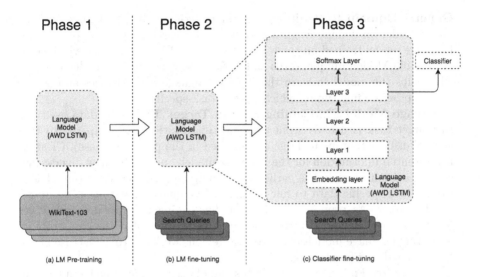

Fig. 1. Inductive Transfer Learning mechanism to identify well-formed search queries

Query Well-formedness Detection Problem: Given a query q, we intend to learn the label C which describes whether the query is a well-formed natural language question or not. We model this task as a binary classification task: $C = 1$ indicates that the query is well-formed while $C = 0$ indicates non-well-formedness.

The ULMFiT Architecture: Previous attempts to use inductive transfer through language modeling have resulted in limited success for NLP tasks [6,16]. However, Howard and Ruder [10] showed that if language models (LMs) are fine-tuned correctly, they would not overfit to small datasets and would enable robust inductive transfer learning. The neural architecture is called the Universal Language Model Fine Tuning architecture. They also proposed novel techniques which prevent catastrophic forgetting during training of the language model. We adapt the ULMFiT model for our inductive transfer learning approach and show that inductive transfer learning is greatly beneficial for identifying well-formed natural language search queries.

The AWD-LSTM model: Our inductive transfer learning mechanism utilizes the state-of-the-art Averaged-SGD Weight-Dropped Long Short Term Memory (AWD-LSTM) networks [13]. It is a variant of the simple LSTM with no short-cut connections, no attention or any other advanced mechanisms, with the same hyperparameters as in typical LSTMs, and no additions other than tuned drop-connect hyperparameters. We use AWD-LSTMs since they have been shown to be effective in learning lower-perplexity language models.

Three Stages of the proposed ITL Framework: The proposed ITL framework for query well-formedness check involves these three important phases:

1. **General Domain Pretraining**: The first phase involves pretraining a language model on a huge English corpus. In our case, we use the pretrained language model trained on the released Wikitext-103 [13] dataset which consists of 103 Million unique words and 28,595 preprocessed Wikipedia articles. This helps the model to learn the general language dependencies and is the first step before fine-tuning which targets task-specific data.

2. **Language Model Fine-tuning for the Target Task**: The data used for the target task is usually from a specific distribution (as compared to the general distribution in the large corpus used in the previous phase). Clearly, it is essential information for the language model – no matter how diverse the general domain data in the earlier pretraining step is. Hence, in this phase, we use task-specific data to fine-tuning our language model in an unsupervised manner. As proposed in [10], our fine-tuning involves discriminative fine-tuning and slanted triangular learning rates to combat the catastrophic forgetting language models exhibited in previous works [6,16] which used language models for fine-tuning.

   Discriminative Fine-tuning (DFT): Instead of keeping the same learning rate for all the layers of the AWD-LSTM, a different learning rate is used for tuning the three different layers. The intuition behind this is that since each of the layers represent a different kind of information [20], they must be fine-tuned to different extents.

   Slanted Triangular Learning Rates (STLR): Using the same learning rate is not the best way to enable the model to converge to a suitable region of the parameter space. Thus we adapt the slanted triangular learning rate [10] which first increases the learning rate and then linearly decays it as the number of training samples increases.

3. **Classifier Fine-tuning for the Target Task**: The weights that we obtain from the second phase are fine-tuned by keeping the same upstream architecture, but also appending 2 fully connected layers for the final classification with the last layer predicting the well-formedness rating. In this phase, we adapt the *gradual unfreezing heuristic* [10] for our task.

   Gradual Unfreezing (GU): All layers are not fine-tuned at the same time, instead the model is gradually unfrozen starting from the last layer, as it contains the least general knowledge [20]. The last layer is first unfrozen and fine-tuned for one epoch. Subsequently, the next frozen layer is unfrozen and all unfrozen layers are fine-tuned. This is repeated until all layers are fine-tuned until convergence is reached.

4   Dataset

For our experiments, we use the recently released benchmark dataset for well-formed natural language questions [8]. It contains a total of 25,100 questions, each labeled with a rating (between 0 and 1) of the query being well-formed. The authors collected these questions by utilizing questions asked by users on WikiAnswers[1], originally published as the Parallex corpus [7]. The compiled dataset primarily contains well-formed questions as queries along with typical constructs of search queries.

A query is annotated as well-formed if the supplied query is *grammatical* in nature, has *perfect spellings* and is an *explicit question*. For each search query, the average of the five scores (over each of the annotator's ratings) is calculated and then documented as the final rating R which indicates the degree of its well-formedness. As suggested in [8], the query is considered well-formed if $R \geq 0.8$ for the query[2]. In our experiments, we make use of the standard train-dev-test split supplied by Faruqui and Das [8], which consists of 17500 training, 3750 development and 3850 test queries. A few queries with their corresponding labels from the dataset are shown in Table 1.

Table 1. Examples from the benchmark well-formedness dataset from [8]

Example	Well-formedness rating
Which form of government is still in place in greece?	1.0
One of Mussolini's goals?	0.0
How many leagues in a mile in the mexican term?	0.4
What is the scotlands longest river?	0.2
How do you get rid of browsing history?	0.8

5   Experiments

Baselines: We compare our proposed method with the following baselines.

- Majority Class Prediction: Classify all queries into the majority class in the test set.
- Question Word Classifier: If the query starts with an interrogative word, classify it as being well-formed.
- Word Bi-Directional LSTM (BiLSTM) Classifier: Use a Bi-LSTM for classification which takes as input a one-hot vector of the input words of the query, and classifies using a softmax for binary classification.

---

[1] http://www.answers.com/Q/.
[2] A rating greater than or equal to 0.8 ensures at least 4 out of 5 annotators marked the query as well-formed.

– Faruqui and Das [8] propose a 2 hidden layer neural architecture with rectified linear unit (ReLU) activations and a final softmax layer for the predictions. For the input, the authors extract word, character and Part-of-Speech (POS) *n-grams*: word-1,2; char-3,4 grams as the lexical features and POS-1,2,3 grams for the syntactic features to form the n-gram embeddings via concatenation.

Table 2. Comparison of various classifiers and ablation study for the ITL Model

Model	Accuracy (%)
Question Word Classifier	54.9
Majority Class Prediction	61.5
Word BiLSTM Classifier	65.8
word-1,2 char-3,4 grams [8]	66.9
word-1,2 POS-1,2,3 grams [8]	70.7
word-1,2 char-3,4 POS-1,2,3 grams [8]	70.2
(Inductive Transfer Learning)	
No pretraining with WikiText-103	68.2
No LM fine-tuning	72.8
Fine-tuning without DFT and STLR	73.0
No gradual unfreezing	72.4
All (Pre-train + Fine-tune with DFT and STLR + Gradual unfreezing)	**75.0**

Hyper-parameter Settings: As suggested in [10], we use the AWD-LSTM language model with 3 layers, 1150 hidden activations per layer and an embedding size of 400. The hidden layer of the classifier is of size 50. A batch size of 30 is used to train the model. The LM and classifier fine-tuning is done with a base learning rate of 0.004 and 0.01 respectively. All experiments are on the standard train-dev-test split as proposed in [8] with the classification results reported on the 3850 sized test data.

Results and Analysis: Table 2 shows the performance of our ITL model as compared with various baselines and the state-of-the-art method [8]. The overall ITL model has an accuracy of 75.03% improving significantly over the previous state-of-the-art (feature-engineered solution of [8]). To assess the impact of each of the three steps involved in ITL, we perform an ablation study as follows.

– **No pretraining**: Train the model without the pretraining step.
– **No LM fine-tuning**: Phase 2 is ignored.
– **Fine-tuning without DFT and STLR**: LM fine-tuning without discriminative fine-tuning and without Slanted Triangular Learning Rates.
– **No gradual unfreezing**: No gradual unfreezing during classifier fine-tuning.

As expected, from Table 2, we observe that not fine-tuning the LM on the target task results in a worse performance versus fine-tuning. Using DFT and STLR is beneficial. Gradual unfreezing helps in increasing the performance. All the three steps in the fine-tuning process contribute towards improving accuracy.

6 Conclusions

In this work, we showed that the idea of using inductive transfer learning by fine-tuning language models aids in identifying whether search queries are well-formed natural language questions. On a large dataset of 25,100 questions, we showed that our method beats the baselines with a significant margin. In the future, we plan to explore the "accuracy versus labeled dataset size" tradeoff for this approach across multiple resource-poor languages.

References

1. Baeza-Yates, R., Calderón-Benavides, L., González-Caro, C.: The intention behind web queries. In: Crestani, F., Ferragina, P., Sanderson, M. (eds.) SPIRE 2006. LNCS, vol. 4209, pp. 98–109. Springer, Heidelberg (2006). https://doi.org/10.1007/11880561_9

2. Barr, C., Jones, R., Regelson, M.: The linguistic structure of English web-search queries. In: Proceedings of the Conference on Empirical Methods in Natural Language Processing, pp. 1021–1030. Association for Computational Linguistics (2008)

3. Bawa, M., Bayardo Jr., R.J., Rajagopalan, S., Shekita, E.J.: Make it fresh, make it quick: searching a network of personal webservers. In: Proceedings of the 12th International Conference on World Wide Web, pp. 577–586. ACM (2003)

4. Bergsma, S., Wang, Q.I.: Learning noun phrase query segmentation. In: Proceedings of the 2007 Joint Conference on Empirical Methods in Natural Language Processing and Computational Natural Language Learning (EMNLP-CoNLL) (2007)

5. Copestake, A.A., Flickinger, D.: An open source grammar development environment and broad-coverage English grammar using HPSG. In: LREC, Athens, Greece, pp. 591–600 (2000)

6. Dai, A.M., Le, Q.V.: Semi-supervised sequence learning. In: Advances in Neural Information Processing Systems, pp. 3079–3087 (2015)

7. Fader, A., Zettlemoyer, L., Etzioni, O.: Paraphrase-driven learning for open question answering. In: Proceedings of the 51st Annual Meeting of the Association for Computational Linguistics (Volume 1: Long Papers), vol. 1, pp. 1608–1618 (2013)

8. Faruqui, M., Das, D.: Identifying well-formed natural language questions. In: EMNLP (2018, to appear)

9. Gupta, M., Bendersky, M., et al.: Information retrieval with verbose queries. Found. Trends® Inf. Retrieval 9(3–4), 209–354 (2015)

10. Howard, J., Ruder, S.: Universal language model fine-tuning for text classification. In: Proceedings of the 56th Annual Meeting of the Association for Computational Linguistics (Volume 1: Long Papers), vol. 1, pp. 328–339 (2018)

11. Manshadi, M., Li, X.: Semantic tagging of web search queries. In: Proceedings of the Joint Conference of the 47th Annual Meeting of the ACL and the 4th International Joint Conference on Natural Language Processing of the AFNLP: Volume 2-Volume 2, pp. 861–869. Association for Computational Linguistics (2009)

12. Markatos, E.P.: On caching search engine query results. Comput. Commun. 24(2), 137–143 (2001)

13. Merity, S., Keskar, N.S., Socher, R.: Regularizing and optimizing LSTM language models. arXiv preprint arXiv:1708.02182 (2017)

14. Mishra, N., Saha Roy, R., Ganguly, N., Laxman, S., Choudhury, M.: Unsupervised query segmentation using only query logs. In: Proceedings of the 20th International Conference Companion on World Wide Web, pp. 91–92. ACM (2011)
15. Mostafazadeh, N., Misra, I., Devlin, J., Mitchell, M., He, X., Vanderwende, L.: Generating natural questions about an image. arXiv preprint arXiv:1603.06059 (2016)
16. Mou, L., et al.: How transferable are neural networks in NLP applications? arXiv preprint arXiv:1603.06111 (2016)
17. Roy, R.S., Choudhury, M., Bali, K.: Are web search queries an evolving protolanguage? In: The Evolution of Language, pp. 304–311. World Scientific (2012)
18. Yang, J., Hauff, C., Bozzon, A., Houben, G.J.: Asking the right question in collaborative Q&A systems. In: Proceedings of the 25th ACM Conference on Hypertext and Social Media, pp. 179–189. ACM (2014)
19. Yannakoudakis, H., Rei, M., Andersen, Ø.E., Yuan, Z.: Neural sequence-labelling models for grammatical error correction. In: Proceedings of the 2017 Conference on Empirical Methods in Natural Language Processing, EMNLP, pp. 2795–2806 (2017)
20. Yosinski, J., Clune, J., Bengio, Y., Lipson, H.: How transferable are features in deep neural networks? In: Advances in Neural Information Processing Systems, pp. 3320–3328 (2014)

Towards Spatial Word Embeddings

Paul Mousset[1,2(✉)], Yoann Pitarch[1], and Lynda Tamine[1]

[1] IRIT, Université de Toulouse, CNRS, Toulouse, France
{paul.mousset,yoann.pitarch,lynda.tamine}@irit.fr
[2] Atos Intégration, Toulouse, France

Abstract. Leveraging textual and spatial data provided in spatio-textual objects (eg., tweets), has become increasingly important in real-world applications, favoured by the increasing rate of their availability these last decades (eg., through smartphones). In this paper, we propose a spatial retrofitting method of word embeddings that could reveal the localised similarity of word pairs as well as the diversity of their localised meanings. Experiments based on the semantic location prediction task show that our method achieves significant improvement over strong baselines.

Keywords: Word embeddings · Retrofitting · Spatial

1 Introduction

The last decades have witnessed an impressive increase of geo-tagged content known as spatio-textual data or geo-texts. Spatio-textual data includes Places Of Interest (POI) with textual descriptions, geotagged posts (eg., tweets), geo-tagged photos with textual tags (eg., Instagram photos) and check-ins from location-based services (eg., Foursquare). The interplay between text and location provides relevant opportunities for a wide range of applications such as crisis management [11] and tourism assistance [5]. This prominence gives also rise to considerable research issues underlying the matching of spatio-textual objects which is the key step in diverse tasks such as querying geo-texts [24], location mention [6,9] and semantic location prediction [3,25]. Existing solutions for matching spatio-textual objects are mainly based on using a combination of textual and spatial features either for building scalable object representations [24] or for designing effective object-object matching models [3,25]. The goal of our work is to explore the idea of jointly leveraging spatial and textual knowledge to build enhanced representations of textual units (namely words) that could be used at either object representation and matching levels. The central thesis of our work is driven by two main intuitions: (1) co-occurrences of word pairs within spatio-textual objects reveal localised word similarities. For instance *dinosaur* and *museum* are semantically related near a natural history museum, but less related near an art museum; (2) As a corollary of intuition 1, distinct

© Springer Nature Switzerland AG 2019
L. Azzopardi et al. (Eds.): ECIR 2019, LNCS 11438, pp. 53–61, 2019.
https://doi.org/10.1007/978-3-030-15719-7_7

meanings of the same word could be conveyed using the spatial word distribution as source of evidence. For instance *dinosaur* can refer to a prehistoric animal or to a restaurant chain specifically in New York. Thus, we exploit the spatial distribution of words to jointly identify semantically related word pairs as well as localised word meanings. To conceptualise our intuitions, we propose a retrofitting strategy [7,20] as means of refining pre-trained word embeddings using spatial knowledge. We empirically validate our research intuitions and then show the effectiveness of our proposed spatial word embeddings within semantic location prediction as the downstream task.

2   Preliminaries

2.1   Definitions and Intuitions

Definition 1 (Spatio-textual object). A spatio-textual object o is a geo-tagged text (eg., a POI with a descriptive text). The geotag is represented by its coordinates *(lat, lon)* referring to the geographic location l denoted $o.l$ (eg., the physical location of a POI). We adopt a word-based vectorial representation of object o including all its textual attributes (eg., POI description) $o = [w_1^{(o)}, \ldots, w_m^{(o)}]$ where each word $w_i^{(o)}$ is drawn from a vocabulary $\mathcal{W}$.

Definition 2 (Spatial distance). The spatial distance between spatio-textual objects o_i, o_j refers to the geographic distance, under a distance metric, between locations $o_i.l$ and $o_j.l$. The spatial distance between words w_i, w_j refers to an aggregated (eg., average) spatial object-object distance over the sets of spatio-textual objects O_i, O_j they respectively belong to.

Intuition 1. Words that occur in close spatio-textual objects tend to have similar meanings. Basically, the spatially closer the words are, regarding the distance between their associated objects, the closer are their meanings (eg., intuitively *cup* is semantically closer to *football* in Europe than in the USA).

Intuition 2. Let us consider a localised meaning of a word as being represented by the set of spatially similar words with respect to *intuition 1*. A word could convey different localised meanings depending on the geographical area where it is spatially dense (eg., *football* in Europe does not refer to the same sport as in the USA).

2.2   Problem Definition

Based on *intuition 1*, we conjecture that spatial signals could contribute to the building of distributed representations of word vectors. As previously suggested [7,20], one relevant way is to inject external knowledge into initial learned word embeddings. However different meanings of the same word are conflated into a single embedding [10,13]. Thus, from *intuition 2*, we build for each word a set of embedding vectors based on its occurrence statistics over the associated spatio-textual objects. Formally, given a set of word vector representations

$\widehat{\mathbf{W}} = \{\widehat{\mathbf{w}}_1, \ldots, \widehat{\mathbf{w}}_n\}$, where $\widehat{\mathbf{w}}_i$ is the k-dimensional embedding vector built for target word $w_i \in \mathcal{W}$, using a standard neural language model (eg., Skip-gram model [14]), the problem is how to build for each word w_i the set of associated spatial word embeddings $\widehat{\mathbf{w}}_i^s = \{\widehat{\mathbf{w}}_{i,1}^s, \ldots, \widehat{\mathbf{w}}_{i,j}^s, \ldots, \widehat{\mathbf{w}}_{i,n_i}^s\}$. Each spatial word vector $\widehat{\mathbf{w}}_{i,j}^s$, derived from an initial embedding $\widehat{\mathbf{w}}_i$, refers to the localised distributional representation of word w_i over a dense spatial area, and n_i is the number of distinct localised meanings of word w_i derived from its spatial distribution over the spatio-textual objects O_i it belongs to.

3   Methodology

3.1   Overview

Our algorithm for building the spatial word embeddings is described in Algorithm 1. For each word w_i, we first identify the spatio-textual objects O_i it belongs to. To identify dense spatial areas of word w_i, we perform a K-Means clustering [12]. More formally, for each word w_i, we determine n_i spatial clusters represented with their respective barycenters $\mathcal{B}_i = \{\mathcal{B}_{i,1}, \cdots, \mathcal{B}_{i,n_i}\}$, where $\mathcal{B}_{i,j}$ is the j-th barycenter of word w_i and n_i the optimal number of clusters for word w_i determined using the silhouette analysis [19]. Each barycenter $\mathcal{B}_{i,j}$ can be seen as a spatial representative of the area that gives rise to a local word meanings of word w_i represented by the distributed vector $\mathbf{w}_{i,j}^s$. We detail in the following section the key step of building the spatial embedding $\mathbf{w}_{i,j}^s$ based on a retrofitting process from word embedding $\widehat{\mathbf{w}}_i$ and considering both spatially neighbour words $W_{i,j}^+$ and distant words $W_{i,j}^-$ with respect to barycenter $\mathcal{B}_{i,j}$.

Algorithm 1. Algorithm for building spatial word embeddings

Input: Vocabulary $\mathcal{W}$; Set of word embeddings $\widehat{\mathbf{W}} = \{\widehat{\mathbf{w}}_1, \ldots; \widehat{\mathbf{w}}_{|\mathcal{W}|}\}$; Set of spatio-textual objects O

Output: Set of spatial word embeddings $\mathbf{W}^s = \{\mathbf{w}_{1,1}^s, \ldots, \mathbf{w}_{1,n_1}^s, \ldots, \mathbf{w}_{|\mathcal{W}|,n_k}^s\}$

  for $i \in \{1, .., |\mathcal{W}|\}$ **do**

1 | $O_i = \text{ExtractObjects}(w_i, O)$

2 | $\text{SpatialClustering}(O_i, \mathcal{B}_i, n_i)$

  end

  repeat

  | **for** $i \in \{1, .., |\mathcal{W}|\}$ **do**

  | | **for** $j \in \{1, .., n_i\}$ **do**

3 | | | $W_{i,j}^+ = \text{Neighbours}(w_i, \mathcal{B}_{i,j})$

4 | | | $W_{i,j}^- = \text{Distant}(w_i, \mathcal{B}_{i,j})$

5 | | | $\mathbf{w}_{ij}^s = \text{Retrofit}\,(\widehat{w}_i, W_{i,j}^+, W_{i,j}^-)$ (see Sect. 4.2)

  | | **end**

  | **end**

  until *Convergence*;

3.2 Spatially Constrained Word Embedding

Our objective here is to learn the set of spatial word embeddings $\mathbf{W}^s$. We want the inferred word vector $\mathbf{w}_{i,j}^s$ (i) to be semantically close (under a distance metric) to the associated word embedding $\widehat{\mathbf{w}}_i$, (ii) to be semantically close to its spatial neighbour words $W_{i,j}^+$ and (iii) to be semantically unrelated to the spatially distant words $W_{i,j}^-$. Thus, the objective function to be minimised is given by:

$$\Psi(\mathbf{W}^s) = \sum_{i=1}^{|W|} \sum_{j=1}^{n_i} \left[\alpha \, d(\mathbf{w}_{i,j}^s, \widehat{\mathbf{w}}_i) + \beta \sum_{w_k \in W_{i,j}^+} d(\mathbf{w}_{i,j}^s, \widehat{\mathbf{w}}_k) + \gamma \sum_{w_k \in W_{i,j}^-} 1 - d(\mathbf{w}_{i,j}^s, \widehat{\mathbf{w}}_k) \right]$$

where $d(w_i, w_j) = 1 - sim(w_i, w_j)$ is a distance derived from a similarity measure (eg., cosinus), $\mathbf{W}_{i,j}^+$ (resp. $\mathbf{W}_{i,j}^-$) is the set of words spatially **close to** (resp. **distant from**) the word $w_{i,j}$, ie., words within (resp. beyond) a radius r^+ (resp. r^-) around its barycenter $\mathcal{B}_{i,j}$, and $\alpha, \beta, \gamma \geq 0$ are hyperparameters that control the relative importance of each term. In our experimental setting, r^+ and r^- are set to 100 and 500 meters and $\alpha = \beta = \gamma = 1$.

4 Evaluation

4.1 Experimental Setup

Evaluation Task and Dataset. We consider the *semantic location prediction task* [3, 25]. Given the tweet t, the task consists in identifying, if any, the POI p that the tweet t semantically focuses on (ie., reviews about). Formally, semantic location identifies a single POI p which is the topmost $p* \in \mathcal{P}$ of a ranked list of candidate POIs returned by a semantic matching function. We employ a *dataset* of English geotagged tweets released by Zhao et al. [25]. The dataset, consists of 74K POI-related tweets, collected from 09.2010 to 01.2015 in New York (NY) and Singapore (SG). Using the Foursquare API, we collected 800K POIs located in NY and SG cities including user-published reviews. The entire dataset consists of 238,369 distinct words, on which we applied K-Means clustering (see Sect. 3.1). As result of clustering, we found 630,732 spatial word clusters with around 2.6 local word meaning $w_{i,j}^s$ created per word w_i. We notice that 166,139 (69.7%) words have only one local meaning.

Baselines, Scenarios and Metrics. We compare our approach with a set of stat-of-the-art matching baseline models: (1) DIST [4]: the Haversine distance Tweet-POI; (2) BM25 [18] ; (3) CLASS [25]: a POI ranking model that combines spatial distance with a text-based language model. To evaluate the effectiveness of our approach, we inject the embedding into the CLASS model as follows: (a) CLASS-MATCH (CM): we compute the cosine similarity of a pair (t,p) instead of the language model score. (b) CLASS-EXPAND (CE): we expand the tweet with the top likely similar words following the approach proposed by Zamani and Croft [23]. For the two above scenarios we consider either the traditional or the spatial word embeddings. Practically, for scenarios using spatial word embeddings, we use the closest local word $w_i^{(t)}$ (resp. $w_i^{(p)}$) by minimising the Haversine distance between tweet (resp. POI) location $t.l$ (resp. $p.l$) and word barycenters $\mathcal{B}_{i,j}$. We exploit two well-known evaluation metrics, namely $Acc@k$ [17] and $Mean$ $Reciprocal\ Rank\ (MRR)$ [2]. Given the semantic location task description, it is worth to mention that low values of k are particularly considered.

4.2 Analysis of Spatial Driven Word Similarities

To validate the intuitions presented in Sect. 2.1, we first build as shown in Fig. 1, the heat-map of the similarity values between the embedding vectors of a sample of insightful words where the darker the cell, the more similar the pair of words. To exhibit the localised meanings of the words, we partition the dataset in two distinct subsets depending on the city the tweets were emitted from (ie., either in NY or SG). For each subset, cosine similarities are then damped by a spatial factor $f_s(w_i, w_j)$ which conveys how spatially close are the word w_i and w_j. Formally, $f_s(w_i, w_j)$ is defined as $f_s(w_i, w_j) = \exp\{-\frac{dist(\mathcal{B}_i, \mathcal{B}_j) - \mu}{\sigma}\}$ where $dist(\mathcal{B}_i, \mathcal{B}_j)$ is the Haversine distance between the barycenters of w_i and w_j and μ (resp. σ) is the average distance (resp. standard deviation) between all word pairs that describe the POIs located in the city. For simplicity purposes, we consider one barycenter per word for each subset. The heat-map of these weighted matrices are shown in Figs. 1b and c for NY and SG respectively. We can see for instance, that the cell ($restaurants$, $dinosaur$) is darker in Fig. 1b than in Fig. 1a while the cell is lighter in Fig. 1c than in Fig. 1a for the same word pair. Generally speaking, there is no objective obvious reason about why the words $restaurants$ and $dinosaur$ should be related to each other, as outlined by the similarity of their word embeddings in Fig. 1a. However, some restaurants in NY are named $Dinosaur\ Bar\text{-}B\text{-}Que$ leading to an over-representativeness of tweets where these two terms co-occur in NY, leading to a local stronger semantic relation within this word pair in NY as revealed by Fig. 1b. This fits with our $intuition\ 1$. Besides, cross-looking at Fig. 1a and its spatial variants Figs. 1b and c provides some clues on why our $intuition\ 2$ is well-founded. Indeed, we can see that words $dinosaur$ and $museum$ are similar regardless of the location. By relating this observation with the previous one, we can infer that $dinosaur$ could refer to both $museum$ and $restaurant$ specifically in NY as revealed by the strength of its similarity with words such as $burger$ and $cheese$ in Fig. 1b which is clearly less pronounced in Fig. 1c.

4.3 Effectiveness

Table 1 summarises the effectiveness results obtained based on the semantic location prediction task. We compute relative changes (R-Chg) using the ratio of the geometric means of the MRR and compute the relative improvements suited for non aggregated measures for $Acc@k$. Overall, we can see that the scenarios involving matching with spatial embeddings (CM-$\mathbf{W}^s$ and CE-$\mathbf{W}^s$) significantly

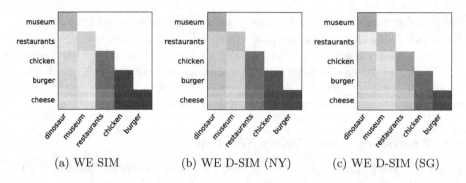

(a) WE SIM   (b) WE D-SIM (NY)   (c) WE D-SIM (SG)

Fig. 1. Cosine similarities of traditional WE SIM (a), WE SIM damped by word-word barycenter distances in NY dataset (b) and in SG dataset (c)

overpass all the compared models. For instance, CE-$\mathbf{W}^s$ displays better results in terms of MRR with relative changes ranging between 140.7% and 161.3% compared to DIST, BM25 and CLASS models. More precisely, CE-$\mathbf{W}^s$ allows a more effective mapping tweet-POI: more than 48% of the tweets are associated with the relevant POI based on the top-1 result, against 43% for DIST. In addition, we can observe that while injecting embeddings (either traditional or spatial) allows to improve the effectiveness of the CLASS model, the spatial embeddings allow the achievement of significant better performance. For instance, the MRR of the scenario CE significantly increases by 119%. Specifically looking at the two scenarios involving spatial embeddings, we can notice that CE-$\mathbf{W}^s$ improves MRR by 128.2% and $Acc@1$ by 5.05% compared to CM-$\widehat{\mathbf{W}}$. These results could be explained by the approach used to inject the embeddings. While in CE-$\mathbf{W}^s$, spatial embedding vectors are intrinsically used to expand the tweet description before the matching, they are rather used in the scenario CM-$\widehat{\mathbf{W}}$ to build tweet and POI embeddings using an IDF weighted average of embeddings which might generate biases in their representations. This observation clearly shows the positive impact of the intrinsic use of the spatial embeddings.

5   Related Work

A standard approach for improving traditional word embeddings is to inject external knowledge, mainly lexical resource constraints, using either an *online* or *offline* approach [14,16]. The online approach exploits external knowledge during the learning step [8,21,22]. For instance, Yu et al. and Xu et al. [21,22] propose the RCM model which extends the skip-gram objective function with semantic relation between word pairs, as provided by a lexical resource, based on the assumption that related words yield similar contexts. The offline approach, also called *retrofitting*, uses external resources outside the learning step [7,15, 20]. For instance, Faruqui et al. [7] propose a method for refining vector space representations by favouring related words, as provided by a lexical resource (eg.,

Table 1. Effectiveness evaluation. R-Chg: CE-$\mathbf{W}^s$ relative changes. R-Imp: CE-$\mathbf{W}^s$ relative improvements. Significant Student's t-test $*$: $p < 0.05$.

		MRR		Acc@1		Acc@5	
		Value	R-Chg	Value	R-Imp	Value	R-Imp
Dist. based	DIST	0.514	+140.7 $*$	0.430	+19.61 $*$	0.605	+15.45 $*$
Text based	BM25	0.423	+161.3 $*$	0.307	+64.68 $*$	0.668	+4.49 $*$
Text-Dist. based	CLASS	0.507	+159.9 $*$	0.401	+25.85 $*$	0.624	+11.79 $*$
Traditional	CM-$\widehat{\mathbf{W}}$	0.521	+128.0 $*$	0.413	+24.52 $*$	0.640	+9.06 $*$
Embeddings	CE-$\widehat{\mathbf{W}}$	0.563	+119.0 $*$	0.470	+9.41 $*$	0.659	+5.94 $*$
Spatial	CM-$\mathbf{W}^s$	0.577	+128.2 $*$	0.489	+5.05 $*$	0.675	+3.36 $*$
Embeddings	CE-$\mathbf{W}^s$	0.604	$-$	0.515	$-$	0.698	$-$

WordNet, FramNet), to have similar vector representations. To the best of our knowledge, our work is the first attempt for retrofitting word embeddings using spatial knowledge. To tackle the meaning conflation deficiency issue of word embeddings [1,10,13], the general approach is to jointly learn the words and their senses. For instance, Iacobacci et al. [10] first disambiguate words using the *Babelfy* resource, and then revise the continuous bag of words (CBOW) objective function to learn both word and sense embeddings.

6 Conclusion

In this paper, we introduced spatial word embeddings as a result of *retrofitting* traditional word embeddings. The retrofitting method leverages spatial knowledge toward revealing localised semantic similarities of word pairs, as well as localised meanings of words. The experimental evaluation shows that our proposed method successfully refines pre-trained word embeddings and allows achieving significant results over the semantic location prediction task. As future work, we plan to evaluate the effectiveness of our proposed spatial word embeddings within other location-sensitive tasks including spatial summarization of streaming objects such as tweets.

Acknowledgments. This research was supported by IRIT and ATOS Intégration research program under ANRT CIFRE grant agreement #2016/403.

References

1. Cheng, J., Wang, Z., Wen, J.R., Yan, J., Chen, Z.: Contextual text understanding in distributional semantic space. In: Proceedings of CIKM 2015, pp. 133–142 (2015)
2. Craswell, N.: Mean reciprocal rank. In: Liu, L., Özsu, M.T. (eds.) Encyclopedia of Database Systems, p. 1703. Springer, Boston (2009). https://doi.org/10.1007/978-0-387-39940-9
3. Dalvi, N., Kumar, R., Pang, B., Tomkins, A.: A translation model for matching reviews to objects. In: Proceedings of CIKM 2009, pp. 167–176 (2009)
4. De Smith, M., Goodchild, M.F.: Geospatial Analysis: A Comprehensive Guide to Principles, Techniques and Software Tools. Metador (2007)
5. Deveaud, R., Albakour, M.D., Macdonald, C., Ounis, I.: Experiments with a venue-centric model for personalised and time-aware venue suggestion. In: Proceedings of CIKM 2015, pp. 53–62 (2015)
6. Fang, Y., Chang, M.W.: Entity linking on microblogs with spatial and temporal signals. Trans. Assoc. Comput. Linguist. **2**, 259–272 (2014)
7. Faruqui, M., Dodge, J., Jauhar, S.K., Dyer, C., Hovy, E., Smith, N.A.: Retrofitting word vectors to semantic lexicons. In: Proceedings of NAACL 2015, pp. 1606–1615 (2015)
8. Glavaš, G., Vulić, I.: Explicit retrofitting of distributional word vectors. In: Proceedings of ACL 2018, pp. 34–45 (2018)
9. Han, J., Sun, A., Cong, G., Zhao, W.X., Ji, Z., Phan, M.C.: Linking fine-grained locations in user comments. Trans. Knowl. Data Eng. **30**(1), 59–72 (2018)
10. Iacobacci, I., Pilehvar, M.T., Navigli, R.: SensEmbed: learning sense embeddings for word and relational similarity. In: Proceedings of ACL and IJCNLP 2017, pp. 95–105 (2017)
11. Imran, M., Castillo, C., Diaz, F., Vieweg, S.: Processing social media messages in mass emergency: a survey. ACM Comput. Surv. **47**(4), 67:1–67:38 (2015)
12. MacQueen, J., et al.: Some methods for classification and analysis of multivariate observations. In: Proceedings of BSMSP 1967, pp. 281–297 (1967)
13. Mancini, M., Camacho-Collados, J., Iacobacci, I., Navigli, R.: Embedding words and senses together via joint knowledge-enhanced training. In: Proceedings of CoNLL 2017, pp. 100–111 (2017)
14. Mikolov, T., Sutskever, I., Chen, K., Corrado, G.S., Dean, J.: Distributed representations of words and phrases and their compositionality. In: Proceedings of NIPS 2013, pp. 3111–3119 (2013)
15. Mrkšić, N., et al.: Counter-fitting word vectors to linguistic constraints. arXiv preprint (2016)
16. Pennington, J., Socher, R., Manning, C.D.: Glove: global vectors for word representation. In: Proceedings of EMNLP 2014, pp. 1532–1543 (2014)
17. Powers, D.M.: Evaluation: from precision, recall and f-measure to ROC, informedness, markedness & correlation. J. Mach. Learn. Technol. **2**, 37–63 (2011)
18. Robertson, S.E., Jones, K.S.: Relevance weighting of search terms. J. Am. Soc. Inf. Sci. **27**(3), 129–146 (1976)
19. Rousseeuw, P.J.: Silhouettes: a graphical aid to the interpretation and validation of cluster analysis. J. Comput. Appl. Math. **20**, 53–65 (1987)
20. Vulić, I., Mrkšić, N.: Specialising word vectors for lexical entailment. In: Proceedings of NAACL-HLT 2018, pp. 1134–1145 (2018)
21. Xu, C., et al.: RC-NET: a general framework for incorporating knowledge into word representations. In: Proceedings of CIKM 2014, pp. 1219–1228 (2014)

22. Yu, M., Dredze, M.: Improving lexical embeddings with semantic knowledge. In: Proceedings of ACL 2014, pp. 545–550 (2014)
23. Zamani, H., Croft, W.B.: Estimating embedding vectors for queries. In: Proceedings of ICTIR 2016, pp. 123–132 (2016)
24. Zhang, D., Chan, C.Y., Tan, K.L.: Processing spatial keyword query as a top-k aggregation query. In: Proceedings of SIGIR 2014, pp. 355–364 (2014)
25. Zhao, K., Cong, G., Sun, A.: Annotating points of interest with geo-tagged tweets. In: Proceedings of CIKM 2016, pp. 417–426 (2016)

Asymmetry Sensitive Architecture
for Neural Text Matching

Thiziri Belkacem[✉], Jose G. Moreno, Taoufiq Dkaki, and Mohand Boughanem

IRIT UMR 5505 CNRS, University of Toulouse, Toulouse, France
{thiziri.belkacem,jose.moreno,taoufiq.dkaki,mohand.boughanem}@irit.fr

Abstract. Question-answer matching can be viewed as a puzzle where missing pieces of information are provided by the answer. To solve this puzzle, one must understand the question to find out a correct answer. Semantic-based matching models rely mainly in semantic relatedness the input text words. We show that beyond the semantic similarities, matching models must focus on the most important words to find the correct answer. We use attention-based models to take into account the word saliency and propose an asymmetric architecture that focuses on the most important words of the question or the possible answers. We extended several state-of-the-art models with an attention-based layer. Experimental results, carried out on two QA datasets, show that our asymmetric architecture improves the performances of well-known neural matching algorithms.

Keywords: Asymmetric · Attention models · Relevance matching

1 Introduction

Short text matching in general and Question Answering (QA) in particular include several problems that can be grouped into two main classes. The first one consists in identifying whether two texts are semantically similar. Current solutions are based on syntactic and semantic relatedness of the inputs. We refer to this as the *symmetric matching problem*, it includes tasks such as sentence completion [14,19] and question pairs identification [1,2]. The second is whether an input text provides the information sought in another text. In this case, the nature of input texts is not the same and their association is determined not only by the semantic relationship but also by complementarity. We refer to this as *the asymmetric matching problem* and it mainly includes question-answer matching [11,15], where the question contains some of the requested information but not the information itself. Several matching models based on convolutional neural networks (CNN) [7,9] and Long Short-Term Memory Models (LSTM) [13,14] have recently been proposed. In these models, the input sentences are first mapped to a set of word vectors, then processed in a symmetric architecture through different layers. To take into account the asymmetric aspect of the question-answer matching task, we believe that existing QA models must go

© Springer Nature Switzerland AG 2019
L. Azzopardi et al. (Eds.): ECIR 2019, LNCS 11438, pp. 62–69, 2019.
https://doi.org/10.1007/978-3-030-15719-7_8

beyond the classical symmetric text matching architecture. An ideal model must focus more on the most important words in the question to better address it. Attention-based models [3,18] can provide a way to fit this requirement. Based on the importance of words in the text, these models learn *attention* coefficients that allow subsequent processes to focus on the most important words of an input text. In this paper, we propose an attention-based architecture that allows to better handle the asymmetric aspect of the question-answer matching, such that the model gives more focus to the most important words of the question. Our main contributions are as follow:

1. We propose an *asymmetric* matching architecture to handle *asymmetric* matching problems.
2. We extend several state-of-the-art models using the proposed architecture.
3. We conduct a comparative experimental study of existing models against extended ones using our architecture.

2   Related Work

Deep neural architectures in recent text matching models are based on a siamese-like architecture [5], where both inputs undergo the same[1] type of processing. This architecture is adopted in several existing models. In [12], Shen et al. propose a C-DSSM model using a convolutional network. The C-DSSM architecture uses a *word hashing* layer, as a common function to construct embedded representations of the inputs, then stacked layers map the representations to low dimensional vectors for the matching process. In [7], the authors proposed two convolutional architectures: ARC-I and ARC-II. The first constructs a sentence representation using a sequence of convolution and pooling layers, then computes the matching score of the input sentences. ARC-II applies a series of convolution and pooling layers to a matching matrix of input word vectors. Experimental results show that ARC-II outperforms ARC-I. In [9], Pang et al. proposed the MatchPyramid model. The architecture of MatchPyramid is also symmetric: first, inputs are represented using embedded vectors then a matching tensor is computed and fed to a sequence of convolution and pooling layers, in order to extract high level interaction signals. In [14], Wan et al. propose the MV-LSTM model, a position-based model for question answering. The symmetry of MV-LSTM consists in a bidirectional LSTM (bi-LSTM) layer that constructs a position-aware representation for both input sentences. Interaction matrices are then computed and passed through a pooling and fully connected layers to compute the final matching score. Mitra et al. [8] proposed a duet architecture to match documents and queries. The duet architecture is composed of the local model that uses the interaction matrix of query and document words, and the distributed model that learns embeddings of the query and the document text before matching. The parameters of both models are optimized jointly during

---

[1] Some differences may exist but they are only related to the input size which is considered as a non-architectural difference.

training and the final matching score is provided by the duet architecture. Both the local and the distributed models are based on a symmetric architecture, since both the inputs are dealt with in the same way.

All the mentioned models use symmetric architectures regardless the nature of the addressed task. In both symmetric [7,8] and asymmetric [9,14] tasks, only symmetric architectures were adopted. The asymmetric aspect of inputs processing is already discussed in [4], where Bordes et al. devised an architecture to learn more than one relation at a time from a knowledge base. However, none of the previous work provide an asymmetric architecture for text matching.

3   Asymmetric Matching Architecture

Motivation

Expressed in a natural language, the question describes a specific user's information need. It is like a puzzle whose missing pieces must be found and put together, in a logical way, to solve the problem. A pattern of each gap in this puzzle describes the corresponding missing part. It is all the same for the question-answer matching: a pattern must be filled by one or more answers, not all answers can be suitable, only those that conform to the pattern (the question) that describes the missing part are able to correctly fill it (give the sought information). The example in Fig. 1 from the WikiQA dataset shows this perception.

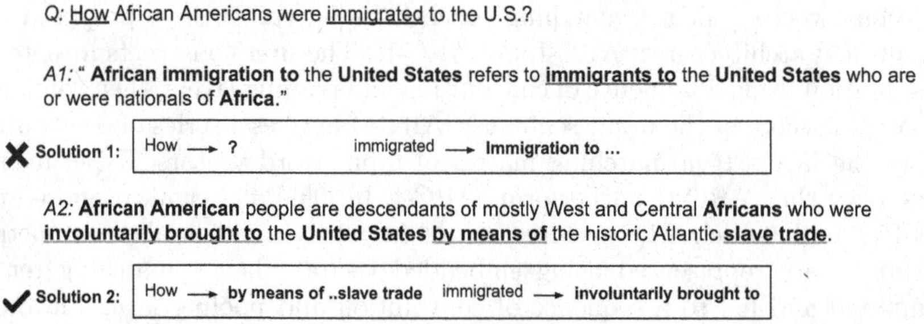

> Q: How African Americans were immigrated to the U.S.?
>
> A1: " **African immigration to** the **United States** refers to **immigrants to** the **United States** who are or were nationals of **Africa**."
>
> ✘ Solution 1: | How ⟶ ?   immigrated ⟶ Immigration to ... |
>
> A2: **African American** people are descendants of mostly West and Central **Africans** who were **involuntarily brought to** the **United States by means of** the historic Atlantic **slave trade**.
>
> ✔ Solution 2: | How ⟶ by means of ..slave trade   immigrated ⟶ involuntarily brought to |

Fig. 1. In the answers *A1* and *A2*, words in **bold** represent semantic and syntactic relatedness. Underlined words are: the key words (pattern clues) in *Q* and corresponding matches in *A1* and *A2*. Solution 2 fills completely the missing part described in *Q*.

In this figure, to correctly answer the question *Q*, we have to focus on the words *"How"* and *"immigrated to"* describing the missing part. *Q* is about how the immigration process was done. We have two different answers *A1* and *A2* with corresponding solutions 1 and 2 respectively, as illustrated in Fig. 1. Based on semantic and syntactic relatedness, both the answers contain several words corresponding to the question. However, one notice that solution 2 better solves

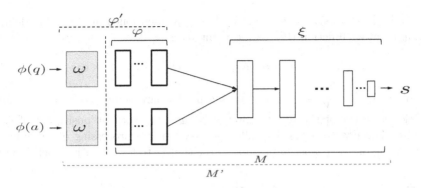

Fig. 2. The asymmetric architecture M' extends a model M. φ processes the inputs in parallel. ξ a sequence of processing layers to compute the matching score s. The attention-based layers ω are activated according to the configuration of Eq. 1.

the problem than solution 1 does. Literally, the answer $A2$ contains the corresponding information, solution 2 represents the missing piece of this puzzle. This example, highlights the asymmetry aspect of the question-answer matching task. Notice that syntactic and semantic based similarities are not sufficient to solve the problem completely. The question has some words that require a particular attention to retrieve the correct answer. Attention-based models, used in machine translation [3], sentiment classification [10,17] and paraphrase identification [18], enable to identify the kernel information to be considered in a given sequence and focus on discriminating elements. In the previous example, words *"How"* and *"immigrated to"* of the question Q, require more focus. Formally, given a word sequence S, the attention-based model learns a coefficient vector α that determines how much attention should be given to each element of S according to the task to be performed. We propose an asymmetric model architecture using attention-based layers, in order to focus on most important words of the different inputs. Let us consider a question q, an answer a, a matching model M and the embedding function ϕ. Most of state-of-the-art neural-based models can be summarized in Fig. 2. Where $\phi(q)$ and $\phi(a)$ are the embedded representations of q and a respectively. M can be viewed as a sequence of two main blocks of different layers: φ is a sequence of layers used to learn input representations simultaneously. The block ξ is another sequence of different processing layers, used to compute the final matching score s. We define the asymmetric model M' that extends the model M by adding layer ω that can be applied differently for asymmetric inputs, as highlighted in the Fig. 2. We define a function φ' as in Eq. 1 to handle asymmetric inputs in the model M'. For a sequence S of l words with $S \in \{q,a\}$, we define a parameter $e_S \in \{0,1\}$ to set up the asymmetric processing as follows: $e_S = 1$ activates the shaded layer ω in Fig. 2 for the corresponding input. If $e_S = 0$ for both the inputs then $M' = M$.

$$\varphi'(\phi(S)) = \begin{cases} \varphi \circ \omega(\phi(S)) & \text{if } e_S = 1 \\ \varphi(\phi(S)) & \text{otherwise} \end{cases} \tag{1}$$

where ω is the extension attention-based layer as mentioned in Fig. 2. We define ω using a gating function [15], as represented in Eq. 2.

$$\omega(\phi(S)) = [w_1 \times \alpha_1, w_2 \times \alpha_2, ..., w_t \times \alpha_t, ..., w_l \times \alpha_l] \tag{2}$$

with $\alpha_t = \frac{exp(V^T.w_t)}{\sum_{j=1}^{l} exp(V^T.w_j)}$, where V is a model parameter. It is the attention coefficients vector of the input sentence S. w_t is a word at position t of the sentence. The layer ω is used to allow an asymmetric processing of the inputs and focusing on their most important words thanks to the attention coefficients.

4   Experiments and Results

4.1   Experimental Protocol

Experiments were performed[2] using the MatchZoo [6] framework for neural text matching models. We used two datasets. First, WikiQA Corpus [16] composed of 3047 questions from Bing query logs and 1473 candidate answers from Wikipedia. Second, QuoraQP dataset composed of 404351 question pairs. We adopted a cross-validation with 80% to train, 10% to test and 10% to validate the different models. We used a public pre-trained 300-dimensional word vectors of GloVe[3], which are trained in a Common crawl dataset. Existing and proposed models were trained using ranking hinge loss function during 400 epochs, on the WikiQA dataset and categorical cross entropy as loss function during 500 epochs, on the QuoraQP[4] dataset. We reported performances at the end of all training epochs. In both *Symmetric* and *Asymmetric* architectures, we opted by the recommended hyper-parameters configuration, either on the corresponding paper or in Matchzoo. For the C-DSSM model, we used embedded word vectors rather than the tri-letter hashing method [12] in order to compare the symmetric and asymmetric version.

4.2   Results and Discussion

Table 1 shows the performance results, in WikiQA and QuoraQP datasets, of the different models with two architecture configurations: the symmetric configuration includes the *Original* architecture of the corresponding models and their respective architecture *(Q+A)* where the attention layer ω is applied at both inputs simultaneously. The asymmetric architecture refers to the extended model where layer ω is added to one input at a time: question input *(Q)* or answer input *(A)*. Superscripts ▲ and ▼ show respectively the significance[5] of

---

[2] The corresponding code will be available on MatchZoo and public to allow the reproducibility of the results we show in this paper.

[3] http://nlp.stanford.edu/data/glove.840B.300d.zip.

[4] The loss values of some of the models converged after more than 400 epochs in QuoraQP dataset.

[5] We performed Student's test with $P = 0.05$.

Table 1. Comparison of the symmetric and asymmetric architectures using several text matching models, in WikiQA and QuoraQP datasets. The ".ω" refers to application of layer ω with the corresponding model, as described in Fig. 2. Values in **Bold** indicates the best performances. Superscripts ▲ and ▼ refer to the significance of the results improvement and deterioration respectively.

Performance on WikiQA				MRR	ndcg@3	ndcg@5	MAP
Models				MRR	ndcg@3	ndcg@5	MAP
Classical models	LM			0.5981	0.5841	0.6282	0.5932
	BM25			0.5811	0.5668	0.6203	0.5762
Neural Models	Symmetric	Original	ARC-II	0.5708	0.5410	0.6095	0.5606
			C-DSSM	0.5586	0.5149	0.5902	0.5451
			DUET	0.6259	0.6016	0.6561	0.6113
			MatchPyramid	0.6529	0.6442	0.6902	0.6436
			MV-LSTM	0.6215	0.6101	0.6549	0.6046
		(Q+A)	ARC-II.ω	0.5814	0.5548	0.6194	0.5743
			C-DSSM.ω	0.5622	0.5266	0.5891	0.5523
			DUET.ω	0.5982	0.5589▼	0.6283	0.5801
			MatchPyramid.ω	0.4698▼	0.4272▼	0.5202▼	0.4697▼
			MV-LSTM.ω	0.5904	0.5562	0.6145	0.5562
	Asymmetric	(Q)	ARC-II.ω	0.5748	0.5117	0.5872	0.5465
			C-DSSM.ω	0.5222	0.4973	0.5530▼	0.5134
			DUET.ω	0.6314	0.6116	0.6619	0.6158
			MatchPyramid.ω	**0.6715**	**0.6649**	**0.7068**	**0.6591**
			MV-LSTM.ω	0.6691▲	0.6519▲	0.6948▲	0.6507▲
		(A)	ARC-II.ω	0.5528	0.5627	0.6151	0.5741
			C-DSSM.ω	0.5886	0.5461	0.6190	0.5763
			DUET.ω	0.6383	0.6113	0.6679	0.6251
			MatchPyramid.ω	0.5575▼	0.5360▼	0.5952▼	0.5502▼
			MV-LSTM.ω	0.6174	0.6003	0.6590	0.6165

Accuracy on QuoraQP dataset				
Models	Symmetric		Asymmetric	
	Original	(Q+A)	(Q)	(A)
C-DSSM	0.670969	0.668107	0.751076	0.748548
ARC-II	0.803320	0.785819	0.786159	0.789267
MatchPyramid	**0.818289**	**0.808887**	**0.817010**	**0.815087**
MV-LSTM	0.759707	0.774055	0.790089	0.780036
DUET	0.777509	0.765360	0.767198	0.761784

the improvements and the deteriorations of the models performances. In the WikiQA dataset, the results show that for all models and metrics, at least one of the asymmetric architectures, (Q) or (A), outperforms its symmetric counterparts, including Original and (Q+A). Indeed, the performances obtained with the asymmetric (Q)-MatchPyramid.ω are the best for this dataset. Besides, the (Q)-MV-LSTM.ω outperforms significantly the original model MV-LSTM. Note that these results strongly support our claim about the impact of the asymmetric architectures w.r.t. the question or the answer. Even if there are not significant improvements in several models, the asymmetric architecture enable the neural models, such as CDSSM, to reach results of the classical models such as BM25. In the QuoraQP dataset, the symmetric task does not benefit of the asymmetric architecture given the symmetric nature of the question-question matching. There are no significant improvements over the asymmetric architectures compared to the original. We retained from this analysis that the asymmetric aspect of the inputs has an important impact on the matching process. By consequence, matching models must adapt their architectures to the nature of the task. Note that we carried additional investigations and we figured out that the asymmetric architecture performs differently w.r.t. question type (*what*, *who*, ...). Results are omitted due to paper size limitation.

5 Conclusion

In this paper we proposed an asymmetric architecture for asymmetric matching tasks. We used an attention layer to extend several state-of-the-art models and construct the corresponding asymmetric architectures. Experiments in two different QA datasets showed promising results of the asymmetric architecture as compared to the symmetric one. We conclude that when the model performs an asymmetric matching task, our architecture enables to acknowledge the asymmetric aspect and provide better results. Since there were no significant differences with some experimented models between the original and extended versions, the up coming work will involve the use of additional datasets to confirm the importance of the asymmetric architecture. Our work opens a new perspective for future research and will focus our attention on how to make a neural model automatically adapt to the nature of the task being addressed.

References

1. Abishek, K., Hariharan, B.R., Valliyammai, C.: An enhanced deep learning model for duplicate question pairs recognition. In: Nayak, J., Abraham, A., Krishna, B.M., Chandra Sekhar, G.T., Das, A.K. (eds.) Soft Computing in Data Analytics. AISC, vol. 758, pp. 769–777. Springer, Singapore (2019). https://doi.org/10.1007/978-981-13-0514-6_73
2. Addair, T.: Duplicate question pair detection with deep learning. Stanf. Univ. J. (2017)
3. Bahdanau, D., Cho, K., Bengio, Y.: Neural machine translation by jointly learning to align and translate. arXiv preprint arXiv:1409.0473 (2014)
4. Bordes, A., Weston, J., Collobert, R., Bengio, Y.: Learning structured embeddings of knowledge bases. In: Proceedings of the Twenty-Fifth AAAI Conference on Artificial Intelligence, AAAI 2011, pp. 301–306. AAAI Press (2011). http://dl.acm.org/citation.cfm?id=2900423.2900470
5. Bromley, J., Guyon, I., LeCun, Y., Säckinger, E., Shah, R.: Signature verification using a "Siamese" time delay neural network. In: Advances in Neural Information Processing Systems, pp. 737–744 (1994)
6. Fan, Y., Pang, L., Hou, J., Guo, J., Lan, Y., Cheng, X.: MatchZoo: a toolkit for deep text matching. arXiv preprint arXiv:1707.07270 (2017)
7. Hu, B., Lu, Z., Li, H., Chen, Q.: Convolutional neural network architectures for matching natural language sentences. In: Ghahramani, Z., Welling, M., Cortes, C., Lawrence, N.D., Weinberger, K.Q. (eds.) Advances in Neural Information Processing Systems 27, pp. 2042–2050. Curran Associates, Inc. (2014). http://papers.nips.cc/paper/5550-convolutional-neural-network-architectures-for-matching-natural-language-sentences.pdf

8. Mitra, B., Diaz, F., Craswell, N.: Learning to match using local and distributed representations of text for web search. In: Proceedings of the 26th International Conference on World Wide Web, WWW 2017, pp. 1291–1299. International World Wide Web Conferences Steering Committee, Republic and Canton of Geneva, Switzerland (2017). https://doi.org/10.1145/3038912.3052579

9. Pang, L., Lan, Y., Guo, J., Xu, J., Wan, S., Cheng, X.: Text matching as image recognition. In: AAAI, pp. 2793–2799 (2016)

10. Parikh, A.P., Täckström, O., Das, D., Uszkoreit, J.: A decomposable attention model for natural language inference. arXiv preprint arXiv:1606.01933 (2016)

11. Peng, Y., Liu, B.: Attention-based neural network for short-text question answering. In: Proceedings of the 2018 2nd International Conference on Deep Learning Technologies, ICDLT 2018, pp. 21–26. ACM, New York (2018). https://doi.org/10.1145/3234804.3234813

12. Shen, Y., He, X., Gao, J., Deng, L., Mesnil, G.: Learning semantic representations using convolutional neural networks for web search. In: Proceedings of the 23rd International Conference on World Wide Web, WWW 2014 Companion, pp. 373–374. ACM, New York (2014). https://doi.org/10.1145/2567948.2577348

13. Sutskever, I., Vinyals, O., Le, Q.V.: Sequence to sequence learning with neural networks. In: Ghahramani, Z., Welling, M., Cortes, C., Lawrence, N.D., Weinberger, K.Q. (eds.) Advances in Neural Information Processing Systems 27, pp. 3104–3112. Curran Associates, Inc. (2014). http://papers.nips.cc/paper/5346-sequence-to-sequence-learning-with-neural-networks.pdf

14. Wan, S., Lan, Y., Guo, J., Xu, J., Pang, L., Cheng, X.: A deep architecture for semantic matching with multiple positional sentence representations. AAAI **16**, 2835–2841 (2016)

15. Yang, L., Ai, Q., Guo, J., Croft, W.B.: aNMM: ranking short answer texts with attention-based neural matching model. In: Proceedings of the 25th ACM International on Conference on Information and Knowledge Management, pp. 287–296. ACM (2016)

16. Yang, Y., Yih, W., Meek, C.: WikiQA: a challenge dataset for open-domain question answering. In: Proceedings of the 2015 Conference on Empirical Methods in Natural Language Processing (EMNLP). Association for Computational Linguistics, Lisbon, September 2015

17. Yang, Z., Yang, D., Dyer, C., He, X., Smola, A., Hovy, E.: Hierarchical attention networks for document classification. In: Proceedings of the 2016 Conference of the North American Chapter of the Association for Computational Linguistics: Human Language Technologies, pp. 1480–1489 (2016)

18. Yin, W., Schütze, H., Xiang, B., Zhou, B.: ABCNN: attention-based convolutional neural network for modeling sentence pairs. Trans. Assoc. Comput. Linguist. **4**(1), 259–272 (2016)

19. Zweig, G., Platt, J.C., Meek, C., Burges, C.J.C., Yessenalina, A., Liu, Q.: Computational approaches to sentence completion. In: Proceedings of the 50th Annual Meeting of the Association for Computational Linguistics: Long Papers - Volume 1, ACL 2012, pp. 601–610. Association for Computational Linguistics, Stroudsburg (2012). http://dl.acm.org/citation.cfm?id=2390524.2390609

QGraph: A Quality Assessment Index for Graph Clustering

Maria Halkidi[1(✉)] and Iordanis Koutsopoulos[2]

[1] University of Piraeus, Piraeus, Greece
mhalk@unipi.gr
[2] Athens University of Economics and Business, Athens, Greece
jordan@aueb.gr

Abstract. In this work, we aim to study the cluster validity problem for graph data. We present a new validity index that evaluates structural characteristics of graphs in order to select the clusters that best represent the communities in a graph. Since the work of defining what constitutes cluster in a graph is rather difficult, we exploit concepts of graph theory in order to evaluate the cohesiveness and separation of nodes. More specifically, we use the concept of *degeneracy*, and *graph density* to evaluate the connectivity of nodes *in* and *between* clusters. The effectiveness of our approach is experimentally evaluated using real-world data collections.

Keywords: Cluster validity · Graph clustering · Data analysis

1 Introduction

In recent years, there are many application domains (web applications, biomedicine, social networks) where the available data are represented as graphs and thus the requirement for graph data analysis techniques is stronger than ever. *Graph clustering* is one of the main tasks in graph data analysis and has attracted the interest of data mining research community. A graph clustering can be defined as a set of subgraphs, further referred to as graph clusters or communities, characterized by dense connections between vertices in clusters and low density between vertices of different clusters. The last few decades, a number of methods for graph clustering (community detection) have been proposed [3,4].

Clustering algorithms with different cost functions give different results, and there is no single optimal choice of the algorithm and the cost function for all available data sets. Even the same clustering algorithm under different assumptions and input parameter values could result in different partitionings of a data set. Then a challenging issue is how to evaluate the quality of different clustering results and select the best possible clustering for a data set. This is the well known *cluster validity problem*.

The problem of *cluster validity* has been widely studied and there is a number of indices for evaluating clustering results [6,8]. They measure the compactness

© Springer Nature Switzerland AG 2019
L. Azzopardi et al. (Eds.): ECIR 2019, LNCS 11438, pp. 70–77, 2019.
https://doi.org/10.1007/978-3-030-15719-7_9

and separability of clusters using variance or density analysis methods. The majority of cluster validity indices are applied to Euclidean space while there are only few works on graph data.

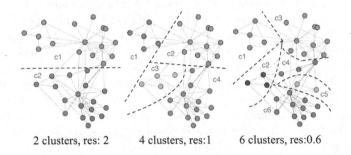

2 clusters, res: 2   4 clusters, res:1   6 clusters, res:0.6

Fig. 1. Zachary karate club data set: partitioning into 2, 4 and 6 clusters.

Since the need for new data analysis techniques that deal with the graph structure increases, the requirement of evaluating the quality of analysis results also arizes. Figure 1 shows the partitioning of graph under different input parameters using a modularity-based clustering algorithm [1]. In most of the cases we are not able to have a visualization of our data and thus it is difficult to identify which is the partitioning that best fits them. Moreover the characteristics and properties of graphs are different from other data types, such as numerical, categorical data, and thus new metrics have to been studied in order to evaluate the structure of graph clusters.

In this work we aim to study the characteristics of graph clusters and develop a new cluster validity approach for evaluating their quality. We exploit concepts of graph theory such as *graph degeneracy* and *density* to evaluate both locally and globally the connectivity in and between graph clusters.

2   Related Work

Clustering algorithms extract clusters from data, which are not known a priori, and thus the final partition of a data set requires some sort of evaluation in most applications [5,6]. The evaluation of clustering results is well known in the research community and a number of research works have been made especially in the area of machine learning, pattern recognition [8]. Since the application of graphs is high, the interest of researchers to develop graph clustering algorithms increases. Fortunato [3] provides an extended overview on community detection methods in graphs while he also discusses issues regarding the significance of clustering and how methods should be tested and compared against each other. However, there is little work on cluster validity approaches for graph data.

Boutin et al. [2], present an overview of validity indices for graph clustering while they also propose some normalized version of the available indices.

There are indices that extend widely used cluster validity indices, such as David Bouldin, Dunn's index, to deal with graph structures. Also there are indices that use number of links and vertices in a graph to evaluate the connectivity within and between graph clusters. A metric that uses the concepts of cohesion and separation to assess the quality of clusters is the Silhouette index. Its definition is based on the distance between vertices within clusters and between clusters. One of the limitation of this index is the calculation cost. Also it presents a tendency of giving better scores for clusterings with many singletons.

Another well-known metric is the conductance of a cut [12]. It compares the number of edges cut (i.e. between clusters) and the number of edges in either of the two clusters induced by the cut. Also the coverage [13] of clustering is a metric that used for evaluating clustering results. It is defined as the fraction of intra-cluster edges with respect of the edges of the whole graph. In [14], another cluster validity index, called *performance* metric, is presented. It counts the number of edges withing clusters along with the edges that do not exist between the cluster vertices and the other vertices in the graph.

3 Problem Statement

We assume a graph $G = (V, E)$, where V is the set of nodes (vertices) and E is the set of edges. Let $SC = \{C_1, \ldots, C_N\}$ be a set of different partitionings (clusterings) of G. The number of clusters (also known as communities) could be different in each partitioning C_i.

We desire to define an index that assigns to each $C_i \in SC$ a value. This value should be indicative of the quality of clustering C_i, i.e. it shows how well C_i captures the structure of clusters in the graph G. The definition of such an index should be compatible with the main idea of clustering that is the extraction of compact and well separated clusters. Among the available partitionings in SC, we expect that C_i that best fits the clusters in graph G, would correspond to the optimal value of the validity index.

In summary, given an index Q, a graph G and a set of its partitionings SC, we aim to find the partitioning $C_i \in SC$ such that

$$max/min_{C_i \in SC} Q(C_i)$$

The selection of *max* or *min* depends on the index definition.

4 Definition of a Cluster Validity Index for Graphs

We consider an undirected graph $G = (V, E)$ comprising a set V of vertices together with a set E of edges and let $SC = \{C_1, \ldots, C_N\}$ be the set of different partitionings (clusterings) of G.

We denote the degree of a vertex $v \in V$ as $d_G(v) = |\{u|(u,v) \in E\}|$. The k-core of a graph is the maximal subgraph of G, $G' = (V', E')$ where $\forall v \in V'$,

$d_{G'}(v) \geq k$. The core number of a vertex v, $core(v)$, is the order of the highest-order core that v belongs to. A vertex has core number k if v belongs to the k-core but not to the $(k+1)$-core.

The *degeneracy* of a graph G, denoted by $deg(G)$, is the largest value k such that it has a k-core. A k-degenerate graph is an undirected graph in which every sub-graph has a vertex of degree at most k. Then the degeneracy of a graph can be defined as the maximum core number of vertices in V: $deg(G) = k_{max}\text{-}core = max_{v \in V} core(v)$. The k_{max}-core is also called *degeneracy*-core.

Evaluating the Compactness of Clusters. The degeneracy of a graph G has been extensively used for evaluating and detecting strongly cohesive communities in real-word graphs [7]. It indicates the existence of sub-graphs in G where each vertex has at least $deg(G)$ neighbors. Then the *degeneracy* can be considered as measure of graph's sparsity.

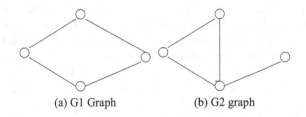

(a) G1 Graph   (b) G2 graph

Fig. 2. Degeneracy vs degeneracy coverage: the graphs G1 and G2 have the same degeneracy $(deg(G1) = deg(G2) = 2)$ while their degeneracy coverage is different $deg\_coverage(G1) = 1$, $deg\_coverage(G2) = \frac{3}{4}$.

A question that arises at this point is what percentage of graph vertices participate to the *degeneracy*-core (i.e. they have degree at least $deg(G)$). In Fig. 2 the graphs $G1$, and $G2$ have the same degeneracy but the *degeneracy*-core of each graph covers different part of the graph. In $G1$ all the vertices are part of the *degeneracy*-core while in $G2$ three of the four vertices participate in *degeneracy*-core. Thus in order to evaluate the connectivity of the clusters (communities) in a graph we have to take into account both the degeneracy of the graph cluster and the coverage of its *degeneracy*-core.

We denote the *degeneracy-core* of a graph G as $dG = (dV, dE)$, $dV \subset V$ and $dE \subset E$. We define the coverage of *degeneracy*-core as: $deg\_coverage(G) = \frac{|dV|}{|V|}$.

We evaluate the *linkage of a graph cluster* c_i based on the concepts of *degeneracy* and the *coverage of its degeneracy-core*. More specifically, we define the *intra_linkage* of a graph cluster as: $intra\_linkage(c_i) = deg(c_i) \cdot deg\_coverage(c_i)$.

Considering a clustering of G into m clusters $C = \{c_1, \ldots, c_m\}$, the linkage within C is defined as average *intra_linkage* of all clusters in C:

$$intraLink(C) = \frac{1}{m} \sum_{c_i \in C} intra\_linkage(c_i)$$

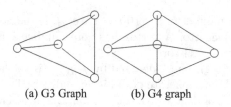

(a) G3 Graph   (b) G4 graph

Fig. 3. Intra-linkage vs density: the graphs G3 and G4 have the same intra-linkage($inta\_linkage(G3) = intra\_linkage(G4) = 3$) while their *intra_density* is different $intra\_dens(G3) = 1$, $intra\_dens(G4) = \frac{4}{5}$.

Another metric that we use to evaluate the cohesion of a graph is the *density* defined as the percentage of the expected edges that exists in the graph. Then the *density of a graph cluster* $c_i = (V_i, E_i)$ is given by: $dens(c_i) = \frac{2 \cdot |E_i|}{|V_i|(|V_i|-1)}$.

There are cases that graphs have similar *intra_linkage* but the density of graphs is different. Figure 3 shows two graphs $G3$, $G4$ that have the same *intra_linkage* but the $G3$ is densest than $G4$. Moreover Fig. 2 shows the existence of graphs with the same number of nodes and edges (i.e. same density) but their *intra_linkage* is different ($intra\_linkage(G1) = 1$, $intra\_linkage(G2) = \frac{3}{2}$). Then both *density* and *intra-linkage* should be taken into account in order to evaluate the quality of a graph cluster.

The *density of a cluster*, $dens(c_i)$, measures the connectivity among all the vertices in the cluster while *intra_linkage* concentrates its evaluation at the densest parts of a cluster.

Evaluating the Separation of Clusters. Assume a pair of graph clusters c_i and c_j with vertices VC_i, VC_j, respectively. We denote $G(c_i, c_j) = (VC_i \cup VC_j, E(c_j, c_j))$ the graph that contains the vertices of the clusters c_i and c_j and the set of edges $E(c_j, c_j)$ which connect the vertices in VC_i with the vertices in VC_j.

The *inter_linkage* of a pair of clusters (c_i, c_j) measures the linkage of two clusters. It is defined based on the degeneracy of the graph $G(c_i, c_j)$ and the coverage of respective *degeneracy-core*. That is:

$$inter\_linkage(c_i, c_j) = deg(G(c_i, c_j)) * deg\_coverage(G(c_i, c_j))$$

Then the *inter-linkage* of a partitioning of G into m clusters is given by:

$$interLink(C) = \frac{2}{m(m-1)} \sum_{i} \sum_{j, j<i} inter\_linkage(G(c_i, c_j))$$

Defining the Validity Index. A partitioning that best fits the graph structure of G is expected to contain graph clusters with high *intra_linkage* and low *inter_linkage*. Then a high difference between *intra-linkage* and *inter-linkage* of a clustering C for graph G indicates that C is a good partitioning of G.

The *inter-linkage* of clusters is mainly focused on the densest parts between clusters. The density of the graph defined by two graph clusters is also used to evaluate the separation of clusters.

The *inter-density* of a pair of clusters c_i, c_j is defined as the percentage of expected edges across the clusters that exist in the graph: $interDens(c_i, c_j) = \frac{|E(c_i,c_j)|}{|V_i||V_j|}$.

The connectivity of two clusters should be evaluated in comparison with the connectivity within the clusters. Given a pair of clusters (c_i, c_j) we define the connectivity of these clusters as follows: $InterCon(c_i, c_j) = \frac{InterDens(c_i,c_j)}{min\{dens(c_i),dens(c_j)\}}$.

A partitioning C whose clusters are well separated are expected to have low inter-connectivity, that is $Separation(C) = \frac{2}{m \cdot (m-1)} \sum_i \sum_j \frac{1}{InterCon(c_i,c_j)}$.

Then we define the following index as indicator of the quality of graph clustering.

$$QGraph(C) = (intraLink(C) - interLink(C)) + Separation(C)$$

The focus of the first part of the proposed index is on the densest areas between and within clusters while the second part refers to a more global evaluation of clusters connectivity. Based on the above definition we could infer that the higher the value of $QGraph$, the better the quality of clustering.

Table 1. The values of cluster validity indices for the data sets: (A) Zachary karate club, (B) Euro-core emails.

(A) Zachary karate club			(B) Euro-core email		
Number of clusters	Modularity	QGraph	Number of clusters	Modularity	QGraph
2	7.23	**8.43**	25	**72.48**	2.82
4	**9.13**	5.38	27	43.0043	2.85
6	9.03	4.85	29	30.596	3.36
			42	11.569	**20.83**

5   Experimental Study

Data sets. We have experimented with the real world dataset *EU-core network* [10] for which there is available the "ground-truth" community memberships of nodes[1]. The network was generated using email data from a large European research institution. It contains 1005 nodes and 25571 edges. The network is organized into 42 communities. The average clustering coefficient of the network is 0.3994. Moreover we used the *Zachary karate club* network [11]. This is a social network of friendships between 34 members of a karate club at a US university.

---

[1] http://snap.stanford.edu/data/.

In this data set we can find two groups of people into which the karate club was split after an argument between two teachers.

Discussion on Experimental Results. Each dataset is partitioned in different number of clusters using the clustering algorithm presented in [1]. We evaluate the clustering results using the *modularity measure* presented in [2] and the proposed *QGraph* index. Table 1 presents in a comparative fashion the values of the validity indices with respect to the number of clusters. The highest value of the indices indicates the partitioning that the index selects as the best one for the considered dataset.

In case of the EU core (see Table 1(B)), the *QGraph* index takes its highest value for the partitioning of 42 communities (corresponding to ground truth) while the *modularity index* select 25 clusters as the best partitioning.

Figure 1 depicts different partitionings of the *Zachary karate club* dataset. The results of cluster validity indices for each of the defined partitionings are presented in Table 1(A). We can observe that *QGraph* takes its highest value for the partitioning of two clusters while the *modularity index* selects 4 clusters as the best partitioning.

The above experimental study shows that *QGraph* achieves in all cases to select the partitioning that best fits the underlying graph data.

6   Conclusion

In this paper we proposed a new validity index *QGraph* for evaluating graph clustering results. The concepts of graph degeneracy and graph density are properly combined to assess the compactness and separation of extracted clusters. As further work, we plan to evaluate the scalability of the approach and its performance using data sets with various structures and sizes.

Acknowledgment. This work has been partly supported by the University of Piraeus Research Center. I. Koutsopoulos acknowledges the support from the AUEB internal project "Original scientific publications".

References

1. Blondel, V.D., Guillaume, J.L., Lambiotte, R., Lefebvre, E.: Fast unfolding of communities in large networks. J. Stat. Mech.: Theory Exp. **2008**, P10008 (2008)
2. Boutin, F., Hascoet, M.: Cluster validity indices for graph partitioning. In: Proceedings of the International Conference of Information Visualisation (2004)
3. Fortunato, S.: Community detection in graphs. Phys. Rep. **486**, 75–174 (2010)
4. Schaeffer, S.E.: Graph clustering. Comput. Sci. Rev. **1**, 27–64 (2007)
5. Halkidi, M., Batistakis, Y., Vazirgiannis, M.: On clustering validation techniques. J. Intell. Inf. Syst. **17**, 107–145 (2001)
6. Halkidi, M., Vazirgiannis, M.: Quality assessment approaches in data mining. In: The Data Mining and Knowledge Discovery Handbook: A Complete Guide for Practitioners and Researchers. Kluwer Academic Publishers (2005)

7. Giatsidis, C.: Graph mining and community evaluation with degeneracy. Ph.D. thesis (2013)
8. Theodoridis, S., Koutroubas, K.: Pattern recognition. Academic Press, Cambridge (1999)
9. Lancichinetti, A., Fortunato, S., Radicchi, F.: Benchmark graphs for testing community detection algorithms. Phys. Rev. **78**, 046110 (2008)
10. Yin, H., Benson, A.R., Leskovec, J., Gleich, D.F.: Local higher-order graph clustering. In: Proceedings of the 23rd ACM SIGKDD International Conference on Knowledge Discovery and Data Mining (2017)
11. Kannan, R., Vempala, S., Vetta, A.: An information flow model for conflict and fission in small groups. J. Anthropol. Res. **33**, 452–473 (1977)
12. Kannan, R., Vempala, S., Vetta, A.: On clusterings: good, bad and spectral. J. ACM **51**(3), 497–515 (2004)
13. Brandes, U., Gaertler, M., Wagner, D.: Engineering graph clustering: models and experimental evaluation. J. Exp. Algorithmics **12**, 1–26 (2008)
14. van Dongen, S.M.: Graph clustering by flow simulation. Ph.D. thesis, University of Utrecht, The Netherlands (2000)

A Neural Approach to Entity Linking on Wikidata

Alberto Cetoli[1]([✉]), Stefano Bragaglia[1], Andrew D. O'Harney[1], Marc Sloan[1], and Mohammad Akbari[2]

[1] Contextscout, London, UK
{alberto,stefano,andy,marc}@contextscout.com
[2] UCL, London, UK
m.akbari@ucl.ac.uk

Abstract. We tackle Named Entity Disambiguation (NED) by comparing entities in short sentences with *Wikidata* graphs. Creating a context
vector from graphs through deep learning is a challenging problem that
has never been applied to NED. Our main contribution is to present an
experimental study of recent neural techniques, as well as a discussion
about which graph features are most important for the disambiguation
task. In addition, a new dataset (Wikidata-Disamb) is created to allow
a clean and scalable evaluation of NED with *Wikidata* entries, and to
be used as a reference in future research. In the end our results show
that a Bi-directional Long Short-TermMemory (Bi-LSTM) encoding of
the graph triplets performs best, improving upon the baseline models
and scoring an F1 value of 91.6% on the Wikidata-Disamb test set (The
dataset and the code (with configurations) for this paper can be found
at https://github.com/contextscout/ned-graphs).

Keywords: Named Entity Disambiguation · Graphs · Wikidata ·
RNN · GCN

1 Introduction and Motivations

A mentioned entity in a text may refer to multiple entities in a knowledge base.
The process of correctly linking a mention to the relevant entity is called Entity
Linking (EL) or NED [5]. Entity disambiguation is different from Named Entity
Recognition (NER), where the system must detect the relevant mention boundaries given a definite set of entity types. In NED, the system must be able to
generate a context for an entity in a text and an entity in a knowledge base, then
correctly link the two [18,19]. NED is a crucial step in web search tasks [1,4,7],
information retrieval [13], data mining [6,9,12], and semantic search [8,15].

A background knowledge base can appear in many forms: as a collection of
texts, as a relational database, or as a collection of graphs in a graph database.
Representing data as ensembles of linked information is an increasing popular form of storage. One example can be found in the successful *Wikidata*
database [21], which aims to mirror the content of *Wikipedia* in a linked format.

L. Azzopardi et al. (Eds.): ECIR 2019, LNCS 11438, pp. 78–86, 2019.
https://doi.org/10.1007/978-3-030-15719-7_10

Both *Wikipedia* and *Wikidata* contain potentially ambiguous entities. For example, when searching for information about *Captain Marvel*, the results should depend on the context in which this entity appears. Indeed, the name *Captain Marvel* is a character from Marvel comics and a nickname for *Michael Jordan*, the basketball player.

The main contributions of this work are two-fold: First, we aim to empirically evaluate different deep learning techniques to create a context vector from graphs, aimed at high-accuracy NED. Secondly, we create a new dataset to help us in our endeavor. Among the datasets available for this task we took inspiration from Wiki-Disamb30 [10]. In that work, Ferragina and Scaiella tackle the problem of cross referencing text fragments with *Wikipedia* pages. Specifically, they deal with *very short* sentences (30–40 words). We build on their work by translating the pointers of *Wikipedia* pages to *Wikidata* items, thereby creating an *ad hoc* dataset based on Wiki-Disamb30. We call our derivative dataset Wikidata-Disamb. This new dataset creates the perfect playground for us to test various models of NED on *Wikidata*.

2   Methodology

All models share three main elements: A *graph*, a *text*, and an *entity* in the text to disambiguate. The disambiguation task is reduced to a consistency test between the input text and the graph. The graph is composed by nodes connected with edges. The node vectors $\{\mathbf{x}_i\}$ are represented by the centroid of the *Glove* word vectors that make up the nodes: For example, a node called "New York" is represented by averaging the word vectors of "New" and "York". An edge $\mathbf{e}_{ij}$ connects node i with node j. The set of edge vectors $\{\mathbf{e}_{ij}\}$ is computed exactly as for the node vectors, by averaging over the word vectors in each of the edge's labels: For example, the vector of "instance of" is the average of the *Glove* vectors "instance" and "of".

The *text* is described as a sequence of word vectors $\{\mathbf{v}_i\}$, represented using the *Glove* embeddings, while the *item* is used to query the *Wikidata* dataset for the corresponding entry. In most models we have an embedding for the input text $\mathbf{y}_{\text{text}}$ and one for the graph $\mathbf{y}_{\text{graph}}$.

All our models receive as an input the node vectors $\{\mathbf{x}_i\}$, the word embeddings $\{\mathbf{v}_i\}$ (and possibly the edge vectors $\{\mathbf{e}_{i,j}\}$). The output of our models is a binary vector, which tells us whether the input graph is consistent with the entity in the text.

3   Models

3.1   Graph Related Models

In all these models the input text is treated in the same way: The text word vectors $\{\mathbf{v}_i\}$ are first fed to a Bi-LSTM, with outputs $\{\mathbf{y}_i\}$. These outputs are then weighted by a mask: A set of scalars $\{a_i\}$ which are 1 where the item is

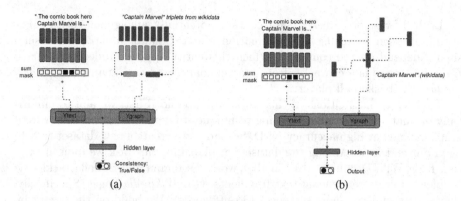

Fig. 1. The *Wikidata* graph can be processed by a Bi-LSTM as a list of triplets (a) or using a GCN approach to disambiguation (b)

supposed to be and 0 otherwise. For example the sentence "The comic book hero Captain Marvel is ..." would have $\{a_i\} = [0,0,0,0,1,1,0,\ldots]$. This mask acts as a "manually induced" attention of the item to disambiguate for. The final output of the Long Short-TermMemory (LSTM) mechanism is the average

$$\mathbf{y}_{\text{text}} = N_{\text{text}}^{-1} \sum_{i=0}^{N_{\text{text}}} a_i \, \mathbf{y}_i, \tag{1}$$

given a sentence with length N_{text}. The following items are our graph-based models:

Text LSTM + Recurrent Neural Network (RNN) of Triplets: In this model, represented in Fig. 1(a), a Bi-LSTM [2] is applied over the sequence of *triplets* in the graph. In the list of input vectors $\mathbf{x}^{\text{triplet}}$ each item is the concatenation of three elements:

$$\mathbf{x}_{i,j}^{\text{triplet}} = \mathbf{x}_i \oplus \mathbf{e}_{i,j} \oplus \mathbf{x}_j \tag{2}$$

where i, j are all the indices between connected nodes in a directed graph. The final states of the Bi-LSTM are then concatenated and then fed to a dense layer, whose output is the graph embedding $\mathbf{y}_{\text{graph}}$.

While this model captures the information of single hops in the graph, it is not suited for capturing the topology of the network. For example, nodes that are topologically close might appear far away in the set of triplets. More importantly, the final embeddings might depend on the specific ordering of the triplets, losing the information about the network shape.

Text LSTM + RNN with Attention: We improve upon the prior model by adding an attention mechanism [3,11,20] after the LSTM for triplets. The output vectors z_i of the LSTM are weighted by an attention coefficient (scalar) b_i and then summed together to create the context vector for the graph.

$$\mathbf{y}_{\text{graph}} = N_{\text{triplets}}^{-1} \sum_{i=0}^{N_{\text{triplets}}} b_i \, \mathbf{z}_i, \tag{3}$$

with

$$\begin{aligned} \mathbf{b} &= \text{softmax}(\mathbf{c}) \\ c_i &= \text{ReLU}(\mathbf{W}_{\text{triplets}} \, \mathbf{z}_i + \mathbf{W}_{\text{text}} \, \mathbf{y}_{\text{text}} + b_{\text{triplets}}), \end{aligned} \tag{4}$$

where the matrices $\mathbf{W}_{\text{triplets}} \in \mathbb{R}^{1 \times dim(z_i)}$ and $\mathbf{W}_{\text{text}} \in \mathbb{R}^{1 \times dim(y_{\text{text}})}$ and the scalar b_{triplets} are learned in training. We expect this attention method to improve the disambiguation task by giving more weight to relevant triplets.

Text LSTM + GCN: We couple the Bi-LSTM with a GCN [14] to compare the sentence to the *Wikidata* graph. The base diagram of the network is in Fig. 1(b). Specifically, the convolutions can be employed to create an embedding vector of the relevant *Wikidata* graph. Ideally, after the graph convolutions, the vector at the position of the central item summarizes the information in the graph.

A graph convolutional network works by stacking convolutional layers based on the topology of the network. Typically, by stacking together N layers the network can propagate the features of nodes that are at most N hops away. The information at the k^{st} layer is propagated to the next one according to the equation

$$\mathbf{h}_v^{k+1} = \text{ReLU}\left(\sum_{u \in \mathcal{N}(v)} \left(\mathbf{W}^k \, \mathbf{h}_u^k + \mathbf{b}^k \right) \right), \tag{5}$$

where u and v are two indices of nodes in the graph. $\mathcal{N}$ is the set of nearest neighbors of node v, plus the node v itself. The vector $\mathbf{h}_u^k$ represents node u's embeddings at the k^{st} layer. The matrix $\mathbf{W}$ and vector $\mathbf{b}$ are learned during training and map the embeddings of node u onto the adjacent nodes in the graph. In this paper the we only consider the outgoing edges from each node[1]. In addition, we reify the relations to appear as additional nodes. In this way the edges become nodes themselves.

Text LSTM + GCN with Attention: We aim to improve upon the prior model by adding an attention mechanism after the GCN layers. This can be obtained by adding a set of weights α_{vu} so that Eq. 5 becomes

$$\mathbf{h}_v^{k+1} = \text{ReLU}\left(\sum_{u \in \mathcal{N}(v)} \alpha_{vu} \left(\mathbf{W}^k \, \mathbf{h}_u^k + \mathbf{b}^k \right) \right). \tag{6}$$

The coefficients α_{uv} signify the attention to be paid to the information being passed from node u to node v. This attention needs to be a function of the vector

---

[1] The *Wikidata* graph is oriented so that the edges direct outwards with respect to the central node. In order to percolate information towards the central node we must consider the outgoing edges.

and edge nodes, as well as a function of the input text. We choose the following method for GCN attention:

$$\begin{aligned}
\boldsymbol{\alpha} &= (\mathbb{1} + \mathbf{A}) \odot \text{softmax}(\mathbf{E}) \\
\mathbf{E} &= \mathbf{B}^\top \mathbf{B} \\
\mathbf{B} &= \text{ReLU}(\mathbf{W}_{\text{graph}}\,\mathbf{H}^k + \mathbf{W}_{\text{text}}\,\mathbf{Q}_{\text{text}} + \mathbf{C}_{\text{graph}}),
\end{aligned} \tag{7}$$

where the softmax function acts on the last dimension of $\mathbf{E}$ and $\mathbf{A}$ is the original adjacency matrix for the graph. The matrix $\mathbf{B}$ models the information propagating from a node in the context of the input text: The columns of the matrix $\mathbf{H}^k \in \mathbb{R}^{m \times n}$ are the layer vectors $\mathbf{h}_u^k$, with n the number of nodes and m the dimension of the layer embeddings; $\mathbf{Q} \in \mathbb{R}^{q \times n}$ is a matrix where all the columns are identical and equal to the input text embeddings $\mathbf{y}_{\text{text}}$, with q the dimension of the text embeddings. In the matrices $\mathbf{W}_{\text{graph}} \in \mathbb{R}^{d \times m}$ and $\mathbf{W}_{\text{text}} \in \mathbb{R}^{d \times q}$ and in the bias matrix $\mathbf{C}_{\text{graph}} \in \mathbb{R}^{d \times n}$ d is an arbitrary intermediate dimension.

To the best of our knowledge, this formulation of GCN attention is original.

3.2 Baseline Models

Feedforward of Averages: We take the average $\bar{\mathbf{v}}$ of the words in the sentence and the average $\bar{\mathbf{x}}$ of all the nodes in the graph, concatenate them, and feed them to a feedforward neural net with one hidden layer. The final output is binary, meaning that the sentence can be either consistent or inconsistent with the *Wikidata* graph.

Text LSTM + Centroid: In this model (and in the following ones) the input text is processed by same Bi-LSTM method of Fig. 1(a) and (b). Here the graph information is instead collapsed onto the average vector $\mathbf{y}_{\text{graph}} = \bar{\mathbf{x}}$ as in the baseline.

Text LSTM + Linear Attention: Instead of using the average $\bar{\mathbf{x}}$ for representing the graph, we employ an attention model over the node vectors. The output of this attention model is

$$\mathbf{y}_{\text{graph}} = N_{\text{nodes}}^{-1} \sum_{i=0}^{N_{\text{nodes}}} b_i \, \mathbf{x}_i, \tag{8}$$

with

$$\begin{aligned}
\mathbf{b} &= \text{softmax}(\mathbf{c}) \\
c_i &= \text{ReLU}(\mathbf{W}_{\text{nodes}}\,\mathbf{x}_i + \mathbf{W}_{\text{text}}\,\mathbf{y}_{\text{text}} + \mathbf{b}_{\text{nodes}}),
\end{aligned} \tag{9}$$

where $\mathbf{W}_{\text{nodes}}$, $\mathbf{W}_{\text{text}}$, and $\mathbf{b}_{\text{nodes}}$ are learned through backpropagation. This attention technique ideally improves the classification task by giving more weight to relevant nodes.

Text LSTM + RNN of Nodes: Instead of taking the average of the nodes in the graph, we use a Bi-LSTM on the nodes and then concatenate the final hidden layers to create the representation vector $\mathbf{y}_{\text{graph}}$ (same structure as Fig. 1(a), but with node vectors in place of triplet vectors).

4   The Dataset

We create a new dataset from the information in the Wiki-Disamb30 set. Originally, the dataset addressed the need of a corpus of very short texts (a few 30–40 words) where a specific entity was linked to the correct *Wikipedia* page. The original dataset contains about 2 million entries and presents three elements for each one: an English sentence, the name of the entity to disambiguate, and the correct *Wikipedia* item corresponding to the entity.

Our dataset provides a conversion from the *Wikipedia* page to a *Wikidata* item, when this conversion exists. If the conversion is not possible (roughly 3% of the cases) the original entry is simply skipped. In order to have a consistent disambiguation task we also select an incorrect *Wikidata* item to pair with the correct one, linked to it by having the same name (or same alias).

Among many alternatives, the incorrect entry is selected to not be trivial, i.e., a disambiguation page or an entry with less than three triplets. The selection among different valid candidates is random. Human supervision has been minimal, limited to checking random items after the set had been generated. In this way we obtain a balanced dataset, where the correct entity appears as many times as the wrong one.

After applying those selection constraints we chose 120000 items, of which 100 thousand in the training set, and 10000 entries each for the development and test sets. Since we measure the consistency with *correct* and *wrong Wikidata* IDs separately, the size of the sets effectively doubles. The reduced size of the dataset allows us for scalable and manageable testing compatible with our computational resources.

In the end, each item in the Wikidata-Disamb dataset contains four items: an English sentence, the name of the entity to disambiguate, the correct *Wikidata* item and an incorrect one.

5   Experimental Results

The results of our experiments are presented in Table 1. We took two evaluations for each model and show the average result. The difference between the lowest and highest score varies between 0.4% and 0.8% for the different models. In absence of more complete statistics, we choose the middle value 0.6% as an estimate for the statistical error to attribute to all our measurements. All results are approximated to the first significant digit of the error.

The modest results of the *linear attention* model are particularly interesting, suggesting that the classification task does not seem to rely on specific easy-to-identify nodes, and that the whole node set information seem to play a role for an accurate result.

The second best results of the paper is given by the *RNN of triplets* model, with $F1 = 91.1\%$ on the test set. This model uses the whole graph information taking as an input the set of triplets that compose the *Wikidata* graph. The *RNN of triplets with attention* seems to perform even better, reaching $F1 = 91.6\%$.

Table 1. Results of our architectures expressed as a percentage (best results in bold).

Description	DEV			TEST		
	prec	rec	F_1	prec	rec	F_1
Feedforwad of averages	82.9	86.9	84.8	82.2	87.1	84.5
Text LSTM + Centroid	87.5	91.4	89.5	87.3	91.8	89.5
Text LSTM + Linear attention	80.4	91.7	85.7	79.7	90.6	84.6
Text LSTM + RNN of nodes	80.4	89.6	84.7	79.6	89.5	84.2
Text LSTM + RNN of triplets	**90.7**	92.2	91.4	90.1	92.0	91.1
Text LSTM + RNN of triplets with attention	90.2	**93.1**	**91.6**	**90.2**	**93.0**	**91.6**
Text LSTM + GCN	71.0	87.6	78.4	70.0	87.8	77.8
Text LSTM + GCN with attention	73.7	91.1	81.5	74.8	88.2	81.0

A straightforward conclusion is that the classification task is mildly helped by paying attention on specific triplets.

Conversely, the *Text LSTM* + GCN model performs poorly, with $F1 = 77.8\%$. The reason of this drop in performance is complex, and we believe it rests in the way GCNs create the final embedding vector. The GCN embeddings sum up information that comes from the graph nodes, edges, and topology of the network. This last piece of information is not considered in the triplet model, and we believe it is what confuses the graph convolutional model: In short, in our experiments the GCN seems to give too much importance to the shape of the graphs in training, ending up being confused when testing.

The *Text LSTM + GCN with attention* model seems to perform about 3% better. The attention model effectively adapts the topology of the network to the input text, alleviating some of the issues with the prior model.

5.1 Comparison with Wiki-Disamb30 Results

The dataset we present in this paper is derived from the Wiki-Disamb30 corpus. A comparison with prior results evaluated on the original dataset seems due, albeit somewhat contrived: In the original dataset the context for disambiguation comes from *Wikipedia* pages, whereas in our work we build an embedding vector from *Wikidata* graphs.

In [17] the authors summarize recent NED results running the Wiki-Disamb30 dataset using the algorithms of the original papers. They report $F1 = 84.6\%$ for [16] and $F1 = 90.9$ for [10]. The state-of-the-art still lies in the work of Raiman and Raiman [17], where they achieve $F1 = 92.4\%$.

6   Concluding Remarks

We have shown that it is possible to disambiguate entities in short sentences by looking at the corresponding entries in *Wikidata*. In order to achieve this result, we created a new dataset Wikidata-Disamb, where we present an equal number of correct and incorrect entity linking candidates.

Our *RNN of triplets with attention* model allows us to achieve the best result $F1 = 91.6\%$ over the test set. This is an improvement from the baseline of the simple *feedforward of averages* model of about 7.1%, where the edges of the *Wikidata* graph are not used. The main contribution of this improvement seems to come from processing the input text with a Bi-LSTM. The second biggest improvement happens when including information about the relation type with the *RNN of triplets*. The GCN based approaches are seen to perform poorly. More interesting graph topologies should make the GCN perform better. We aim to address this issue in following works.

Acknowledgements. This work was partially supported by InnovateUK grant Ref. 103677.

References

1. Artiles, J., Amigó, E., Gonzalo, J.: The role of named entities in web people search, vol. 2, p. 534 (2009). https://doi.org/10.3115/1699571.1699582, http://portal.acm.org/citation.cfm?doid=1699571.1699582
2. Augenstein, I., Rocktäschel, T., Vlachos, A., Bontcheva, K.: Stance detection with bidirectional conditional encoding. In: Proceedings of the 2016 Conference on Empirical Methods in Natural Language Processing, pp. 876–885. Association for Computational Linguistics, Austin, November 2016. https://aclweb.org/anthology/D16-1084
3. Bahdanau, D., Cho, K., Bengio, Y.: Neural machine translation by jointly learning to align and translate. arXiv:1409.0473 [cs, stat], September 2014
4. Blanco, R., Ottaviano, G., Meij, E.: Fast and space-efficient entity linking for queries. In: Proceedings of the Eighth ACM International Conference on Web Search and Data Mining, WSDM 2015, pp. 179–188. ACM Press, Shanghai (2015) https://doi.org/10.1145/2684822.2685317, http://dl.acm.org/citation.cfm?doid=2684822.2685317
5. Bunescu, R., Pasca, M.: Using encyclopedic knowledge for named entity disambiguation, p. 8 (2006)
6. Chang, A.X., Spitkovsky, V.I., Manning, C.D., Agirre, E.: A comparison of named-entity disambiguation and word sense disambiguation, p. 8 (2016)
7. Cucerzan, S.: Large-scale named entity disambiguation based on Wikipedia data, p. 9 (2007)
8. Dietz, L., Kotov, A., Meij, E.: Utilizing knowledge graphs in text-centric information retrieval. In: Proceedings of the Tenth ACM International Conference on Web Search and Data Mining, WSDM 2017, pp. 815–816. ACM, New York (2017). https://doi.org/10.1145/3018661.3022756
9. Dorssers, F., de Vries, A.P., Alink, W.: Ranking triples using entity links in a large web crawl. In: Triple Scorer Task at WSDM Cup 2017, p. 5 (2017)
10. Ferragina, P., Scaiella, U.: TAGME: on-the-fly annotation of short text fragments (by Wikipedia entities), p. 1625. ACM Press (2010). https://doi.org/10.1145/1871437.1871689, http://portal.acm.org/citation.cfm?doid=1871437.1871689

11. Globerson, A., Lazic, N., Chakrabarti, S., Subramanya, A., Ringaard, M., Pereira, F.: Collective entity resolution with multi-focal attention. In: Proceedings of the 54th Annual Meeting of the Association for Computational Linguistics (Long Papers), vol. 1, pp. 621–631. Association for Computational Linguistics, Berlin, August 2016. http://www.aclweb.org/anthology/P16-1059

12. Hoffart, J., et al.: Robust disambiguation of named entities in text, p. 11 (2011)

13. Khalid, M.A., Jijkoun, V., de Rijke, M.: The impact of named entity normalization on information retrieval for question answering. In: Macdonald, C., Ounis, I., Plachouras, V., Ruthven, I., White, R.W. (eds.) ECIR 2008. LNCS, vol. 4956, pp. 705–710. Springer, Heidelberg (2008). https://doi.org/10.1007/978-3-540-78646-7_83. http://dl.acm.org/citation.cfm?id=1793274.1793371

14. Kipf, T.N., Welling, M.: Semi-supervised classification with graph convolutional networks. In: ICLR 2017 (2017). http://arxiv.org/abs/1609.02907

15. Meij, E., Balog, K., Odijk, D.: Entity linking and retrieval for semantic search. In: Proceedings of the Seventh ACM International Conference on Web Search and Data Mining, WSDM 2014, pp. 683–684, February 2014

16. Milne, D., Witten, I.H.: Learning to link with Wikipedia, p. 509. ACM Press (2008). https://doi.org/10.1145/1458082.1458150, http://portal.acm.org/citation.cfm?doid=1458082.1458150

17. Raiman, J., Raiman, O.: DeepType: multilingual entity linking by neural type system evolution. arXiv:1802.01021 [cs], February 2018

18. Ren, X., et al.: CoType: joint extraction of typed entities and relations with knowledge bases, pp. 1015–1024. ACM Press (2017). https://doi.org/10.1145/3038912.3052708, http://dl.acm.org/citation.cfm?doid=3038912.3052708

19. Usbeck, R., et al.: AGDISTIS - graph-based disambiguation of named entities using linked data. In: Mika, P., et al. (eds.) ISWC 2014. LNCS, vol. 8796, pp. 457–471. Springer, Cham (2014). https://doi.org/10.1007/978-3-319-11964-9_29

20. Vaswani, A., et al.: Attention is all you need. In: Guyon, I., et al. (eds.) Advances in Neural Information Processing Systems, vol. 30, pp. 5998–6008. Curran Associates, Inc. (2017). http://papers.nips.cc/paper/7181-attention-is-all-you-need.pdf

21. Vrandečić, D.: Wikidata: a new platform for collaborative data collection. In: Proceedings of the 21st International Conference on World Wide Web, WWW 2012 Companion, pp. 1063–1064. ACM, New York (2012). https://doi.org/10.1145/2187980.2188242

Self-attentive Model for Headline Generation

Daniil Gavrilov, Pavel Kalaidin$^{(\boxtimes)}$, and Valentin Malykh

VK, Nevsky ave., 28, 191023 Saint-Petersburg, Russia
{daniil.gavrilov,pavel.kalaidin,valentin.malykh}@vk.com

Abstract. Headline generation is a special type of text summarization task. While the amount of available training data for this task is almost unlimited, it still remains challenging, as learning to generate headlines for news articles implies that the model has strong reasoning about natural language. To overcome this issue, we applied recent Universal Transformer architecture paired with byte-pair encoding technique and achieved new state-of-the-art results on the New York Times Annotated corpus with ROUGE-L F1-score 24.84 and ROUGE-2 F1-score 13.48. We also present the new RIA corpus and reach ROUGE-L F1-score 36.81 and ROUGE-2 F1-score 22.15 on it.

Keywords: Universal Transformer · Headline generation · BPE · Summarization

1 Introduction

Headline writing style has broader applications than those used purely within the journalism community. So-called naming is one of the arts of journalism. Just as natural language processing techniques help people with tasks such as incoming message classification (see [5] or [6]), the naming problem could also be solved using modern machine learning and, in particular, deep learning techniques. In the field of machine learning, the naming problem is formulated as headline generation, i.e. given the text it is needed to generate a title.

Headline generation can also be seen as a special type of text summarization. The aim of summarization is to produce a shorter version of the text that captures the main idea of the source version. We focus on abstractive summarization when the summary is generated on the fly, conditioned on the source sentence, possibly containing novel words not used in the original text.

The downside of traditional summarization is that finding a source of summaries for a large number of texts is rather costly. The advantage of headline generation over the traditional approach is that we have an endless supply of news articles since they are available in every major language and almost always have a title.

This task could be considered language-independent due to the absence of the necessity of native speakers for markup and/or model development.

© Springer Nature Switzerland AG 2019
L. Azzopardi et al. (Eds.): ECIR 2019, LNCS 11438, pp. 87–93, 2019.
https://doi.org/10.1007/978-3-030-15719-7_11

While the task of learning to generate article headlines may seem to be easier than generating full summaries, it still requires that the learning algorithm be able to catch structure dependencies in natural language and therefore could be an interesting benchmark for testing various approaches.

In this paper, we present a new approach to headline generation based on Universal Transformer architecture which explicitly learns non-local representations of the text and seems to be necessary to train summarization model. We also present the test results of our model on the New York Times Annotated corpus and the RIA corpus.

2   Related Work

Rush et al. [11] were the first to apply an attention mechanism to abstractive text summarization.

In the recent work of Hayashi [4], an encoder-decoder approach was presented, where the first sentence was reformulated to a headline. Our Encoder-Decoder baseline (see Sect. 5.1) follows their setup.

The related approach was presented in [10], where the approach of the first sentence was expanded with a so-called topic sentence. The topic sentence is chosen to be the first sentence containing the most important information from a news article (so called 5W1H information, where 5W1H stands for who, what, where, when, why, how). Our Encoder-Decoder baseline could be considered to implement their approach in OF (trained On First sentence) setup.

Tan et al. in [15] present an encoder-decoder approach based on a pregenerated summary of the article. The summary is generated using a statistical summarization approach. The authors mention that the first sentence approach is not enough for New York Times corpora, but they only use a summary for their approach instead of the whole text, thus relying on external tools of summarization.

3   Background

Consider that we have dataset $D = \{(title_i, fulltext_i)\}_i^N$ of news articles and their titles. An approach for learning summarization is to define a conditional probability $P(y_t|\{y_1, \ldots, y_{t-1}\}, X, \theta)$ of some token $y_t \in V$ at time step $t \in \mathbb{N}$, with respect to article text $X = \{x_1, \ldots, x_N\}$ ($x_i \in V$ too) and previous tokens of the title $\{y_1, \ldots, y_{t-1}\}$, parameterized by a neural network with parameters θ.

Then model parameters are found as $\theta_{MLE} = argmax_\theta \prod_i^N P(Y_i|X_i, \theta)$.

We can then apply two methods for finding the most probable sentence under trained model: *greedy*, decoding token-by-token by finding the most probable token at each time step, and *beam-search*, where we find the top-k most probable tokens at each step. The latter method yields better results though it is more computationally expensive.

Sutskever et al. [14] proposed a model that defines $P(y_t|\{y_1, \ldots, y_{t-1}\}, X, \theta)$ by propagating initial sequence X through a Recurrent Neural Network (RNN). Then last hidden state of RNN is used as context vector c and is then passed to the second RNN with $y_1, \ldots, y_{t-1}$ to obtain distribution over y_t.

RNNs have a commonly known flaw. They rapidly forget earlier timesteps, e.g. see [2]. To mitigate this issue, attention [1] was introduced to the Encoder-Decoder architecture. The attention mechanism makes a model able to obtain a new context vector at every decoding iteration from different parts of an encoded sequence. It helps capture all the relevant information from the input sequence, removing the bottleneck of the fixed size hidden vector of the decoder's RNN.

4 Our Approach

4.1 Universal Transformer

While RNNs could be easily used to define the Encoder-Decoder model, learning the recurrent model is very expensive from a computation perspective. The other drawback is that they use only local information while omitting a sequence of hidden states $H = \{h_1, \ldots, h_N\}$. I.e. any two vectors from hidden state h_i and h_j are connected with $j - i$ RNN computations that makes it hard to catch all the dependencies in them due to limited capacity. To train a rich model that would learn complex text structure, we have to define a model that relies on non-local dependencies in the data.

In this work, we adopt the Universal Transformer model architecture [3], which is a modified version of Transformer [16]. This approach has several benefits over RNNs. First of all, it could be trained in parallel. Furthermore, all input vectors are connected to every other via the attention mechanism. It implies that Transformer architecture learns non-local dependencies between tokens regardless of the distance between them, and thus it is able to learn a more complex representation of the text in the article, which proves to be necessary to effectively solve the task of summarization. Also, unlike [4,15], our model is trained end-to-end using the text and title of each news article.

4.2 Byte Pair Encoding

We also adopt byte-pair encoding (BPE), introduced by Sennrich for the machine translation task in [13]. BPE is a data compression technique where often encountered pairs of bytes are replaced by additional extra-alphabet symbols. In the case of texts, like in the machine translation field, the most frequent words are kept in the vocabulary, while less frequent words are replaced by a sequence of (typically two) tokens. E.g., for morphologically rich languages, the word endings could be detached since each word form is definitely less frequent than its stem. BPE encoding allows us to represent all words, including the ones unseen during training, with a fixed vocabulary.

5   Experiments

In our experiments, we consider two corpora: one in Russian and another in English. It is important to mention that we have not done any additional pre-processing other than lower casing, unlike other approaches [4,10]. We apply BPE encoding, which allows us to avoid usage of the $<UNK>$ token for out-of-vocabulary words. For our experiments, we withheld 20,000 random articles to form the test set. We have repeated our experiments 5 times with different random seeds and report mean values.

English Dataset. We use the New York Times Annotated Corpus (NYT) as presented by the Linguistic Data Consortium in [12]. This dataset contains 1.8 million news articles from the New York Times news agency, written between the years 1987 and 2006. For our experiments, we filtered out news articles containing titles shorter than 3 words or longer than 15 words. We also filtered articles with a body text shorter than 20 words or longer than 2000 words. In addition, we skipped obituaries in the dataset. After filtering, we had 1444919 news available to us with a mean title length of 7.9 words and mean text length of 707.6 words.

Russian Dataset. Russian news agency "Rossiya Segodnya" provided us with a dataset (RIA) for research purposes[1]. It contains news documents from January, 2010 to December, 2014. In total, there are 1003869 news articles in the provided corpus with a mean title length 9.5 words and mean text length of 315.6 words.

5.1   Baseline Models

First Sentence. This model takes the first sentence of an article and uses it as its hypothesis for an article headline. This is a strong baseline for generating headlines from news articles.

Encoder-Decoder. Following [10], we use the encoder-decoder architecture on the first sentence of an article. The model itself is already described at recent works section as Seq-To-Seq with RNNs of Sutskever et al. [14]. For this approach, we use the same preprocessing as we did for our model, including byte pair encoding.

5.2   Training

For both datasets, NYT and RIA, we used the same set of hyper-parameters for the models, namely 4 layers in the encoder and decoder with 8 heads of attention. In addition, we added a Dropout of $p = 0.3$ before applying Layer Normalization [8].

---

[1] The dataset is available at https://vk.cc/8W0l5P.

The models were trained with the Adam optimizer using a scaled learning rate, as proposed by the authors of the original Transformer with the number of warmout steps equal to 4000 in both cases and $\beta = (0.9, 0.98)$. Both models were trained until convergence.

We trained the BPE tokenizator separately on the datasets. NYT data was tokenized with a vocabulary size of active tokens equal to 40000, while RIA data was tokenized using 50000 token vocabulary. In addition, we have limited length of the documents with 3000 BPE tokens and 2000 BPE tokens for RIA and NYT datasets respectively. Any exceeding tokens were omitted. word2vec [9] embeddings were trained on each dataset with the size of each embedding equal to 512. For headline generation, we adopted beam-search size of 10.

6 Results

In Table 1 we present results based on two corpora: the New York Times Annotated (NYT) corpus for English, and the Rossiya Segodnya (RIA) corpus for Russian. For the NYT corpus, we reached a new state of the art on ROUGE-1, ROUGE-2 and ROUGE-L F_1 scores. For the RIA corpus, since it has no previous art, we present results for the baselines and our model.[2] For our model we also experimented with label smoothing following [7].

In our experiments, we noticed that some of the generated headlines are scored low by ROUGE metrics despite seeming reasonable, e.g. top sample in Table 3. This lead us to a new series of experiments. We conducted human evaluation of obtained results for both NYT and RIA corpora. The results are presented in Table 2. 5 annotators marked up 100 randomly sampled articles from a

Table 1. ROUGE-1, 2, L F_1 and recall scores, on NYT corpus and RIA corpus.

Model	R-1-f	R-1-r	R-2-f	R-2-r	R-L-f	R-L-r
New York Times						
First sentence	11.64	**34.67**	2.28	7.43	7.19	**31.39**
Encoder-decoder	23.02	21.90	11.84	11.44	21.23	21.31
Summ-hieratt [15]	–	29.60	–	8.17	–	26.05
Universal Transformer w/smoothing (ours)	25.60	23.90	12.92	12.42	23.66	25.27
Universal Transformer (ours)	**26.86**	25.33	**13.48**	**13.01**	**24.84**	24.38
Rossiya Segodnya						
First sentence	24.08	**45.58**	10.57	21.30	16.70	**41.67**
Encoder-decoder	39.10	38.31	22.13	**21.75**	36.34	36.34
Universal Transformer w/smoothing (ours)	39.31	37.10	21.82	20.66	36.32	35.37
Universal Transformer (ours)	**39.75**	37.62	**22.15**	21.04	**36.81**	35.91

---

[2] We are providing results from Tan et al. [15], which were achieved using the NYT corpus. Unfortunately, the authors have not published all of their filtering criteria and seed for random sampling for this corpus, so we could not follow their setup completely. Therefore, these results are presented here for reference.

Table 2. Human evaluation results for NYT and RIA datasets.

Dataset	User preference		
	Human	Tie	Machine
New York Times Annotated	57.4	27.4	15.2
Rossiya Segodnya	54.4	30.6	15.0

Table 3. Samples of headlines generated by our model.

Original text, truncated: Unethical and irresponsible as the assertion that antidepressant medication, an excellent treatment for some forms of depression, will turn a man into a fish. It does a disservice to psychoanalysis, which offers rich and valuable insights into the human mind. ... Homosexuality is not an illness by any of the usual criteria in medicine, such as an increased risk of morbidity or mortality, painful symptoms or social, interpersonal or occupational dysfunction as a result of homosexuality itself... **Original headline:** homosexuality, not an illness, can't be cured **Generated headline:** why we can't let gay therapy begin
Original text, truncated: southwest airlines said yesterday that it would add 16 flights a day from chicago midway airport, moving to protect a valuable hub amid the fight breaking out over the assets of ata airlines, the airport's biggest carrier. southwest said that beginning in january, it would add the flights to 13 cities that it already served from midway... **Original headline:** southwest is adding flights to protect its chicago hub **Generated headline:** southwest airlines to add 16 flights from chicago
Original text, truncated: москва, 1 апр - риа новости. количество сделок продажи элитных квартир в москве выросло в первом квартале этого года, по сравнению с аналогичным периодом предыдущего, в два раза, говорится в отчете компании intermarksavill s. при этом, также сообщается в нем, количество заключенных в столице первичных сделок в сегменте бизнес-класса в первом квартале 2010 года оказалось на 20 выше, чем в первом квартале прошлого года... **Original headline:** продажи элитного жилья в москве увеличились в 1 квартале в два раза **Generated headline:** продажи элитных квартир в москве в 1 квартале выросли вдвое

train set of each corpora. Each number shows the percentage of annotator preference over three possible options: original headline (Human), generated headline (Machine), no preference (Tie).

For the both corpora, we could see that our model is not reaching human parity yet, having 42.6% and 45.6% of (Machine + Tie) user preference for NYT and RIA datasets respectively, but this result is already close to human parity and leaves room for improvement.

7   Conclusion

In this paper, we explore the application of Universal Transformer architecture to the task of abstractive headline generation and outperform the abstractive state-of-the-art result on the New York Times Annotated corpus. We also present a newly released Rossiya Segodnya corpus and results achieved by our model applied to it.

Acknowledgments. Authors are thankful to Alexey Samarin for useful discussions, David Prince for proofreading, Madina Kabirova for proofreading and human evaluation organization, Anastasia Semenyuk and Maria Zaharova for help obtaining the New York Times Annotated corpus, and Alexey Filippovskii for providing the Rossiya Segodnya corpus.

References

1. Bahdanau, D., Cho, K., Bengio, Y.: Neural machine translation by jointly learning to align and translate. arXiv preprint arXiv:1409.0473 (2014)
2. Bengio, Y., Simard, P., Frasconi, P.: Learning long-term dependencies with gradient descent is difficult. IEEE Trans. Neural Netw. **5**(2), 157–166 (1994). https://doi.org/10.1109/72.279181
3. Dehghani, M., Gouws, S., Vinyals, O., Uszkoreit, J., Kaiser, Ł.: Universal transformers. arXiv preprint arXiv:1807.03819 (2018)
4. Hayashi, Y., Yanagimoto, H.: Headline generation with recurrent neural network. In: Matsuo, T., Mine, T., Hirokawa, S. (eds.) New Trends in E-service and Smart Computing. SCI, vol. 742, pp. 81–96. Springer, Cham (2018). https://doi.org/10.1007/978-3-319-70636-8_6
5. Howard, J., Ruder, S.: Fine-tuned language models for text classification. arXiv preprint arXiv:1801.06146 (2018)
6. Joulin, A., Grave, E., Bojanowski, P., Mikolov, T.: Bag of tricks for efficient text classification. In: Proceedings of the 15th Conference of the European Chapter of the Association for Computational Linguistics, Short Papers, vol. 2, pp. 427–431 (2017)
7. Kim, B., Kim, H., Kim, G.: Abstractive summarization of Reddit posts with multi-level memory networks. arXiv (2018)
8. Ba, J.L., Kiros, J.R., Hinton, G.: Layer normalization (2016)
9. Mikolov, T., Sutskever, I., Chen, K., Corrado, G., Dean, J.: Distributed representations of words and phrases and their compositionality. In Proceedings of NIPS (2013)
10. Putra, J.W.G., Kobayashi, H., Shimizu, N.: Experiment on using topic sentence for neural news headline generation (2018)
11. Rush, A.M., Chopra, S., Weston, J.: A neural attention model for abstractive sentence summarization. In: Empirical Methods in Natural Language Processing, pp. 379–389 (2015)
12. Sandhaus, E.: The New York Times annotated corpus LDC2008T19. DVD. Linguistic Data Consortium, Philadelphia (2008)
13. Sennrich, R., Haddow, B., Birch, A.: Neural machine translation of rare words with subword units (2015). https://arxiv.org/abs/1508.07909
14. Sutskever, I., Vinyals, O., Le, Q.V.: Sequence to sequence learning with neural networks (2014). https://arxiv.org/abs/1409.3215
15. Tan, J., Wan, X., Xiao, J.: From neural sentence summarization to headline generation: a coarse-to-fine approach. In: Proceedings of the 26th International Joint Conference on Artificial Intelligence, pp. 4109–4115. AAAI Press (2017)
16. Vaswani, A., et al.: Attention is all you need. In: Advances in Neural Information Processing Systems, pp. 5998–6008 (2017)

Can Image Captioning Help Passage Retrieval in Multimodal Question Answering?

Shurong Sheng[(✉)], Katrien Laenen, and Marie-Francine Moens

Department of Computer Science, KU Leuven, 3001 Leuven, Belgium
{shurong.sheng,katrien.laenen,sien.moens}@cs.kuleuven.be

Abstract. Passage retrieval for multimodal question answering, spanning natural language processing and computer vision, is a challenging task, particularly when the documentation to search from contains poor punctuation or obsolete word forms and with little labeled training data. Here, we introduce a novel approach to conducting passage retrieval for multimodal question answering of ancient artworks where the query image caption of the multimodal query is provided as additional evidence to state-of-the-art retrieval models in the cultural heritage domain trained on a small dataset. The query image caption is generated with an advanced image captioning model trained on an external dataset. Consequently, the retrieval model obtains transferred knowledge from the external dataset. Extensive experiments prove the efficiency of this approach on a benchmark dataset compared to state-of-the-art approaches.

Keywords: Multimodal question answering · Passage retrieval · Query image caption · Markov random field

1 Introduction

In this paper, we focus on multimodal question answering (MQA) in a museum room where the multimodal query consists of the pictures of (a part of) an artwork taken by an end-user from different viewpoints and a textual question about this artwork. The answers to the multimodal query are obtained by searching multimodal documents. *Passage retrieval* for MQA in such a setting identifies the top-ranked passages that may contain the answer to a given query, thereby reducing the answer search space from a large document collection to a fixed number of passages. Such an MQA application can facilitate visitors' interaction with cultural heritage in a natural and personalized fashion. Please refer to [7] for more details about this research topic.

Passage retrieval for MQA in the cultural heritage domain is inherently challenging. Firstly, it is not practical to collect a large dataset of question-passage

Electronic supplementary material The online version of this chapter (https://doi.org/10.1007/978-3-030-15719-7_12) contains supplementary material, which is available to authorized users.

L. Azzopardi et al. (Eds.): ECIR 2019, LNCS 11438, pp. 94–101, 2019.
https://doi.org/10.1007/978-3-030-15719-7_12

pairs that could be used to train a supervised model with many learning parameters. Secondly, the information needs of the users in this domain are diverse. The dataset used in this work [6] contains 9 different question types. Thirdly, the multimodal documentation source is noisy with non-standard spelling, poor punctuation, and obsolete grammar and word forms. Lastly, some questions without ground truth answers need to be classified as NIL questions. If the questions are wrongly classified, the errors propagate through the passage retrieval task. To address these issues, we explore the query image caption as additional information and feed it into the passage retrieval models. In this way, the retrieval model obtains transferred knowledge from the query image caption which is generated by a deep learning model trained on an external image captioning dataset. The contributions of our work are:

1. We propose a novel way to conduct passage retrieval for MQA by providing the query image captions as additional evidence to a retrieval model.
2. We explore two kinds of caption formats to better exploit the additional caption information, i.e., a hard-decision caption that retains words from the most probable captions of a query image, and a soft-decision caption consisting of the top-5 most probable captions and their generation probabilities. Our model improves the performance of two state-of-the-art retrieval models by adding the query image captions. In addition, we prove the retrieval efficiency of adding captions by comparing performance with retrieval models integrating a query image type (the artwork type, e.g., stone sculpture) and candidate passage match.
3. We collect a domain-specific image captioning dataset containing 20951 cultural image-caption pairs.

The remainder of this paper is organized as follows. Section 2 reviews related research. Next, Sect. 3 describes our model architecture. Section 4 discusses the experiments and evaluation results. Finally, Sect. 5 concludes this paper and provides directions for future research.

2   Related Work

Recent works often perform image annotation for cultural images [10–12] with a focus on feature engineering, e.g., extracting low-level image features such as edges and textures. These works require expertise to design the models. Current state-of-the-art image captioning models for natural images are data-driven methods using powerful deep neural networks [3,9,13]. Here, we apply an advanced image captioning method for natural images [3] to annotate ancient artwork images.

Unlike previous research performing MQA leveraging on question-answer pairs [1,4,14], we implement MQA from the viewpoint of passage retrieval, which is rarely studied. The state-of-the-art passage retrieval method for MQA in the cultural heritage domain is the Markov random field (MRF) model introduced in [7]. This model is a probabilistic framework that can infer whether a passage is relevant to a question, by encoding the candidate passage and its feature matching information regarding the question in a graph. Here, we extend the MRF with the matching information between the candidate passages and the query image caption. This new matching information is computed according to the word embedding based correlation model proposed in [5].

3   Methodology

Formally, given a multimodal query q_i and the selected documentation $D = \{pg_1, \ldots, pg_j, \ldots, pg_n\}$ consisting of n candidate passages, our model will extract a hard-decision caption qc_hard_i and soft-decision caption qc_soft_i from query q_i. Then, we compute the probability $p(pg_j)$ that passage pg_j contains the answer to query q_i for $j = 1..n$, and rank the candidate passages based on these probabilities. We propose to use the query image caption as additional evidence to the retrieval model in order to offer more informative words than the query itself. For example for the query image in Fig. 1, in the associated multimodal query "who is this man?", "man" is the only descriptive term while the query image caption brings "head", "fragmentary", "statue", etc.

We first describe how we generate query image captions and encode them into the two formats in Sect. 3.1. Next, we elaborate on how we adapt the state-of-the-art passage retrieval models of [7] to take into account the image captions in Sect. 3.2.

3.1   Query Image Captions

We use the state-of-the-art SCA-CNN model proposed by [3] to perform query image caption generation. For an in-depth treatment of this model the reader may refer to prior work [3]. There are two key problems associated with using the query image captions in the retrieval model. Firstly, a multimodal query has several query image variants taken from different viewpoints as shown in Fig. 1. These images may produce captions containing different words and thus should be merged effectively. Secondly, a query image caption is generated with a certain probability in SCA-CNN. This leads to some uncertainties for the retrieval model, i.e., if we only retain the most probable caption, then some relevant words for the query might be filtered out. On the other hand, if more caption sentences are extracted, then these captions bring more irrelevant words to the retrieval model. To tackle these problems, we extract two kinds of captions for a given query q_i. The first one is a **hard-decision caption**, which we obtain by generating captions for the query image variants of query q_i with a beam size of 1 and merging these captions by removing duplicate words. More formally, the hard-decision caption $qc\_hard_i = \{qc\_w_1, qc\_w_2, \ldots qc\_w_k\}$ where k denotes the number of distinct words in the query image captions. The second is a **soft-decision caption**, acquired by generating captions for the query image variants of query q_i with a larger beam size and only retaining the top t most probable captions. Hence, the soft-decision caption $qc\_soft_i = \{(qc_1, prob(qc_1)), \ldots, (qc_t, prob(qc_t))\}$.

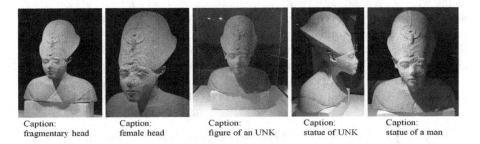

| Caption: | Caption: | Caption: | Caption: | Caption: |
| fragmentary head | female head | figure of an UNK | statue of UNK | statue of a man |

Fig. 1. Example of query image variants of a multimodal query and their most probable captions generated by the image captioning model.

3.2 Retrieval Model

We experiment with the two kinds of query image captions for query q_i, qc_hard_i and qc_soft_i, as additional information for the BF and EF-2 passage retrieval models proposed in [7], the EF-2 model outperforms the BF model by adding the (question type)-(named entity) match information into the BF model. We use the EF-2 model to explain our model architecture; for the BF model we take a similar approach to incorporate the query image captions.

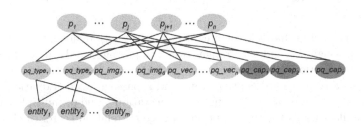

Fig. 2. The retrieval framework integrating caption-passage match information.

We extend the EF-2 model by including a fourth feature match between a candidate passage pg_g and the query q_i: the caption-passage match. As a result, we obtain a new MRF $G_{cap} = \ <\{P^{answer}, PQ^{type}, PQ^{image}, PQ^{vec}, PQ^{cap}\}, E>$ to evaluate whether a passage pg_j contains the answer to query q_i. The structure of MRF G_{cap} is shown in Fig. 2. All nodes in this graph are random variables. More precisely, each $p_j \in P^{answer}$ in the first row is a binary variable denoting whether the corresponding candidate passage pg_j is an answer passage. PQ^{image}, PQ^{type} and PQ^{vec} are represented by the gray nodes in the second row and refer to the three feature matches between a candidate passage pg_g and the query q_i originally defined in [7]: image match, (question type)-(named entity) match and word co-occurrence. The highlighted green nodes $pq\_cap_j \in PQ^{cap}$ define whether a candidate passage pg_j matches the query image caption. Each node and edge in MRF G_{cap} are associated with a potential function. A node potential decides the probability of the binary node being 'True' or 'False'. An edge potential controls the compatibility between two connected nodes. Our goal is to obtain a list of probabilities $S = [p(pg_1), \ldots, p(pg_j), \ldots, p(pg_n)]$ where $p(pg_j)$ is the probability of passage pg_j containing the answer to query q_i and corresponds to the potential value of node p_j. The objective to train MRF G_{cap} is:

$$P = \frac{1}{Z} \prod_{\substack{j \in \{1,2,..,n\} \\ k \in \{1,2,..,g\}}} \varphi_{p_j} * \varphi_{pq\_type_j} * \varphi_{pq\_img_j} * \varphi_{pq\_vec_j} * \varphi_{pq\_cap_j} * \varphi_{entity_k} * \varphi_{edge} \quad (1)$$

with n, the number of candidate passages, g the entity number in the documentation, and φ the potential for a specific node or edge. We exploit a **word embedding correlation based (WEC)** method [5] to initialize $\varphi_{pq\_cap_j}$, which refers to the node potential of $pq\_cap_j \in PQ^{cap}$. This approach integrates word-to-word correlation scores into the caption-passage matching score. Depending on the kind of query image caption, the initial potential value of $\varphi_{pq\_cap_j}$ with the caption-passage match being 'True' is computed differently:

$$\varphi_{pq\_hardcap_j} = \frac{1}{|pg_j|} \sum_{td} \max_{ta} cosine(v_{qc\_hard_{i\_ta}}, v_{pg_j\_td}) \quad (2)$$

$$\varphi_{pq\_softcap_j} = \sum_{qc\_soft_m} p(qc\_soft_m) \frac{1}{|pg_j|} \sum_{td} \max_{tb} cosine(v_{qc\_soft_{mi\_tb}}, v_{pg_j\_td}) \quad (3)$$

Equation 2 is used for hard-decision captions. Here, $|pg_j|$ denotes the length of a candidate passage pg_j, index td ranges over the passage words, index ta ranges over the hard caption words and $v_{qc\_hard_{i\_ta}}$ and $v_{pg_j\_td}$ denote the word embedding of respectively a hard caption word and a passage word. Equation 3 deals with the soft-decision caption, using the expectation value of multiple caption-passage similarities as the final caption-passage correlation score. In this equation, qc_soft_m ranges over the soft captions, $|pg_j|$ denotes the length of a candidate passage, index td ranges over the passage words, index tb ranges over the soft caption words and $v_{qc\_soft_{mi\_tb}}$ and $v_{pg_j\_td}$ refer to the embedding of respectively a soft caption word and a passage word. The initialization approaches for the other potential variables are the same as in [7]. We also follow [7] to perform NIL classification and inference.

4   Experiments and Results

4.1   Query Image Caption Generation

We collected an image captioning dataset on ancient Egyptian art from two online digital archives: The Metropolitan Museum of Art[1] and the Brooklyn Museum[2]. This dataset contains 20951 image-caption pairs and is split into an 80%, 10% and 10% partition for training, validation, and testing respectively. Detailed statistics of the dataset can be found in Sect. 1 the Supplementary Materials. The SCA-CNN image captioning model is trained using the default settings of the paper [3]. It achieves a BLEU-1 score of 0.22 on the test set, which is 8% higher than that of the model proposed in [9]. We consider the BLEU-1 score since we ignore the word dependencies of the captions in the retrieval model. For the hard-decision captions using a beam size of 1 during caption generation, the average length of the query image captions is 4, which is quite short. This inspired us to merge captions from images with different viewpoints. For the soft-decision captions generated with a larger beam size, we empirically selected the top $t = 5$ most probable captions for each artwork.

Table 1. Evaluation results for full-image level questions.

Comparison	Model	mAP	MRR	NIL precision	NIL recall
Comparison 1	EF-2 + hard-decision caption	**0.18**	**0.23**	0.34	0.94
	EF-2 + soft-decision caption	0.13	0.17	0.32	**0.95**
	EF-2 (baseline) [7]	0.15	0.19	**0.35**	0.76
Comparison 2	BF + hard-decision caption	0.14	0.18	0.30	0.78
	BF + soft-decision caption	0.13	0.17	0.31	0.77
	BF (baseline) [7]	0.13	0.17	0.31	0.77
Comparison 3	BF + image type-passage match	0.12	0.17	0.30	0.72
	EF-2 + image type-passage match	0.16	0.21	0.32	0.70

4.2   Retrieval Models

The dataset and evaluation metrics are the same as in [7]. We focus on full-image-level questions because additional image processing is needed to generate partial-image captions. In the BF model, the candidate passages are extracted from the related documents, and in the EF-2 model, passages without a corresponding named entity are filtered out for 'who' and 'where' questions. In our WEC method we use 300-dimensional word embeddings obtained from a skip-gram model [2] trained on a word set composed of the words in the multimodal

---

[1] https://www.metmuseum.org/.
[2] https://www.brooklynmuseum.org/.

documentation. The NIL classification threshold value is set based on a validation dataset with 20% of the data and mAP score serves as the main metric to conduct this (More explanation about the evaluation metrics can be found in Sect. 3 Supplementary Materials). Table 1 shows improved performance for both the BF and the EF-2 model when using hard-decision captions.

We also conducted a set of experiments that extend BF and EF-2 with a query image type and candidate passage match similar to the caption-passage match. The query image type refers to the artwork type, e.g., ceramics reliefs. In this experiment, a CNN consisting of layers up to the mix6d-layer in the Inception-v3 [8] framework was used to predict the query image types. We derived an image type classification dataset from the image captioning dataset as explained in Sect. 2 of the Supplementary Materials. We use the same data partition as for image captioning. The image type prediction model achieves an accuracy of 0.97 on a test set of around 8600 images. As shown in Table 1 under 'Comparison 3', image type information cannot improve the retrieval performance as the query image captions do when incorporated into the retrieval models with the same approach. This confirms how difficult it is to add useful evidence into the BF and the EF-2 model, and proves the effectiveness of leveraging query image captions to the retrieval models.

5   Conclusion and Future Work

In this paper, we introduced a passage retrieval method for MQA leveraging the generated query image caption of a multimodal query and which achieved promising improvements. Next, we will explore how to improve the image captioning performance for both full and partial artwork images to make the retrieval improvements more significant.

Acknowledgments. This work is funded by the KU Leuven BOF/IF/RUN/2015. We additionally thank our anonymous reviewers for the helpful comments.

References

1. Anderson, P., et al.: Bottom-up and top-down attention for image captioning and visual question answering. In: CVPR, vol. 3, p. 6 (2018)
2. Bojanowski, P., Grave, E., Joulin, A., Mikolov, T.: Enriching word vectors with subword information. Trans. Assoc. Comput. Linguist. **5**, 135–146 (2017)
3. Chen, L., et al.: SCA-CNN: spatial and channel-wise attention in convolutional networks for image captioning. In: Proceedings of the IEEE Conference on Computer Vision and Pattern Recognition, pp. 6298–6306 (2017)
4. Li, H., Wang, Y., de Melo, G., Tu, C., Chen, B.: Multimodal question answering over structured data with ambiguous entities. In: Proceedings of the 26th International Conference on World Wide Web Companion, pp. 79–88. International World Wide Web Conferences Steering Committee (2017)
5. Shen, Y., Rong, W., Jiang, N., Peng, B., Tang, J., Xiong, Z.: Word embedding based correlation model for question/answer matching. In: Proceedings of the Thirty-First AAAI Conference on Artificial Intelligence, pp. 3511–3517 (2017)

6. Sheng, S., Van Gool, L., Moens, M.F.: A dataset for multimodal question answering in the cultural heritage domain. In: Proceedings of the Workshop on Language Technology Resources and Tools for Digital Humanities (LT4DH), pp. 10–17 (2016)
7. Sheng, S., Venkitasubramanian, A.N., Moens, M.-F.: A Markov network based passage retrieval method for multimodal question answering in the cultural heritage domain. In: Schoeffmann, K., et al. (eds.) MMM 2018. LNCS, vol. 10704, pp. 3–15. Springer, Cham (2018). https://doi.org/10.1007/978-3-319-73603-7_1
8. Szegedy, C., Vanhoucke, V., Ioffe, S., Shlens, J., Wojna, Z.: Rethinking the inception architecture for computer vision. In: Proceedings of the IEEE Conference on Computer Vision and Pattern Recognition, pp. 2818–2826 (2016)
9. Vinyals, O., Toshev, A., Bengio, S., Erhan, D.: Show and tell: a neural image caption generator. In: Proceedings of the IEEE Conference on Computer Vision and Pattern Recognition, pp. 3156–3164 (2015)
10. Wang, J.Z., Grieb, K., Zhang, Y., Chen, C.C., Chen, Y., Li, J.: Machine annotation and retrieval for digital imagery of historical materials. Int. J. Dig. Libr. 6(1), 18–29 (2006)
11. Xu, L., Merono-Penuela, A., Huang, Z., van Harmelen, F.: An ontology model for narrative image annotation in the field of cultural heritage. In: Proceedings of the 2nd Workshop on Humanities in the Semantic Web (WHiSe), pp. 15–26 (2017)
12. Xu, L., Wang, X.: Semantic description of cultural digital images: using a hierarchical model and controlled vocabulary. D-Lib Mag. 21(5/6) (2015)
13. You, Q., Jin, H., Wang, Z., Fang, C., Luo, J.: Image captioning with semantic attention. In: Proceedings of the IEEE Conference on Computer Vision and Pattern Recognition, pp. 4651–4659 (2016)
14. Zhu, Y., Groth, O., Bernstein, M., Fei-Fei, L.: Visual7W: grounded question answering in images. In: Proceedings of the IEEE Conference on Computer Vision and Pattern Recognition, pp. 4995–5004 (2016)

A Simple Neural Approach to Spatial Role Labelling

Nitin Ramrakhiyani[1,2]([✉]), Girish Palshikar[1], and Vasudeva Varma[2]

[1] TCS Research, Tata Consultancy Services Ltd., Pune, India
{nitin.ramrakhiyani,gk.palshikar}@tcs.com
[2] International Institute of Information Technology, Hyderabad, India
vv@iiit.ac.in

Abstract. Spatial Role Labelling involves identification of text segments which emit spatial semantics such as describing an object of interest, a reference point or the object's relative position with the reference. Tasks in SemEval exercises of 2012 and 2013 propose problems and datasets for Spatial Role Labelling. In this paper, we propose a simple two-step neural network based approach to identify static spatial relations along with the three primary roles - Trajector, Landmark and Spatial Indicator. Our approach outperforms the task submission results and other state-of-the-art results on these datasets. We also include a discussion on the explainability of our model.

Keywords: Spatial role labelling ·
Spatial representation and reasoning · Deep learning · BiLSTM

1 Introduction

Spatial Role Labelling (SpRL) is the process of assigning segments of text in a sentence, with roles they perform based on their spatial semantics. In natural language sentences describing spatial information, there is generally an object whose spatial position is being described (the *Trajector* role), a reference object (*Landmark*) and a spatial trigger (*Spatial Indicator*). There are other roles like *Path* and *Motion Indicator* which describe the dynamic position of a *Trajector*. SpRL is similar to Semantic Role Labelling (SRL) on certain counts and dissimilar on various others. It is similar to SRL mainly because both consider a central element whose arguments have to be found. Spatial indicators and motion indicators in SpRL are like verbs in SRL and other roles like Trajector, Landmark and Path are the arguments of these indicators. SpRL is however different from SRL as the central element may not always emit a spatial sense or otherwise can be part of several spatial relations.

Tasks on SpRL were introduced as Task 3 at SemEval 2012 [5], as Task 3 at SemEval 2013 [4] and as Task 8 (SpaceEval) at SemEval 2015 [10]. The tasks saw a moderate participation with organizers also providing baseline systems in some cases. The tasks introduced the various spatial roles and their semantics,

© Springer Nature Switzerland AG 2019
L. Azzopardi et al. (Eds.): ECIR 2019, LNCS 11438, pp. 102–108, 2019.
https://doi.org/10.1007/978-3-030-15719-7_13

while increasing the complexity of the problem each year. In Task 3 at SemEval 2012, the core task of spatial role labelling was introduced involving identification of roles namely *Trajector*, *Landmark*, *Spatial Indicator* and static relations among these roles. In Task 3 at SemEval 2013, apart from the previous year's problem, the task involving identification of dynamic relations was added. In the SpaceEval 2015 task, identification of finer roles along with their attributes was introduced.

In this paper, we focus only on the identification of static spatial relations and the roles *Trajector*, *Landmark* and *Spatial Indicator*. Hence, we do not attempt the dynamic spatial relation identification sub-problem of Task 3 at SemEval 2013. Also, we do not tackle any problems of SpaceEval 2015 due to introduction of new notion of spatial entities, change in relations to MOVELINK, QSLINK and OLINK and change in evaluation of the relation identification sub-problems.

We propose a simple two step neural approach for these tasks. We train a BiLSTM for a sequence labelling task of identifying spatial roles only to develop context vectors for the words. We then use contexts from this pre-trained BiLSTM for a relation classification step and deduce the corresponding roles from identified relations. The proposed neural model outperforms the participating systems and other state-of-the-art approaches on the datasets of the two tasks. As part of the analysis, we also discuss on the semantics of the context embeddings learned by the BiLSTM.

2 Relevant Literature

2.1 SemEval 2012 Task 3

Task 3 at SemEval 2012 [5], introduced the basic task of spatial role labelling which involved two sub-problems: identification of the three roles namely *Trajector*, *Landmark*, *Spatial Indicator* and identification and classification of static spatial relations involving these roles. The task data was a subset of image descriptions available as a part of the IAPR TC-12 image benchmark [3]. The image descriptions described entities in the images and their relative or absolute positions with respect to other entities in the image. As per the task, each spatial relation is formed of a *Trajector*, a *Spatial Indicator* and an optional *Landmark* and the relation type is classified as: *region* (describing topology such as on, inside, etc.), *direction* (describing orientation such as above, to the left of) and *distal* (describing distance such as far, away, etc.).

As an example from the dataset, consider the sentence: a woman and a child are walking over the square. Here, positions of the woman and child are being described and hence are the *Trajectors*. The square is the reference entity working as a *Landmark* and over describes the position of the Trajectors with respect to the Landmark, thus being the *Spatial Indicator*. There are two relations which can be identified: (woman, over, square) and (child, over, square). Both relations are *direction* type of relations.

2.2   SemEval 2013 Task 3

Task 3 at SemEval 2013 extended the task introduced in SemEval 2012 in two directions. Firstly, they considered full phrase spans of the text identified under a role instead of only head words as considered in the SemEval 2012 task. Secondly, they introduced a sub-problem of identification of dynamic spatial relations and corresponding roles like *Path*, *Direction* and *Distance*. The change from head words to full spans prompted the organizers to change from the annotation format used in SemEval 2012 to a different character based offset format.

2.3   Other Work

Another pivotal task - Task 8 (SpaceEval) at SemEval 2015 extended the earlier tasks in multiple directions. The organizers followed the annotation scheme specified in the ISOSpace standard [11] to enhance the granularity of the spatial semantics and used a more fine grained set of annotation tags. The sub-problems introduced covered identification of spatial entities which may be Places or Objects, identification of static and dynamic relations and identification of various attributes of these relations.

Mazalov et al. [8] proposes a CNN based technique similar to a one used for semantic role labelling for the tasks in Spatial Role Labelling. The authors report results on static spatial relation extraction on the SemEval 2013 dataset. Kordjamshidi and Moens [6] propose a structured learning based solution for spatial ontology population from text and report results on the datasets of SemEval 2012 and SemEval 2013 tasks. In another work [7], the authors propose visually informed embedding of words (VIEW) for use in a spatial arrangement prediction task. The paper reports its results on the SemEval 2013 Task 3 dataset.

3   Proposed Approach

We hypothesize that the context of spatial text elements is useful in finding their spatial roles. Also, employing their word level features along with their context can boost this classification further. We propose a two step approach to achieve spatial relation identification and deduction of spatial roles of the involved elements.

As the first step, we train a BiLSTM to get context embeddings for words of the sentences in the dataset. The BiLSTM is trained for a sequence labelling task of identifying spatial indicators and spatial actors. For this step, we denote both trajectors and landmarks as a single type of entity namely "spatial actor". As the only goal of this step is to learn the context of the words it is not necessary to learn a sequence labeller for the three roles separately. Moreover, it is observed that some words play multiple roles in a sentence i.e. a trajector in one relation and landmark in another. For example, in the sentence a man sitting on a bench in front of the wall., the word bench behaves as a landmark in the relation (man, on, bench) but behaves as trajector in the relation (bench, in `front of, benchwall`). This is observed for more than 10% sentences in the datasets which restricts us from posing it as a sequence labelling problem over all the three roles. The output of this network is not utilized in the later step of the approach. This network is formed of an input layer feeding into a BiLSTM layer followed finally by a prediction layer for each time step.

As part of the second step, we first generate a list of candidate relations from each sentence. We develop a dependency parsing based candidate generation logic by analysing the dependency parse of multiple sentences in training data along with their spatial roles. The candidate generation procedure checks for each preposition in the sentence and marks its prepositional child as a possible landmark and its head as a possible trajector, if they are nouns. If they aren't nouns the algorithm continues traversing the dependency path further on each side till a noun is found and is marked as trajector or landmark. If a trajector found is connected to another noun by a conjunction dependency relation that noun is also added as a possible trajector. Landmarks are also expanded similarly. The candidate generation procedure returns a list of relation triples of trajector, spatial indicator and landmark for each sentence.

We then add the class information to each candidate relation. If the candidate relation is present in the true relations for the sentence, the relation's type (region, direction or distal) is added as its class. Otherwise the relation is assigned a null type. A second neural network is developed and trained for the task of classifying these candidate relations. The network has a input layer with three input vectors each corresponding to trajector, spatial indicator and landmark, followed by a hidden layer and finally a prediction layer. As input to the network, we propose two configurations - (i) only the context of trajector, spatial indicator and landmark of the candidate relation or (ii) context concatenated with the original (not retrained) word embeddings of the trajector, spatial indicator and landmark. The context of each word is obtained by passing the sentence through the pre-trained BiLSTM of the first step and collecting the hidden layer output at each word from both directions.

While testing, each sentence gets its candidate relations generated. For each candidate relation, context of its trajector, landmark and spatial indicator is obtained from the BiLSTM trained on training data. The candidate relation is input to the trained relation identification network and if predicted as true the corresponding trajector, landmark and spatial indicator are also marked as true. The network output and deduced roles are processed to generate annotations as per the evaluation scripts provided by the organizers.

Changes for Task 3 of SemEval 2013. For this task, it was required to predict the full span of the spatial role and not just the head word as in the earlier task. The training and testing learned from the earlier task are re-used as the data remains the same and annotations change marginally. While making predictions on the testing data the network predicts only head words as part of spatial roles. These predicted head words are expanded using another dependency based procedure. For each trajector and landmark all their determiner, compound, adjective modifier, numerical modifier and adverb modifier children are included to form the complete span. For spatial indicators, constructions of the form "on the left of" are developed starting from the predicted indicator "on".

4   Experimentation, Evaluation and Analysis

Experimentation: We use keras [2] to implement the neural networks. We use 300 dimensional Glove word embeddings [9] trained on a common web crawl of 42 billion words. For arriving at the right neural network parameters, we use five fold cross validation on the training data. For the BiLSTM network, we find the best results for 10 epochs, batch size of 32, dropout of 0.1 and 300 LSTM units. For the relation identification network, the best parameters are 10 epochs, batch size of 32, dropout of 0.3 and 600 hidden units for the context based network and 900 hidden units for the context + embeddings based network. To account for randomness in network weight initialization, we carry out the training (and testing) 10 times and report averaged results over the runs.

Evaluation: We use the evaluation scripts provided by the task organizers as part of the released dataset, to compute the results. We produce the output of our approaches as desired by the evaluation jar files, run the jar files and report results thus obtained. This puts our approach at par in comparison to the participating systems and other state-of-the-art.

For Task 3 at SemEval 2012, along with the baseline results provided by the organizers, we use the results reported in the best run [12] submitted by the single participating team, as a baseline for comparison. The results from the best approach proposed in Kordjamshidi and Moens [6] on the IAPR TC-12 dataset is also included as a baseline for comparison. In the interest of space, we request readers to refer to [6] for details on the approaches.

For Task 3 at SemEval 2013, we use the results reported in the best run [12] submitted by the single participating team, as a baseline for comparison. We only report results under the relaxed evaluation criteria as specified by the organizers. We also compare with the best results proposed in [7]. We however, do not compare with the approach in Mazalov et al. [8] as it is not clear from the paper whether the authors evaluate using the organizer provided scripts. The authors mention changing the format of the data leading to the possibility of final evaluation being carried out differently.

Analysis: The pre-trained BiLSTM when tested using five fold cross validation on the training data, showed superior results of F1 greater than 0.9 for both classes - spatial indicator and spatial actor.

It can be observed from Tables 1 and 2 that our hypothesis stating use of context and embeddings for text elements to predict spatial roles gets established. For Task 3 of SemEval 2012, our approaches outperform all baselines by substantial margins. It is important to note here that the UTD submission [12] relies on a fixed list of prepositions as spatial indicators which though a curated list, can be limiting in many cases. Our relation candidate generation logic does not rely on a fixed list and considers each preposition for a possible relation. Also the

Table 1. F1 scores of various systems for the SemEval 2012 tasks

Approach	TR	LM	SP	Relation	Relation type
Organizer baseline [5]	0.646	0.756	**0.900**	0.500	NA
UTD best run [12]	0.707	0.772	0.823	0.573	0.566
EtoE-IBT-CLCP [6]	0.673	0.797	0.869	0.617	NA
Context	0.835	0.856	0.883	0.775	0.706
Context + Embeddings	**0.848**	**0.875**	**0.900**	**0.794**	**0.741**

Table 2. F1 scores of various systems for the SemEval 2013 static spatial relation tasks

Approach	TR	LM	SP	Relation
UNITOR best run [1]	0.682	0.785	**0.926**	0.458
VIEW [7]	0.732	0.678	0.749	0.235
Context	**0.823**	**0.814**	0.901	**0.562**
Context + Embeddings	0.808	0.8	0.878	0.556

context and embedding approach is seen to perform better than the only context based approach. For the SemEval 2013 task, our context based approach shows improvement in relation identification performance over the baselines. Analysis reveals that the change from head words to full span of the roles has lead to certain inconsistencies in the annotations. This is also highlighted by authors in [8]. The lower relation identification performance on this task can be attributed to these changes.

To understand the semantics captured in the context of spatial indicators in the BiLSTM network we perform an experiment. We check whether the context of spatial indicators from different sentences shows similar semantics. If so, we can conclude that this context representation of a spatial indicator does represent its true spatial function. To perform this check, we collect the context vectors of the spatial indicators from all training sentences and cluster them using average linkage clustering with an empirically decided distance threshold of 0.7. A manual observation of the clusters shows that the same indicator from different sentences lies mostly in a single cluster and different spatial indicators lie in their respective clusters, thus validating the proposed understanding.

5   Conclusion

In this paper, we attempt to solve the problem of identification of spatial roles and static spatial relations in text. We show that context of words learned from a BiLSTM trained for a sequence labelling task can help in the identification process. We show that our two-step approach of generating context vectors and relation identification based on the learned context vectors, outperforms the state-of-the-art results on tasks of SemEval 2012 and SemEval 2013.

References

1. Bastianelli, E., Croce, D., Basili, R., Nardi, D.: UNITOR-HMM-TK: structured kernel-based learning for spatial role labeling. In: Second Joint Conference on Lexical and Computational Semantics (* SEM), Volume 2: Proceedings of the Seventh International Workshop on Semantic Evaluation (SemEval 2013), vol. 2, pp. 573–579 (2013)
2. Chollet, F.: Keras. GitHub repository. https://github.com/fchollet/keras. Accessed 26 Oct 2018
3. Grubinger, M., Clough, P., Müller, H., Deselaers, T.: The IAPR TC-12 benchmark: a new evaluation resource for visual information systems. In: International Workshop OntoImage, vol. 5 (2006)
4. Kolomiyets, O., Kordjamshidi, P., Moens, M.F., Bethard, S.: SemEval-2013 task 3: spatial role labeling. In: Second Joint Conference on Lexical and Computational Semantics (* SEM), Volume 2: Proceedings of the Seventh International Workshop on Semantic Evaluation (SemEval 2013), vol. 2, pp. 255–262 (2013)
5. Kordjamshidi, P., Bethard, S., Moens, M.F.: SemEval-2012 task 3: spatial role labeling. In: Proceedings of the First Joint Conference on Lexical and Computational Semantics-Volume 1: Proceedings of the Main Conference and the Shared Task, and Volume 2: Proceedings of the Sixth International Workshop on Semantic Evaluation, pp. 365–373. Association for Computational Linguistics (2012)
6. Kordjamshidi, P., Moens, M.F.: Global machine learning for spatial ontology population. Web Semant.: Sci. Serv. Agents World Wide Web **30**, 3–21 (2015)
7. Ludwig, O., Liu, X., Kordjamshidi, P., Moens, M.F.: Deep embedding for spatial role labeling. arXiv preprint arXiv:1603.08474 (2016)
8. Mazalov, A., Martins, B., Matos, D.: Spatial role labeling with convolutional neural networks. In: Proceedings of the 9th Workshop on Geographic Information Retrieval, p. 12. ACM (2015)
9. Pennington, J., Socher, R., Manning, C.: Glove: global vectors for word representation. In: Proceedings of the 2014 Conference on Empirical Methods in Natural Language Processing (EMNLP), pp. 1532–1543 (2014)
10. Pustejovsky, J., Kordjamshidi, P., Moens, M.F., Levine, A., Dworman, S., Yocum, Z.: SemEval-2015 task 8: SpaceEval. In: Proceedings of the 9th International Workshop on Semantic Evaluation (SemEval 2015), pp. 884–894 (2015)
11. Pustejovsky, J., Moszkowicz, J.L., Verhagen, M.: ISO-space: the annotation of spatial information in language. In: Proceedings of the Sixth Joint ISO-ACL SIGSEM Workshop on Interoperable Semantic Annotation, vol. 6, pp. 1–9 (2011)
12. Roberts, K., Harabagiu, S.M.: UTD-SPRL: a joint approach to spatial role labeling. In: Proceedings of the First Joint Conference on Lexical and Computational Semantics-Volume 1: Proceedings of the Main Conference and the Shared Task, and Volume 2: Proceedings of the Sixth International Workshop on Semantic Evaluation, pp. 419–424. Association for Computational Linguistics (2012)

Neural Diverse Abstractive Sentence Compression Generation

Mir Tafseer Nayeem$^{(\boxtimes)}$, Tanvir Ahmed Fuad, and Yllias Chali

University of Lethbridge, Lethbridge, AB, Canada
mir.nayeem@alumni.uleth.ca, t.fuad@uleth.ca, chali@cs.uleth.ca

Abstract. In this work, we have contributed a novel abstractive sentence compression model which generates diverse compressed sentence with paraphrase using a neural **seq2seq** encoder decoder model. We impose several operations in order to generate diverse abstractive compressions at the sentence level which was not addressed in the past research works. Our model jointly improves the information coverage and abstractiveness of the generated sentences. We conduct our experiments on the human-generated abstractive sentence compression datasets and evaluate our system on several newly proposed Machine Translation (**MT**) evaluation metrics. Our experiments demonstrate that the methods bring significant improvements over the state-of-the-art methods across different metrics.

Keywords: Abstractive summarization · Diverse sentence compression

1 Introduction

The task of automatic text summarization aims at finding the most relevant information in a text and presenting them in a condensed form. A good summary should retain the most important contents of the original text, while being non-redundant and grammatically readable [5,15]. Summarization on the sentence level is called sentence compression. Sentence compression approaches can be classified into two categories: extractive and abstractive sentence compression. Most sentence compression models follow extractive approaches that select the most relevant information from the source sentence and generate a shorter representation of the sentence by deleting unimportant fragments which is still grammatical. On the other hand, abstractive methods, which are still a growing field, are highly complex as they need extensive natural language generation to rewrite the sentences from scratch based on the understanding of the sentences [17]. The abstractive techniques which we traditionally use are sentence compression, fusion and lexical paraphrasing [16].

2 Related Works

Recent success of neural sequence-to-sequence (**seq2seq**) models provide an effective way for text generation which achieved huge success in the case of

© Springer Nature Switzerland AG 2019
L. Azzopardi et al. (Eds.): ECIR 2019, LNCS 11438, pp. 109–116, 2019.
https://doi.org/10.1007/978-3-030-15719-7_14

abstractive sentence summarization. These systems have adopted techniques such as encoder-decoder with attention [2,12] models from the field of machine translation to model the sentence summarization task [8,13,19,23]. The deep neural network architectures are completely data driven hence more training data will produce good quality output sequences. Therefore, almost all the past works on sentence summarization using neural networks [4,8,19,22,26] made use of the English Gigaword dataset [14].

Unfortunately, this line of research under the term sentence compression, which can generate deletion based compressive sentences, somewhat misleadingly called abstractive summarization in some follow-up research works [13,23,26]. Our experimental results clearly demonstrate the fact that they are producing compressions by copying the source sentence words with morphological variations, no paraphrasing is involved in the process.

3   Diverse Abstractive Sentence Compression Model

Our neural **D**iverse **P**araphrastic **C**ompression model is based on Neural Machine Translation (NMT). **DPC** uses **NMT** to translate from a source sentence to an abstractive compression. Given a source sentence $\mathbf{X} = (x_1, x_2, \ldots, x_N)$, our model learns to predict its abstractive compression target $\mathbf{Y} = (y_1, y_2, \ldots, y_M)$ with diversity, where $M < N$. Inferring the target Y given the source X is a typical sequence to sequence learning problem, which can be modeled with attention-based encoder-decoder models [2,12]. As the name suggests, the basic form of an encoder-decoder model consists of two components.

Encoder. The encoder in our case is a bi-directional GRU (Bi-GRU), unlike [12] which uses uni-directional LSTM [11]. Another important modification we can do to the Bi-GRUs following [12] is stacking multiple layers on top of each other. They can extract more abstract features of the current words or sentences. However, stacking RNNs suffer from the vanishing gradient problem in the vertical direction from the output layer ($\mathbf{GRU_3}$) to the layer close to the input ($\mathbf{GRU_1}$), just as the standard RNN suffers in the horizontal direction. This causes the earlier layers of the network to be under-trained. A simple solution to this problem is to add residual connections, which has been shown to be extremely useful for the image recognition task [10]. The idea behind these networks is simply to add the output of the previous layer directly to the result of the next layer. For example, in a 3-layer stacked GRU with residual connections, the calculation at time step t would look as follows,

$$h_{1,t} = \mathbf{BiGRU_1}(e(x_t), h_{1,t-1}) + e(x_t)$$

$$h_{2,t} = \mathbf{BiGRU_2}(h_{1,t}, h_{2,t-1}) + h_{1,t}$$

$$h_{3,t} = \mathbf{BiGRU_3}(h_{2,t}, h_{3,t-1}) + h_{2,t}$$

where, the $h_{1,t} \in \mathbb{R}^n$ encodes all content seen so far at time t from layer 1, which is computed from h_{t-1} and $e(x_t)$, where $e(x_t) \in \mathbb{R}^m$ is the m-dimensional embedding of the current word x_t. Therefore, we use the idea of residual networks for building our encoder decoder model **DPC** to perform the abstractive compression generation task which is illustrated in Fig. 1. The initial hidden states of the encoder are set to zero vectors, i.e., $\overrightarrow{h^S_{1,1}} = 0$, $\overleftarrow{h^S_{1,N}} = 0$. In our **DPC** model, the encoder transforms the source sentence **X** into a sequence of hidden states $(\mathbf{h}^S_{3,1}, \mathbf{h}^S_{3,2}, \ldots, \mathbf{h}^S_{3,N})$ with a stacked residual network.

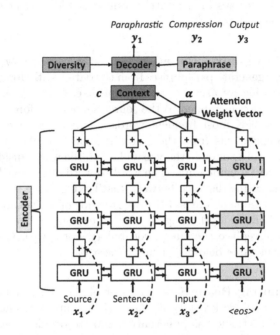

Fig. 1. Neural **D**iverse **P**araphrastic **C**ompression Generation Model

Decoder and Attender. The decoder uses a simple GRU with attention to generate one word y_{t+1} at a time in the target sentence **Y**.

$$P(\mathbf{Y}|\mathbf{X}) = \prod_{t=1}^{M} P(y_t|y_{1:t-1}, \mathbf{X}) \tag{1}$$

We use the ($\cdot$) *dot* attention mechanism [12] due to its efficiency and which is simple to implement. The dot attention mechanism is actually the dot product between two hidden vectors.

Decoder with Copying Mechanism. At each generation step of the decoder, the output word is selected according to the probability distribution over the whole target vocabulary in the softmax layer, which is the most time and capacity-consuming part of the system. Therefore, we limit our vocabularies to be the top 60K most frequent words. The infrequent words were removed from the vocabulary and were replaced with the symbol UNK, meaning unknown word. However, it has been observed that the infrequent words are usually proper nouns or named-entities that have an impact on the meaning of the sentence. Therefore, we used the **COPYNET** model proposed by [9] which can integrate the regular way of word generation in the decoder with the new copying mechanism which can choose words or subsequences in the input sequence and put them at appropriate places in the output sequence. Please refer to the original paper by [9] for more detail.

Paraphrasing in Context. Our model implicitly learned how to paraphrase and can eventually generate paraphrases from the data itself. Moreover, to ensure complete paraphrasing we also impose an explicit edit operation. The pre-edit paraphrasing operation is applied to the source sentence before giving it to the model. We use the 60K most frequent words as our model vocabulary, out of almost 300K unique words from the whole training set. We create an alignment table for the words outside the vocabulary to the words inside the vocabulary using **fastText** [3] embedding. The word-to-word alignment has been done by calculating the cosine distance between **fastText** average word vectors. We found an alignment table of almost 8K words outside of the vocabulary to words inside with $CosDistance \geq 0.7$ (e.g., pricey $\Rightarrow$ expensive, detested $\Rightarrow$ hated). Our model tries to replace the out of vocabulary words with the words inside using the alignment table before sequence generation.

Diversity-Promoting Beam Search. Most of the generated outputs using standard beam search are lexically similar and they are different at only some small parts of the text, such as punctuation and stopwords. A solution to this problem is to force the beam-search decoder to generate more diverse outputs. In this work, we use a technique proposed by [1] for enforcing the diversity between beams. This work has shown to be effective for creating diverse image captions, machine translation and visual question generation. The authors divide K beams into G groups and control the diversity between these groups while expanding each beam. They modify the log probability of each predicted word on group G_i (except the first group G_1) as follows:

$$score(w_t^i \mid \mathbf{X}, W_{t-1}^i) = logP(w_t^i \mid \mathbf{X}, W_{t-1}^i) + \lambda \cdot \Delta(P_{i-1}) \qquad (2)$$

where, w_t^i is the candidate word of group G_i at time-step t, P_{i-1} is the list of last generated tokens from previous groups, λ is the diversity factor and Δ is the dissimilarity of current group with previous groups[1]. The beam size of the decoder was set to be 10. We present our N-best ($N = 5$) model generated output in Table 1.

---

[1] We use Hamming Diversity due to its simplicity and efficiency as *Delta* function.

Table 1. Our **DPC** model generated output (**CR** means Compression Ratio and highlighted words indicate paraphrasing in context).

Source sentence	It is the right message, sent while it is still early enough to do something constructive about the disappointing quality of the work so far
Reference (*Best*)	It is the right message to send to correct the disappointing quality of work so far. (**CR: 0.36**)
Output#1	*This* message is the right message. (**CR: 0.76**)
Output#2	It is the right message, sent while it is still early enough to do something *suitable*. (**CR: 0.44**)
Output#3	It is the right message, sent while it is still early enough to do something *faster* about the work. (**CR: 0.24**)
Output#4	*This* message is the right message, sent while it is still early enough to do something *useful* about the work so far. (**CR: 0.12**)
Output#5	It is the right message, sent while it is still early enough to do something *faster* about the work so far. (**CR: 0.16**)

4   Experiments

4.1   Datasets

For training set, we use a parallel corpus which was constructed from the Annotated English Gigaword dataset [14]. We use the script released by [19] to generate **3.8M** sentence-summary pairs as training set. For validation and test set, we use MSR-ATC dataset [24]. We filtered out the compressions which involve multiple source sentences. The final validation and test set contains 271 and 459 pairs of single sentence abstractive compression with maximum of five human rewrite variations.

4.2   Evaluation Metric

We evaluate our system automatically using various automatic metrics such as **BLEU** [18], **SARI** [25] and **METEOR-E** [21]. **Compression Ratio (CR)** is a measure of how terse a compression. We define **Copy Rate** as how many tokens are copied to the abstract sentence from the source sentence without paraphrasing. Lower copy rate score means more paraphrasing is involved in the output abstract sentence. Copy rate of 100% means no paraphrasing.

4.3   Performance Comparison and Discussion

We compare our model with the systems which include both deletion-based and near abstractive models. **ILP**, an integer linear programing approach for

Table 2. Performance of different systems compare to our proposed model.

Model	Information coverage		Abstractiveness		
	BLEU	SARI	METEOR-E	CR	Copy rate
T3 [7]	11.1	25.7	0.22	**0.75**	90.6
ILP [6]	54.7	38.1	0.36	0.29	99.5
Seq2Seq [8]	53.8	35.5	0.34	0.39	99.7
NAMAS [19]	38.7	36.6	0.31	0.24	99.8
PG + C [20]	45.5	37.3	0.37	0.21	99.3
SEASS [26]	44.6	38.5	0.35	0.34	99.6
DPC (ours)	**54.9**	**39.3**	**0.41**	0.47	**84.5**

sentence compression which involves word deletion [6]; **T3**, a tree-to-tree transduction model for abstractive sentence compression [7]; **seq2seq**, a neural model for deletion-based compression [8]; and **NAMAS**, a neural model for abstractive compression and sentence summarization [19]. The output generated by the above mentioned systems were collected from [24]. Moreover, we also compare our system with [20] which uses Pointer Generator Networks and Coverage Mechanism and with [26] which uses a selective gate network and an attention equipped decoder to tackle sentence summarization task.

We take the identical test set of [24] for comparison. We use the generated output directly from the baseline models using their settings to compare with our model across the metrics discussed earlier. For fair comparison, we add all the top ($N = 5$) candidates in the evaluation process. The results of different baseline systems across different evaluation metrics are presented in Table 2. Our model balances the information coverage (BLUE, SARI) and complete abstractiveness (METEOR-E, Copy Rate), instead of over compressing the generated sentences (Compression Ratio (**CR**)). As our model is generating diverse paraphrastic compression, we obtain a higher BLEU score compare to all the models presented in Table 2. We get a slightly higher score in terms of SARI because of the multiple human references. The Copy Rate scores of the baseline systems other than T3 clearly indicates that they are doing completely compression, no paraphrasing is involved. Lower copy rate means more new words were generated in the output sentences. We also get a higher score in METEOR-E metric because of the lexical substitution operations.

5 Conclusion and Future Work

In this paper, we have designed a new abstractive compression generation model at the sentence level which jointly performs diverse sentence compression and paraphrasing. We have imposed several operations to this architecture to reduce the extractiveness of abstractive sentence level output summaries.

Acknowledgements. The research reported in this paper was supported by the Natural Sciences and Engineering Research Council (**NSERC**) of Canada - discovery grant and the University of Lethbridge.

References

1. Vijayakumar, A.K., et al.: Diverse beam search: decoding diverse solutions from neural sequence models. In: AAAI 2018, February 2018
2. Bahdanau, D., Cho, K., Bengio, Y.: Neural machine translation by jointly learning to align and translate. In: ICLR 2015 (2015)
3. Bojanowski, P., Grave, E., Joulin, A., Mikolov, T.: Enriching word vectors with subword information. Trans. Assoc. Comput. Linguist. **5**, 135–146 (2017)
4. Cao, Z., Li, W., Li, S., Wei, F.: Retrieve, rerank and rewrite: soft template based neural summarization. In: Proceedings of the 56th Annual Meeting of the Association for Computational Linguistics (Volume 1: Long Papers), pp. 152–161. Association for Computational Linguistics (2018)
5. Chali, Y., Tanvee, M., Nayeem, M.T.: Towards abstractive multi-document summarization using submodular function-based framework, sentence compression and merging. In: Proceedings of the Eighth International Joint Conference on Natural Language Processing, IJCNLP 2017, Taipei, Taiwan, 27 November–1 December 2017, Volume 2: Short Papers, pp. 418–424 (2017)
6. Clarke, J., Lapata, M.: Global inference for sentence compression: an integer linear programming approach. JAIR **31**, 399–429 (2008)
7. Cohn, T., Lapata, M.: Sentence compression as tree transduction. JAIR **34**(1), 637–674 (2009)
8. Filippova, K., Alfonseca, E., Colmenares, C., Kaiser, L., Vinyals, O.: Sentence compression by deletion with LSTMs. In: Proceedings of the 2015 Conference on Empirical Methods in Natural Language Processing (2015)
9. Gu, J., Lu, Z., Li, H., Li, V.O.: Incorporating copying mechanism in sequence-to-sequence learning. In: Proceedings of the 54th Annual Meeting of the Association for Computational Linguistics (Volume 1: Long Papers), Berlin, Germany, pp. 1631–1640, August 2016
10. He, K., Zhang, X., Ren, S., Sun, J.: Deep residual learning for image recognition. In: CVPR, pp. 770–778. IEEE Computer Society (2016)
11. Hochreiter, S., Schmidhuber, J.: Long short-term memory. Neural Comput. **9**(8), 1735–1780 (1997)
12. Luong, M.T., Pham, H., Manning, C.D.: Effective approaches to attention-based neural machine translation. In: Empirical Methods in Natural Language Processing (EMNLP), Lisbon, Portugal, pp. 1412–1421, September 2015
13. Nallapati, R., Zhou, B., dos Santos, C., glar Gulçehre, Ç., Xiang, B.: Abstractive text summarization using sequence-to-sequence RNNs and beyond. CoNLL 2016, p. 280 (2016)
14. Napoles, C., Gormley, M., Van Durme, B.: Annotated gigaword. In: Proceedings of the Joint Workshop on Automatic Knowledge Base Construction and Web-Scale Knowledge Extraction, AKBC-WEKEX 2012, Stroudsburg, PA, USA, pp. 95–100 (2012)
15. Nayeem, M.T., Chali, Y.: Extract with order for coherent multi-document summarization. In: Proceedings of TextGraphs@ACL 2017: The 11th Workshop on Graph-based Methods for Natural Language Processing, Vancouver, Canada, 3 August 2017, pp. 51–56 (2017)

16. Nayeem, M.T., Chali, Y.: Paraphrastic fusion for abstractive multi-sentence compression generation. In: Proceedings of the 2017 ACM on Conference on Information and Knowledge Management, CIKM 2017, Singapore, 06–10 November 2017, pp. 2223–2226 (2017)
17. Nayeem, M.T., Fuad, T.A., Chali, Y.: Abstractive unsupervised multi-document summarization using paraphrastic sentence fusion. In: Proceedings of the 27th International Conference on Computational Linguistics, pp. 1191–1204. Association for Computational Linguistics (2018)
18. Papineni, K., Roukos, S., Ward, T., Zhu, W.J.: Bleu: a method for automatic evaluation of machine translation. In: Proceedings of the 40th Annual Meeting on Association for Computational Linguistics, ACL 2002, Stroudsburg, PA, USA, pp. 311–318 (2002)
19. Rush, A.M., Chopra, S., Weston, J.: A neural attention model for abstractive sentence summarization. In: Proceedings of the 2015 Conference on Empirical Methods in Natural Language Processing, pp. 379–389, September 2015
20. See, A., Liu, P.J., Manning, C.D.: Get to the point: summarization with pointer-generator networks. In: Proceedings of the 55th Annual Meeting of the Association for Computational Linguistics (Volume 1: Long Papers), Vancouver, Canada, pp. 1073–1083, July 2017
21. Servan, C., Berard, A., Elloumi, Z., Blanchon, H., Besacier, L.: Word2Vec vs DBnary: augmenting METEOR using vector representations or lexical resources? In: Proceedings of COLING 2016, the 26th International Conference on Computational Linguistics: Technical Papers, Osaka, Japan, pp. 1159–1168, December 2016
22. Song, K., Zhao, L., Liu, F.: Structure-infused copy mechanisms for abstractive summarization. In: Proceedings of the 27th International Conference on Computational Linguistics, pp. 1717–1729. Association for Computational Linguistics (2018)
23. Suzuki, J., Nagata, M.: Cutting-off redundant repeating generations for neural abstractive summarization. In: Proceedings of the 15th Conference of the European Chapter of the Association for Computational Linguistics: Volume 2, Short Papers, Valencia, Spain, pp. 291–297, April 2017
24. Toutanova, K., Brockett, C., Tran, K.M., Amershi, S.: A dataset and evaluation metrics for abstractive compression of sentences and short paragraphs. In: Proceedings of the 2016 Conference on Empirical Methods in Natural Language Processing, Austin, Texas, pp. 340–350, November 2016
25. Xu, W., Napoles, C., Pavlick, E., Chen, Q., Callison-Burch, C.: Optimizing statistical machine translation for text simplification. Trans. Assoc. Comput. Linguist. **4**, 401–415 (2016)
26. Zhou, Q., Yang, N., Wei, F., Zhou, M.: Selective encoding for abstractive sentence summarization. In: Proceedings of the 55th Annual Meeting of the Association for Computational Linguistics (Volume 1: Long Papers), Vancouver, Canada, pp. 1095–1104, July 2017

Fully Contextualized Biomedical NER

Ashim Gupta[1]([✉]), Pawan Goyal[1], Sudeshna Sarkar[1],
and Mahanandeeshwar Gattu[2]

[1] Indian Institute of Technology Kharagpur, Kharagpur, India
ashimgupta95@gmail.com, {pawang,sudeshna}@cse.iitkgp.ac.in
[2] Excelra Knowledge Solutions, Hyderabad, India
nandu.gattu@excelra.com

Abstract. Recently, neural network architectures have outperformed traditional methods in biomedical named entity recognition. Borrowed from innovations in general text NER, these models fail to address two important problems of polysemy and usage of acronyms across biomedical text. We hypothesize that using a fully-contextualized model that uses contextualized representations along with context dependent transition scores in CRF can alleviate this issue and help further boost the tagger's performance. Our experiments with this architecture have shown to improve state-of-the-art F1 score on 3 widely used biomedical corpora for NER. We also perform analysis to understand the specific cases where our contextualized model is superior to a strong baseline.

1 Introduction

Biomedical Named Entity Recognition (NER) is a fundamental step in several downstream biomedical text mining and information extraction tasks like relation classification, co-reference resolution etc. Traditional Biomedical NER systems [7,8] have often relied on task specific hand crafted features. Recent neural network based architectures in biomedical domain [15] have shown that comparable results can be achieved without making use of these hand engineered features although the performance is still dependent on the quality of learned word representations [16]. Character embeddings and pre-trained distributed embeddings have been used to model complex syntactic and semantic characteristics of words. But these complementary embedding models fail to capture different word uses across different linguistic contexts (i.e., *polysemy*). This problem is compounded in biomedical text due to ambiguous usage of words from general text [14] (ex: *column* in general English means *an upright pillar* while in medical context can be taken to mean *the spine*). Word representations obtained from training on biomedical corpora do not solve this problem because both forms, general English and biomedical, are generally present in the training text.

Electronic supplementary material The online version of this chapter (https://doi.org/10.1007/978-3-030-15719-7_15) contains supplementary material, which is available to authorized users.

L. Azzopardi et al. (Eds.): ECIR 2019, LNCS 11438, pp. 117–124, 2019.
https://doi.org/10.1007/978-3-030-15719-7_15

Another issue specific to biomedical domain is the generous usage of abbreviations (ex: gene/protein names like *ALA, MEN 1*) without explicit mention of their full forms. Neither character embeddings nor distributed word embeddings are effective in solving this issue. Character embeddings do not help as these abbreviations are mostly acronyms, where all characters are capitalized irrespective of the entity type. Word embeddings generally fail as most of these acronyms fall outside their vocabulary.

In order to ameliorate these two issues, we look at contextualization as an alternative. Current biomedical NER systems make use of context with the help of Bi-directional LSTMs that sequentially process a sentence [15,16,18]. Our model captures context more effectively in two additional ways: First, we make use of contextualized word representations based on [13], which have shown to improve sequence tagging with general English text [12,13]. Following the earlier discussion on issues with current biomedical NER systems, we find that these contextualized word representations are especially helpful in biomedical domain. Second, for the CRF layer we use a context dependent transition matrix [3] which is conditioned on token as well as its immediate context. We model this transition matrix non-linearly with the help of different neural networks. Experiments on four widely used biomedical datasets show that we are able to obtain state-of-the-art performance using this fully contextualized NER tagger.

Our main contributions are summarized as: (1) We show that using contextualized word embeddings for Biomedical NER leads to better performance in comparison to the baseline system. (2) We explore the use of different neural networks to model pairwise transition scores for CRF which further improves the tagging performance. (3) Our proposed model provides an improvement over current state-of-the-art on 3 out of 4 standard biomedical NER datasets.

2   Proposed Method

Our overall neural network architecture is shown in Fig. 1, which uses a Bi-directional LSTM with a CRF [5] sequence layer stacked on top of it. The input to this model is obtained by concatenating pre-trained distributed word representations with contextualized embeddings from a bi-directional language model. Following [2,6], a score over the output sequence $\mathbf{y} = (y_1, y_2, \ldots, y_n)$ is computed by summation of unary scores and pairwise transition scores as:

$$s(\mathbf{X}, \mathbf{y}) = \sum_{i=0}^{n} A_{y_i, y_{i+1}} + \sum_{i=1}^{n} P_{i, y_i} \tag{1}$$

where $\mathbf{P}$, an $n \times k$ matrix is used to model the unary scores and $\mathbf{A}$, a $k \times k$ matrix is used for transition scores, with n being the sequence length and k being the size of the tag set. The training is done with backpropagation by minimizing the negative log-likelihood of the correct label sequence $\hat{\mathbf{y}}$ for input $\mathbf{X}$ as follows:

$$-\log(p(\hat{\mathbf{y}}|\mathbf{X})) = -\log\left(\frac{e^{s(\mathbf{X}, \hat{\mathbf{y}})}}{\sum_{\mathbf{y} \in Y} e^{s(\mathbf{X}, \mathbf{y})}}\right) \tag{2}$$

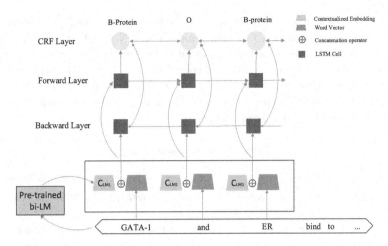

Fig. 1. Overview of the sequence tagging architecture. The word vector from pre-trained skip-gram model is concatenated with contextualized embeddings from a 2 layer bi-directional language model. See text for details.

During inference, Viterbi algorithm is used for determining the final label sequence.

Contextualized Word Representations: Recently, contextualized word vectors have shown to improve performance in many downstream tasks [11–13]. [13] show that contextualized word vectors obtained from a bi-directional language model achieve state of the art results on NER in english domain. They use a CNN with varying filter sizes over characters and use a 2 layer Bi-directional model. Finally, they compute a linear combination over hidden states stacked on each token to get a final word representation, which they call as ELMo.

In order to deal with issue of polysemy and acronyms in biomedical text, we use contextualized embeddings from an architecture similar to ELMo by training it on biomedical text with 1 Billion tokens obtained from PubMed and PMC. A key difference is that we do not take linear combination of the hidden state vectors from each layer (as is done with ELMo) and rather simply concatenate them as we observe that this results in a slight increase in performance on the development set.

Pairwise Modeling with Neural Networks: Most neural linear-chain CRF models for sequence tagging fix the transition matrix, $\mathbf{A}$, after the training process has concluded. In case of rare medical entities, this parameter matrix might not effectively model the transitions between the labels [3]. We, therefore, study two different neural-network based methods to non-linearly model these transition parameters conditioned on the current token and its context.

We keep unary scores for the Bi-LSTM-CRF unchanged and denote the pairwise scores modeled by a neural network by $\phi_{nn}(\mathbf{X})_{i,i+1}$, where i denotes the label position in the sequence. For modeling these scores, we consider two different types of neural networks, namely, a Fully Connected (FC) Feed Forward Neural Network and a Convolutional Neural Network (CNN). For each transition, the neural network takes as input the feature representation from Bi-LSTM of the two neighboring tokens involved in transition and outputs a $k \times k$ matrix of transition scores. So for a sentence with length n, the neural network outputs a tensor of shape $n \times k \times k$. The training objective remains the same and is trained end-to-end corresponding to Eq. (2).

Table 1. Dataset statistics

Dataset	Entity types	Sentences (Train/Dev/Test)	# Mentions
JNLPBA [4]	Protein, DNA, RNA Cell Line, Cell Type	24,806 (18,607/1,939/4,260)	59,973
BC2GM [17]	Protein/Gene	20,000 (10,000/5,000/5,000)	24,550
BC5CDR [9]	Disease, Chemical	13,938 (4,560/4,581/4,797)	28,785
NCBI-Disease [1]	Disease	7,295 (5,432/923/940)	6,892

3   Datasets and Experimental Setup

We use 4 widely used biomedical NER datasets to validate our method. Across these 4 datasets, all important biomedical entity types are covered. Statistics regarding mentions in corpora are mentioned in Table 1. We perform *exact match* (both entity type and entity boundary should be correctly predicted) evaluation based on macro-averaged F1 scores on all of these datasets. We provide the necessary training and implementation details in the supplementary material[1].

4   Results and Discussions

4.1   Evaluation of Architectures for Pairwise Modeling

We explore two prominent neural network architectures, namely CNN and Fully Connected NN (FCNN) for modeling pairwise transition scores. In case of CNN, we perform one dimensional convolution with filter width 2 along with tanh non-linearity. In case of FCNN, we element wise multiply the feature representation for the two tokens involved in the transition. We experiment with a single layer FCNN and a multi layer FCNN with depth 2. We observe that using a CNN for modeling transitions generally performs slightly better than FCNN. We find that FCNN with 2 layers can provide slightly better precision. For each of the datasets, we choose the pairwise modeling technique that performs the best on development set. We find that a CNN with depth 1 works the best for all datasets except for BC2GM, where CNN with depth 2 performs better. Implementation details and more results are provided in the supplementary material.

---

[1] https://www.dropbox.com/s/zc53mw8n77aop27/SupplementaryMaterial.pdf? dl=0.

4.2 Comparison with State of the Art Methods

We compare our results with other neural network methods known to perform well on these datasets. As a baseline, we implement the Bi-LSTM-CRF tagger described in [6], which makes use of pre-trained embeddings and incorporates subword features using another Bi-LSTM. We implement another baseline that instead uses CNN for calculating character embeddings [10]. Finally, we compare our method with a deep multi-task model, which provides state of the art performance on these datasets [18]. For the baselines, we performed hyper-parameter tuning with LSTM cell size and dimension of pre-trained word vectors. No such hyper-parameter tuning is performed for our model. Please refer the supplementary material for more details.

Table 2. Comparison with baseline and state of the art method, based on F1-score. *: Our implementation. **: From paper

Model	NCBI-Dis	BC2GM	BC5CDR	JNLPBA
Lample et al. [6]*	85.59	79.35	86.12	73.75
Ma et al. [10]*	83.70	77.92	85.95	72.72
Deep Multi-Task [18]**	86.14	80.74	**88.78**	73.52
Proposed model	**88.31**	**82.06**	88.64	**76.20**

We observe that our proposed model outperforms the two baselines significantly on all 4 datasets (Table 2). In comparison to the Multi-Task model (MTM), our model is superior in performance by more than 1% on 3 out of 4 datasets. Likely reason the MTM slightly outperforms our model on BC5CDR dataset is that MTM uses 3 different datasets that have either one of the Chemical and Disease entities, in which case using a MTM seems to have helped.

Table 3. Ablations with F1-score on test set. *: Instead use CharLSTM (See Text) to make a fair comparison

Model	NCBI-Dis	BC2GM	BC5CDR	JNLPBA
Full model	88.31	82.06	88.64	76.20
- NN transition scores	87.05	80.96	88.21	75.53
- Contextualized*	86.12	79.92	86.67	74.23

4.3 Ablations

To highlight the importance of the two important components in our model, we perform ablation analysis on the final test set without changing any hyper-parameters (see Table 3). First, we only remove the contextualized embeddings from our model but incorporate LSTM based sub-word features (like the baseline). In the second case, we only replace the context dependent transition matrix with a fixed (after training) transition matrix. We observe that: (1) The contribution of contextualized word embeddings is more prominent and leads to an increase of almost 2% in all cases. (2) When we do not use the context dependent transition matrix, in 3 out of 4 cases, F1 score drops by more than 1%.

4.4 Understanding the Effect of Contextualized Representations

To gain more insights into our proposed model, in particular the importance of contextualized representations in biomedical text, we select some examples from the test set. In Fig. 2, two recurring cases where using contextualized word embeddings have helped are shown. Example 1 exhibits the behavior of the baseline [6] and our proposed model in case of an acronym, the usage of which is very common in scientific texts. Using character representations, which might capture the information that the token is capitalized, alone does not help as the token *FNCAT* is out of vocabulary for pre-trained embeddings. It is interesting to note that a bi-directional LSTM, which is supposed to make use of the context, has not helped here either. This mistake is rectified by using a fully contextualized model, which looks at the neighboring token *Expression*, and infers that the entity involved (*FNCAT*) is a gene/protein.

1	O	O	O	O	O	O	O	O	O
	O	O	B-GENE	O	O	O	O	O	O
	O	O	B-GENE	O	O	O	O	O	O
	Expression	of	FNCAT	increased	on	serum	treatment	indicating	that

Context : ... the region of the FN gene between positions +69 and -510 bp mediated serum responsiveness.

Context : Thus, oxy R mutants are locked on for Ag 43 expression, ...

2	O	O	O	O	O	O	O	B-GENE	I-GENE	O
	O	B-GENE	I-GENE	O	O	O	O	B-GENE	I-GENE	O
	O	B-GENE	I-GENE	O	O	O	O	B-GENE	I-GENE	O
	whereas	dam	mutants	are	locked	off	for	Ag	43	expression

Fig. 2. Example outputs of our proposed model in cases where context helps determine the tags. Example 1 shows a case where the token is an acronym and Example 2 is a case of *polysemy* (for token *dam*). Entity tags highlighted with gold are the gold-standard tags, with red are the ones from baseline, and with blue, are from our proposed model. Related context is also shown. (Color figure online)

Looking at Example 2, we understand how using contextualized embeddings help to deal with polysemy. The token *dam*, which is more commonly associated with a reservoir structure, is incorrectly labeled by the baseline model. The baseline system correctly labeled *oxy R mutants* as a gene/protein entity but did not recognize *dam mutants*. Looking at the context suggests that like *oxy R mutants*, *dam mutants* should also be a protein. Again in such cases, looking at the larger context might have helped our model.

Finally, on analysis, we find that a fully contextualized model like ours also does much better on longer entities. We find that for entities with size greater than 5, our proposed model outperforms the baseline by a bigger margin. Detailed plots are available in the supplementary material.

5   Conclusions

In this paper, we proposed a fully contextualized NER architecture that makes use of context more effectively by using contextualized representations along with context conditioned transition scores. Our proposed model significantly outperformed the baseline on all the datasets. In addition, our experiments have shown to beat the current state-of-the-art results on 3 out of 4 datasets. All of our results were achieved without tuning hyper-parameters to specific datasets. Our detailed analysis in Sect. 4.4 indicates why in the case of acronyms and polysemy, using a fully contextualized model might have helped.

Acknowledgements. This work was sponsored by Ministry of Human Resource Development (MHRD), and Excelra Knowledge Solutions under a UAY project.

References

1. Doğan, R.I., Leaman, R., Lu, Z.: Ncbi disease corpus: a resource for disease name recognition and concept normalization. J. Biomed. Inform. **47**, 1–10 (2014)
2. Huang, Z., Xu, W., Yu, K.: Bidirectional LSTM-CRF models for sequence tagging. arXiv preprint arXiv:1508.01991 (2015)
3. Jagannatha, A.N., Yu, H.: Structured prediction models for RNN based sequence labeling in clinical text. In: Proceedings of the Conference on Empirical Methods in Natural Language Processing, vol. 2016, p. 856. NIH Public Access (2016)
4. Kim, J.D., Ohta, T., Tsuruoka, Y., Tateisi, Y., Collier, N.: Introduction to the bio-entity recognition task at JNLPBA. In: Proceedings of the International Joint Workshop on Natural Language Processing in Biomedicine and its Applications, pp. 70–75. Association for Computational Linguistics (2004)
5. Lafferty, J., McCallum, A., Pereira, F.C.: Conditional random fields: probabilistic models for segmenting and labeling sequence data (2001)
6. Lample, G., Ballesteros, M., Subramanian, S., Kawakami, K., Dyer, C.: Neural architectures for named entity recognition. In: Proceedings of NAACL-HLT, pp. 260–270 (2016)
7. Leaman, R., Gonzalez, G.: Banner: an executable survey of advances in biomedical named entity recognition. In: Biocomputing 2008, pp. 652–663. World Scientific (2008)
8. Leaman, R., Islamaj Doğan, R., Lu, Z.: DNorm: disease name normalization with pairwise learning to rank. Bioinformatics **29**(22), 2909–2917 (2013)
9. Li, J., et al.: BioCreative V CDR task corpus: a resource for chemical disease relation extraction. Database **2016** (2016)
10. Ma, X., Hovy, E.: End-to-end sequence labeling via bi-directional LSTM-CNNs-CRF. In: Proceedings of the 54th Annual Meeting of the Association for Computational Linguistics (Volume 1: Long Papers), vol. 1, pp. 1064–1074 (2016)

11. McCann, B., Bradbury, J., Xiong, C., Socher, R.: Learned in translation: contextualized word vectors. In: Advances in Neural Information Processing Systems, pp. 6294–6305 (2017)
12. Peters, M., Ammar, W., Bhagavatula, C., Power, R.: Semi-supervised sequence tagging with bidirectional language models. In: Proceedings of the 55th Annual Meeting of the Association for Computational Linguistics (Volume 1: Long Papers), vol. 1, pp. 1756–1765 (2017)
13. Peters, M., et al.: Deep contextualized word representations. In: Proceedings of the 2018 Conference of the North American Chapter of the Association for Computational Linguistics: Human Language Technologies, (Volume 1: Long Papers), vol. 1, pp. 2227–2237 (2018)
14. Pisanelli, D.M., Gangemi, A., Battaglia, M., Catenacci, C.: Coping with medical polysemy in the semantic web: the role of ontologies. In: Medinfo, pp. 416–419 (2004)
15. Sahu, S., Anand, A.: Recurrent neural network models for disease name recognition using domain invariant features. In: Proceedings of the 54th Annual Meeting of the Association for Computational Linguistics (Volume 1: Long Papers), vol. 1, pp. 2216–2225 (2016)
16. Sahu, S.K., Anand, A.: Unified neural architecture for drug, disease and clinical entity recognition. arXiv preprint arXiv:1708.03447 (2017)
17. Smith, L., et al.: Overview of biocreative ii gene mention recognition. Genome Biol. **9**(2), S2 (2008)
18. Wang, X., et al.: Cross-type biomedical named entity recognition with deep multi-task learning. arXiv preprint arXiv:1801.09851 (2018)

DeepTagRec: A Content-cum-User Based Tag Recommendation Framework for Stack Overflow

Suman Kalyan Maity[1(✉)], Abhishek Panigrahi[2], Sayan Ghosh[2], Arundhati Banerjee[2], Pawan Goyal[2], and Animesh Mukherjee[2]

[1] Northwestern University, Evanston, IL, USA
suman.maity@kellogg.northwestern.edu
[2] Department of CSE, IIT Kharagpur, Kharagpur, WB, India

Abstract. In this paper, we develop a *content-cum-user* based deep learning framework *DeepTagRec* to recommend appropriate question tags on Stack Overflow. The proposed system learns the content representation from question title and body. Subsequently, the learnt representation from heterogeneous relationship between user and tags is fused with the content representation for the final tag prediction. On a very large-scale dataset comprising half a million question posts, *DeepTagRec* beats all the baselines; in particular, it significantly outperforms the best performing baseline *TagCombine* achieving an overall gain of 60.8% and 36.8% in *precision@3* and *recall@10* respectively. *DeepTagRec* also achieves 63% and 33.14% maximum improvement in *exact-k accuracy* and *top-k accuracy* respectively over *TagCombine*.

Keywords: Tag recommendation · Deep learning · Stack Overflow

1 Introduction

In community based question answering (CQA) websites like Yahoo! Answers, Stack Overflow, Ask.com, Quora etc., users generate content in the form of questions and answers, facilitating the knowledge gathering through collaboration and contributions in the Q&A community. These questions are annotated with a set of tags by users in order to topically organize them across various subject areas. The tags are a form of metadata for the questions that help in indexing, categorization, and search for particular content based on a few keywords. Hashtags in social media or CQA tags are precursor to folksonomy or social/collaborative tagging (Del.icio.us[1], Flickr[2]). The tagging mechanism in

---

[1] http://del.icio.us.
[2] http://www.flickr.com.

S. K. Maity—Most of the work was done when all the authors were at IIT Kharagpur, India. We also acknowledge Prithwish Mukherjee, Shubham Saxena, Robin Singh, Chandra Bhanu Jha for helping us in various stages of this project.

© Springer Nature Switzerland AG 2019
L. Azzopardi et al. (Eds.): ECIR 2019, LNCS 11438, pp. 125–131, 2019.
https://doi.org/10.1007/978-3-030-15719-7_16

folksonomy is fully crowd-sourced and unsupervised and hence the annotation and the overall organization of tags suffers from uncontrolled use of vocabulary resulting in wide variety of tags that can be redundant, ambiguous or entirely idiosyncratic. Tag ambiguity arises when users apply the same tag in different contexts which gives the false impression that resources are similar when they are in fact unrelated. Tag redundancy, on the other hand, arises when several tags bearing the same meaning are used for the same concept. Redundant tags can hinder algorithms that depend on identifying similarities between resources. Further, manual error and malicious intent of users could also lead to improperly tagged questions, thereby, jeopardizing the whole topical organization of the website. A tag recommendation system can help the users with a set of tags from where they can choose tags which they feel best describe the question, thus facilitating faster annotations. Moreover, tag recommendation decreases the possibility of introducing synonymous tags into the tag list due to human error, thereby, reducing tag redundancy in the system.

The problem of tag recommendation has been studied by various researchers from various different perspectives [1–4,6,9–12,16,19]. [12] proposes a content-based method that incorporates the idea of tag/term coverage while [19] proposes a two-way Poisson mixture model for real-time prediction of tags. [17] proposes a user-based vocabulary evolution model. [7] present a framework of personalized tag recommendation in Flickr using social contacts. [9] propose a LDA based method for extracting a shared topical structure from the collaborative tagging effort of multiple users for recommending tags. [16] presents a factorization model for efficient tag recommendation. [10] build a word trigger method to recommend tags and further use this framework for keyphrase extraction [11]. [15] leverage similar questions to suggest tags for new questions. A recent paper by Wu et al. [22] exploits question similarity, tag similarity and tag importance and learns them in a supervised random walk framework for tag recommendation in Quora. Further, Joulin et al. [8] proposes a general text classification which we have adapted for comparison with our model. [3] develops an unsupervised content-based hashtag recommendation for tweets while [18] proposes a supervised topic model based on LDA. [4] uses Dirichlet process mixture model for hashtag recommendation. [13] proposes a PLSA-based topic model for hashtag recommendation. Weston et al. [21] proposes a CNN-based model for hashtag prediction.

In this paper, we employ a *content-cum-user* based deep learning framework for tag recommendation model which takes advantage of the content of the question text and is further enhanced by the rich relationships among the users and tags in Stack Overflow. We compare our method with existing state-of-the-art methods like Xia et al. [20], Krestel et al. [9], Wu et al. [22], Lu et al. [12], Joulin et al.'s *fastText* [8] and Weston's *#TAGSPACE* method [21] and observe that our method performs manifold better.

2   Tag Recommendation Framework

In this section, we describe in detail the working principles of our proposed recommendation system *DeepTagRec*[1]. The basic architecture of the model is shown in Fig. 1. The whole framework consists of three major components: (a) content representation extraction from question title and body, (b) learning user representation from heterogeneous user-tag network using node2vec [5], (c) tag prediction using representation aggregation. We formulate the tag prediction model as a function of the content and the user information.

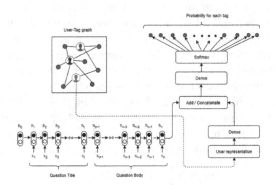

Fig. 1. *DeepTagRec* tag recommendation framework

2.1   Content Representation

To obtain the representation of body and title, we use Gated Recurrent Unit (GRU) model to encode the content as a sequence of words. Given the title of the question T and main body of the question B, we first run a GRU to learn representation of T, denoted by c_T. In the next step, we learn the representation of B denoted by c_B using a GRU, having c_T as the initial hidden state. We then describe the inner mechanism of a GRU. The GRU hidden vector output at step t, h_t, for the input sequence $X = (x_1, \ldots, x_t, \ldots, x_n)$ is given by:

$$
\begin{aligned}
z_t &= \sigma(W_z x_t + U_z h_{t-1}) \\
r_t &= \sigma(W_r x_t + U_r h_{t-1}) \\
\tilde{h}_t &= tanh(W_h x_t + U_h(r_t \odot h_{t-1})) \\
h_t &= z_t \odot \tilde{h}_t + (1 - z_t) \odot h_{t-1}
\end{aligned}
\tag{1}
$$

where, $W_z, W_r, W_h \in R^{m \times d}$ and $U_z, U_r, U_h \in R^{d \times d}$ are the weight vectors, m and d denote the word2vec dimension of word x_t and hidden state h_t respectively, z_t and r_t denote the update gate and reset gate in the GRU. The initial state h_0 is either vector 0 or is given as an input. We shall denote the entire GRU model by G(X, h_0) for future references.

---

[1] The codes and data are available at https://bit.ly/2HsVhWC.

Let T and B denote the sequence of words in the title and body of the question, respectively. Each word present in the sequence has its word2vec [14] representation. In case the predefined word vector does not exist, we consider a 300 dimension zero vector for the word. So, the content representation can be summarized as $c_T = G(T, 0); c_B = G(B, c_T)$.

2.2 User-Tag Network Representation

We construct a heterogeneous network containing nodes corresponding to users and tags. Let the graph be denoted by $G(V, E)$ where $V = V_U \cup V_T$, V_U are nodes corresponding to users and V_T are nodes corresponding to tags. We add an edge between a user and a tag, if the user has posted some question with the tag present in the question's tagset. The basic idea is to create a network and understand the tag usage pattern of each user. Given this graph $G(V, E)$, we use node2vec (a semi-supervised algorithm for learning feature representation of nodes in the network using random-walk based neighborhood search) to learn the representation of each node present in the graph. Let $f: V \to R^d$ be the mapping function from nodes to feature representations. Node2vec optimizes the following objective function, which maximizes the log-probability of observing a network neighborhood $N_S(u)$ for a node u conditioned on its feature representation,

$$\max_{f} \sum_{u \in V} log(P(N_S(u)|u))$$

where $P(N_S(u)|u)$ is given by

$$P(N_S(u)|u) = \prod_{n_i \in N_S(u)} \frac{\exp(f(n_i)f(n_u))}{\sum_{v \in V} \exp(f(v)f(n_u))}$$

Node2vec starts with a random function that maps each node to an embedding in R^d. This function is refined in an iterative way, so that the conditional probability of occurrence of the neighborhood of a node increases. The conditional probability of a node in the neighborhood of another node is proportional to cosine similarity of their embeddings. This is the idea that we use in our model, a user's representation should have high similarity with his/her adjacent tag nodes in the graph.

2.3 Representation Aggregation and Tag Prediction

Once we obtain both the word2vec representation (Q_w) of the question data and the node2vec representation (U_n) of users, we can aggregate these embeddings into final heterogeneous embedding (f_{agg}) by specific aggregation function $g(.,.)$ as follows.

- Addition: $f_{agg} = Q_w + U_n$
- Concatenation: $f_{agg} = [Q_w, U_n]$.

Following this layer, we have a dense or fully connected layer and finally a sigmoid activation is applied in order to get a probability distribution over the 38,196 tags.

3   Evaluation

In this section, we discuss in detail the performance of *DeepTagRec* and compare it with six other recent and popular recommendation systems – (i) Xia et al.'s *TagCombine* [20], (ii) Krestel et al.'s [9] LDA based approach, (iii) Lu et al.'s [12] content based approach, (iv) Wu et al.'s [22] question-tag similarity driven approach, (v) Weston et al.'s CNN-based model (#TAGSPACE) and (vi) Joulin et al.'s fastText model [8].

Training and Testing: We have 0.5 million training questions. Each question has different number of tags associated with it. Maximum length of the question is fixed as 300 words[2]. Each word is represented as a 300(m) dimension vector by using the predefined word2vec embeddings. The tags are represented as one-hot vectors. For a training example with t tags we add these t one hot vectors as the output for that training example. The number of GRU units is taken as 1000(d). The learning rate is taken to be 0.001 and the dropout as 0.5. For testing the model, we use 10K questions and perform the same initial steps of concatenating the title and body and then representing the words in 300 dimension vector form. For learning user representation, we create a user-tag graph over the training examples. We use node2vec over this graph to learn a 128 vector user embedding for all the users present in the training dataset. The output of the joint model is a probability distribution over 38,196 tags. We take the k tags with the highest probability for further evaluation.

Experiments and Results
In this section, we discuss in detail the performance of our proposed model *DeepTagRec* and compare it against the baselines. To understand the effect of content and user information separately, we also experiment with a variant of the proposed model – $DeepTagRec_{content}$ (i.e, only the content representation module of *DeepTagRec*). For evaluation purpose, we have used the following metrics.

Precision@k: Suppose there are q questions and for each question i, let $TagU_i$ be the set of tags given by the asker to the question and $TagR_i$ be the set of top-k ranked tags recommended by the algorithm, then $Precision@k = \frac{1}{q} \sum_{i=1}^{q} \frac{|TagU_i \cap TagR_i|}{|TagR_i|}$.

Recall@k: Similarly as *precision@k*, it can be formally defined as $Recall@k = \frac{1}{q} \sum_{i=1}^{q} \frac{|TagU_i \cap TagR_i|}{|TagU_i|}$ where k is a tunable parameter that determines how many tags the system recommends for each question.

   DeepTagRec significantly outperforms all the baselines (see Table 1) obtaining a *precision@*3 of ∼0.51 and a *recall@*10 of ∼0.76. Comparing the proposed variants, we observe that while most of the improvement of our model comes from the content representation, user information consistently helps in improving the performance. Since *Wu et al.*, *Lu et al.*, *Krestal et al.*, *#TAGSPACE* and *fastText*

---

[2] Avg. length of questions is 129 words. For question length <300, we pad them with zero vectors.

methods perform significantly worse, we have not considered them for subsequent analysis in this paper. We also do not consider $DeepTagRec_{content}$ further, and only compare $DeepTagRec$ with the best performing baseline, $TagCombine$.

Table 1. Precision (P) and Recall (R) in for $DeepTagRec$ and the other baselines.

Model	P@3	P@5	P@10	R@3	R@5	R@10
Krestel et al. [2009]	0.0707	0.0603	0.0476	0.0766	0.1097	0.1738
Lu et al.[2009]	0.1767	0.1351	0.0922	0.1952	0.2477	0.3362
Wu et al. [2016]	0.21	0.16	0.106	0.2325	0.2962	0.3788
#TAGSPACE [2014]	0.105	0.087	0.063	0.111	0.162	0.511
fastText [2016]	0.102	0.0783	0.149	0.0388	0.149	0.227
TagCombine	0.3194	0.2422	0.1535	0.3587	0.4460	0.5565
$DeepTagRec_{content}$	0.4442	0.3183	0.184	0.5076	0.591	0.6702
DeepTagRec	**0.5135**	**0.3684**	**0.2125**	**0.5792**	**0.6736**	**0.7613**

Top-k and Exact-k Accuracy: Apart from precision and recall, we also define the following evaluation metrics for further comparison.

Top-k accuracy: This metric is defined as the fraction of questions correctly annotated by at least one of the *top-k* tags recommended by the algorithm. *Exact-k accuracy*: This metric is defined as the fraction of questions correctly annotated by the kth recommended tag.

Table 2 shows the top-k and exact-k accuracy for both the models and we can observe that $DeepTagRec$ outperforms $TagCombine$ by 33.14%, 22.89% and 13.5% for $k = 3$, 5 and 10 respectively w.r.t. top-k accuracy. $DeepTagRec$ also performs better in exact-k accuracy than $TagCombine$ by achieving maximum and minimum gains of 63% and 10% respectively.

Table 2. Top-k and exact-k accuracy. Values at first 3 columns are for top-k accuracy and rest are for exact-k accuracy.

Model	k = 3	k = 5	k = 10	k = 1	k = 2	k = 3	k = 4	k = 5
TagCombine	0.688	0.769	0.851	0.481	0.289	0.188	0.145	0.108
DeepTagRec	**0.916**	**0.945**	**0.966**	**0.784**	**0.468**	**0.289**	**0.184**	**0.118**

4 Conclusions

In this paper, we propose a neural network based model ($DeepTagRec$) that leverages both the textual content (i.e., title and body) of the questions and the user-tag network for recommending tag. Our model outperforms the most competitive baseline $TagCombine$ significantly. We improve – *precision*@3 by 60.8%, *precision*@5 by 52.1%, *precision*@10 by 38.4%, *recall*@3 by 61.5%, *recall*@5 by 51.03%, *recall*@10 by 36.8% – over $TagCombine$. $DeepTagRec$ also performs better in terms of other metrices where it achieves 63% and 33.14% overall improvement in *exact-k accuracy* and *top-k accuracy* respectively over $TagCombine$.

References

1. Ding, Z., Qiu, X., Zhang, Q., Huang, X.: Learning topical translation model for microblog hashtag suggestion. In: IJCAI (2013)
2. Fu, W.T.: The microstructures of social tagging: a rational model. In: CSCW, pp. 229–238 (2008)
3. Godin, F., Slavkovikj, V., De Neve, W., Schrauwen, B., Van de Walle, R.: Using topic models for Twitter hashtag recommendation, pp. 593–596 (2013)
4. Gong, Y., Zhang, Q., Huang, X.: Hashtag recommendation using Dirichlet process mixture models incorporating types of hashtags, pp. 401–410 (2015)
5. Grover, A., Leskovec, J.: node2vec: scalable feature learning for networks. In: KDD (2016)
6. Heymann, P., Ramage, D., Garcia-Molina, H.: Social tag prediction. In: SIGIR, pp. 531–538 (2008)
7. Hu, J., Wang, B., Tao, Z.: Personalized tag recommendation using social contacts. In: Proceedings of SRS 2011, in Conjunction with CSCW, pp. 33–40 (2011)
8. Joulin, A., Grave, E., Bojanowski, P., Mikolov, T.: Bag of tricks for efficient text classification. arXiv preprint arXiv:1607.01759 (2016)
9. Krestel, R., Fankhauser, P., Nejdl, W.: Latent Dirichlet allocation for tag recommendation. In: RecSys, pp. 61–68 (2009)
10. Liu, Z., Chen, X., Sun, M.: A simple word trigger method for social tag suggestion. In: EMNLP, pp. 1577–1588 (2011)
11. Liu, Z., Liang, C., Sun, M.: Topical word trigger model for keyphrase extraction. In: COLING, pp. 1715–1730 (2012)
12. Lu, Y.T., Yu, S.I., Chang, T.C., Hsu, J.Y.J.: A content-based method to enhance tag recommendation. In: IJCAI, vol. 9, pp. 2064–2069 (2009)
13. Ma, Z., Sun, A., Yuan, Q., Cong, G.: Tagging your tweets: a probabilistic modeling of hashtag annotation in Twitter. In: CIKM, pp. 999–1008 (2014)
14. Mikolov, T., Chen, K., Corrado, G., Dean, J.: Efficient estimation of word representations in vector space. arXiv preprint arXiv:1301.3781 (2013)
15. Nie, L., Zhao, Y.L., Wang, X., Shen, J., Chua, T.S.: Learning to recommend descriptive tags for questions in social forums. ACM TOIS **32**(1), 5 (2014)
16. Rendle, S., Schmidt-Thieme, L.: Pairwise interaction tensor factorization for personalized tag recommendation. In: WSDM, pp. 81–90 (2010)
17. Sen, S., et al.: Tagging, communities, vocabulary, evolution, pp. 181–190 (2006)
18. She, J., Chen, L.: TOMOHA: topic model-based hashtag recommendation on Twitter. In: WWW Companion, pp. 371–372 (2014)
19. Song, Y., et al.: Real-time automatic tag recommendation. In: SIGIR, pp. 515–522 (2008)
20. Wang, X.Y., Xia, X., Lo, D.: TagCombine: recommending tags to contents in software information sites. J. Comput. Sci. **30**(5), 1017–1035 (2015)
21. Weston, J., Chopra, S., Adams, K.: #TagSpace: semantic embeddings from hashtags. In: EMNLP, pp. 1822–1827 (2014)
22. Wu, Y., Wu, W., Li, Z., Zhou, M.: Improving recommendation of tail tags for questions in community question answering. In: AAAI (2016)

Document Performance Prediction for Automatic Text Classification

Gustavo Penha[1], Raphael Campos[2], Sérgio Canuto[2],
Marcos André Gonçalves[2], and Rodrygo L. T. Santos[2(✉)]

[1] Delft University of Technology, Delft, The Netherlands
g.penha-1@tudelft.nl
[2] Computer Science Department, Universidade Federal de Minas Gerais,
Belo Horizonte, Brazil
{rcampos,sergiodaniel,mgoncalv,rodrygo}@dcc.ufmg.br

Abstract. Query performance prediction (QPP) is a fundamental task in information retrieval, which concerns predicting the effectiveness of a ranking model for a given query in the absence of relevance information. Despite being an active research area, this task has not yet been explored in the context of automatic text classification. In this paper, we study the task of predicting the effectiveness of a classifier for a given document, which we refer to as document performance prediction (DPP). Our experiments on several text classification datasets for both categorization and sentiment analysis attest the effectiveness and complementarity of several DPP inspired by related QPP approaches. Finally, we also explore the usefulness of DPP for improving the classification itself, by using them as additional features in a classification ensemble.

Keywords: Performance prediction · Automatic text classification

1 Introduction

Query performance prediction (QPP) is a challenging and fundamental problem in information retrieval. It concerns predicting the effectiveness of a ranking model when there is no relevance information available. Applications for QPP include selecting the best model depending on query features, combining multiple ranking models and requesting more information for potentially poorly formulated queries. QPP approaches have been divided into pre-retrieval [8,12,14,15,20,34] and post-retrieval [5,17,22,23,29,30,32], depending on the information used by the method. Inspired by QPP, in this paper, we derive and adapt methods for document performance prediction (DPP), which aim at predicting the performance of automatic text classifiers.

2 Document Performance Prediction

The performance of automatic text classifiers is usually measured by their average effectiveness over test documents. However, this performance can vary

L. Azzopardi et al. (Eds.): ECIR 2019, LNCS 11438, pp. 132–139, 2019.
https://doi.org/10.1007/978-3-030-15719-7_17

depending on the specific document in question. Inspired by query performance prediction, we define the task of document performance prediction as predicting the effectiveness of a text classifier for a given document, when labeled data is not available. Formally, a document performance predictor π can be defined as a function $\pi : D \times M \to Y$, where D and M denote the space of all documents and classifiers probability outputs, respectively, and Y denotes the space of possible effectiveness assessments given a pair $\langle d, m \rangle$, $d \in D$, $m \in M$. An effective predictor $\pi(d, m)$ approximates the true effectiveness $\Delta(\hat{y}_{dm}, y_{dm})$ as accurately as possible, where Δ is any classification effectiveness metric defined over the classification output $\hat{y}_{dm} = m(d)$ and the label y_{dm}. In our experiments in Sect. 3, we use cross-entropy as a representative evaluation metric Δ.

Depending on the information used by a document performance predictor π, it may fall into one of two categories: pre-classification and post-classification. In particular, a *pre-classification* DPP relies solely on the contents of document d to make its performance prediction. In contrast, inspired by post-retrieval query performance predictors, which leverage the ranked list produced by a target ranking model, a *post-classification* DPP uses the classification output $\hat{y}_{dm}$ in addition to the contents of document d. In the remainder of this section, we propose several pre- and post-classification approaches for DPP.

2.1   Pre-classification DPP

Inspired by prior work on ad-hoc retrieval, we adapted pre-retrieval query performance predictors for DPP. Instead of applying these methods on the query q, we apply them to document d. Some of our proposed DPP also require statistics from a corpus T, comprising documents used for training the classifier.

dT-Stats. Our first category of pre-classification DPP, denoted dT-Stats, includes predictors that rely only on the document d and the training corpus T. These predictors are independent of the classifier and were inspired by several pre-retrieval QPP methods [12]. In particular, *tokenCount* and *termCount* are the total number of tokens and the number of unique tokens in the document, respectively. *AvQL* is the average character size of the tokens in the document. {*Av,Max,Dev*}-*IDF* are the average, maximum and standard deviation of the inverse document frequency of the document terms. *AvICTF* is the average inverse collection term frequency of the document terms, defined as $AvICTF = \frac{1}{n} \sum_{i=1}^{n} [\log_2(cf_i) - \log_2(tf_{i,d})]$, where n is the number of terms in the document, cf_i is the collection frequency of the i-th term and $tf_{i,d}$ is its term frequency in d. *SCS* is the simplified clarity score of document terms, i.e., $SCS \approx \log_2 \frac{1}{n} + \frac{1}{n} \sum_{i=1}^{n} [\log_2(cf_i) - \log_2(tf_{i,d})]$. {*Av,Sum,Max*}-*SCQ* are the average, maximum and standard deviation of the collection document similarity. *AvP* is the average number of senses for document terms, using WordNet function `wordnet.synsets`, and *AvNP* is the average number of noun senses among these. *Av-*{*Path,LCH,WUP*} are the relatedness of a sample of 50 terms from the document by calculating all their pairwise similarities, using three similarity functions provided by WordNet: Path, Leacock-Chodorow, and Wu-Palmer.

d-Latent. Our second category of pre-classification DPP, denoted d-Latent, includes two predictors that are based on a latent representation of document d. In particular, $\{Max, Avg\}$-$PoolingGlove$ denote the maximum and average of each of 50 Glove dimensions from "glove.6B.zip"[1] for the document terms.

2.2   Post-classification DPP

Recent work has shown that post-retrieval query performance predictors are state-of-the-art in ad-hoc retrieval [17,25,26,32]. Unlike QPP, we do not have access to a list of documents retrieved for a query. Instead, we have a probability distribution $\hat{y}_{dm}$ of the classes predicted by a classifier m for document d.

DistBased. Predictors from this category assign the relevance of a document d to each class by calculating distances between a document d and each class centroid or between d and its k nearest neighbors (10 in our experiments) from each class. Here, we use the distance scores (Cosine, Euclidean and Manhattan) themselves as predictors that exploit the combination of global and local information about the distribution of documents in each class, as described in prior works on document classification with distance-based features [11,21].

BaggBased. Predictors from this category relate to the approach of Roitman et al. [26] and other approaches that estimate the variance of the retrieved lists [9, 22,29,31]. Here we bootstrap the estimators from the bagging-based models and use the variance of their predictions for document classes instead of the scores of top-retrieved documents. $BaggCVariance$ is the standard deviation of each class predicted probability for n (20 in our experiments) random base estimators sampled j (50 in our experiments) times for each classification bagging model m from $\{$RF [3], Bert [4], Broof [28]$\}$ and $n\_estimators = 200$ (which is the number of base models included in the bagging model). $BaggQ\{25,50,75\}C$ is similar to $BaggCVariance$, but instead of the standard deviation, we calculate the 25, 50 and 75 quantiles from the class prediction probabilities. $PredEntropy$ is a vector containing the entropy of the base estimators predictions probability distribution for each bagging classification model m from $\{$RF,Bert,Broof$\}$. $NumPredC$ is a vector containing the number of distinct classes (estimated probability not zero) in the base estimators predictions for each bagging classification model m from $\{$RF,Bert,Broof$\}$.

ProbPBased. DPP in this category use the prediction of any classifier, being agnostic to their inductive biases. $ProbPred$ are the probability predictions $\hat{y}_{dm}$ of each class for each classification model m, resulting in a vector of dimensionality $|M| \times |C|$, where M are all the classification models the performances of which are being predicted and C is the target set of classes. $ProbPredVar$ is the standard deviation of probability predictions of each class for each classification model, resulting in a vector of dimensionality $|M| \times |C|$. $ProbPBased$ encompasses the 25, 50 and 75 quantiles of probability predictions of each class for each classification model, resulting in a vector of dimensionality $|M| \times |C| \times 3$.

---

[1] http://nlp.stanford.edu/data/glove.6B.zip.

3   Evaluation

In this section, we aim to answer the following research questions:

Q1. How effective are the proposed DPP?
Q2. How complementary are the proposed DPP?
Q3. How effective are DPP for enhancing a classification ensemble?

3.1   Experimental Setup

We explored two categorization datasets, 20Newsgroups (20NG) and 4Universities (4UNI, aka WEBKB) with about 20,000 and 8,200 documents respectively. We also evaluate our approaches for the sentiment analysis task. We considered four data sets of messages labeled as positive or negative from distinct domains: Amazon, BBC, NYT, YouTube [4]. Inspired by prior work on QPP [8,26,29,32], we compute the correlation between the predicted performance $\pi(d, m)$ and the actual performance $\Delta(\hat{y}_{dm}, y_{dm})$ for document d and classifier m. In particular, we measure the actual performance of m as the cross-entropy between the predicted class distribution $\hat{y}_{dm}$ and the true distribution y_{dm}. The higher the cross-entropy Δ, the more distant the two distributions.

We predict the performance of several classifiers, $M = \{$XGBoost [6], KNN [1], NaiveBayes [33], Bert [4], Broof [28], RandomForest [3], SVM [16], MLP [10]$\}$. Except for Bert and Broof,[2] we used scikit-learn v0.18 implementations and their default hyperparameters, with TF-IDF document representations as input. To evaluate our proposed DPP, we perform a 5-fold cross-validation. In each round, four folds serve as the training corpus T and the remaining fold is used to calculate the correlation between the predicted and actual performance of each classifier m, averaged across all models in M. Accordingly, we report the mean of the average correlation obtained by each DPP across the five test folds.

3.2   DPP Effectiveness

To address $Q1$, Table 1 shows the mean average correlation coefficient (Pearson's ρ and Kendall's τ) attained by the best-performing DPP in each of the categories described in Sects. 2.1 and 2.2. The most successful predictors are from the categories *BaggBased* and *ProbPBased*, which comprise post-classification predictors. This result is somewhat expected given that post-retrieval QPP are state-of-the art. *dT-Stats* and *d-Latent* predictors are not effective for the sentiment analysis datasets. However, they achieve higher correlations in the categorization datasets (4UNI and 20NG). We believe this happens because sentiment analysis datasets are smaller in number of documents as well as in document length, hurting statistics taken on the document and corpus. Finally, *DistBased* predictors were ineffective in our experiments. We attribute this to the fact that neighbor information is used only by one of the eight classifiers whose performance we are predicting (KNN). For the other seven classifiers, this inductive bias does not hold, hence it is not a good predictor of their performance.

---

[2] https://github.com/raphaelcampos/stacking-bagged-boosted-forests.

Table 1. Effectiveness of document performance prediction strategies in terms of Pearson's ρ and Kendall's τ correlation with the cross-entropy loss.

Method	4UNI		20NG		Amazon		BBC		NYT		YouTube	
	K-τ	P-ρ	K-τ	P-ρ	K-τ	P-ρ	K-τ	P-ρ	K-τ	P-ρ	K-τ	P-ρ
BaggBased	.129	**.195**	.205	**.302**	.163	.203	.207	**.488**	0.09	.114	.241	.303
ProbPBased	**.250**	.187	**.408**	.157	**.240**	**.274**	**0.39**	.397	**.113**	**.123**	**.421**	**.448**
DistBased	.017	.024	.030	.026	.019	.026	.040	.059	.020	.030	.017	.026
d-Latent	.121	.149	.081	.092	.020	.029	.046	.076	.020	.030	.039	.047
dT-Stats	.175	.159	.265	.263	.019	.032	.038	.058	.017	.027	.016	.028

Table 2. Effectiveness of the combination of document performance prediction using different groups of methods as input space, in terms of Pearson's ρ and Kendall's τ correlation with the cross-entropy loss. Superscripts [†]/[‡] denote statistically significant improvements over the best raw DPP at 95%/99% confidence intervals.

Input space	4UNI		20NG		Amazon		BBC		NYT		YouTube	
	K-τ	P-ρ	K-τ	P-ρ	K-τ	P-ρ	K-τ	P-ρ	K-τ	P-ρ	K-τ	P-ρ
Best raw DPP	.250	.195	.408	.302	.240	.274	.390	.488	.113	.123	.421	.448
BaggBased	.347[‡]	.471[†]	.386	.516[†]	.304[†]	.400[†]	.257	.536[†]	.145[†]	.203[†]	.401	.512[†]
ProbPBased	.439[†]	.556[†]	.554[‡]	**.720**[†]	.250[‡]	.325[†]	.553[†]	.613[†]	.112	.148[†]	**.444**[†]	.468[†]
TF-IDF	.230	.296[‡]	.334	.276	.243	.314[†]	.242	.260	.103	.158[†]	.325	.363
d-Latent	.229	.291	.099	.120	.015	.022	.033	.045	.014	.022	.021	.032
DistBased	.030	.051	.029	.017	.010	.014	.030	.041	.015	.022	.029	.045
dT-Stats	.220	.258	.297	.360	.008	.012	.022	.038	.010	.012	.018	.026
All-pre-clf	.241	.301	.300	.367	.014	.019	.029	.041	.012	.018	.016	.026
All-post-clf	.478[†]	**.631**[†]	.538	.382	**.317**[†]	**.423**[†]	**.563**[†]	**.703**[†]	**.159**[†]	**.218**[†]	.440[‡]	**.531**[†]
All	**.479**[†]	.628[†]	**.582**[†]	.437	.288[†]	.381[†]	.464	.624[†]	.132[‡]	.183[†]	.396	.488[†]

3.3 DPP Complementarity

The combination of predictors through machine learning has been explored for improving query performance predictors, with the assumption that they capture complementary information [2,7,13,19,30,35]. To address $Q2$, we assess the complementarity of the proposed DPP for a given classifier m when used as input features for a machine-learned DPP (ML-DPP) aimed to predict m's actual performance. Table 2 shows the effectiveness of several different groups of DPP used as input features for a ML-DPP based on a random forest regressor. The single best-performing DPP is included as a baseline. For all datasets, we can significantly improve upon the single best DPP by combining multiple DPP. Therefore, recalling $Q2$, we conclude that the proposed DPP have a degree of complementarity and capture different types of information.

Table 3. Average macro F1 score for an ensemble of eight classifiers with DPP as additional meta-features. Superscripts [†]/[‡] denote statistically significant improvements compared to not using additional meta-features at 95%/99% confidence intervals.

	20NG	4UNI	Amazon	BBC	NYT	YouTube
Stacking	.939	.779	.759	.769	.630	.759
+ BaggBased	**.949**[†] (1.1%)	**.851**[†] (9.2%)	**.836**[†] (10.1%)	**.878**[†] (14.2%)	**.761**[†] (20.8%)	**.858**[†] (13.0%)
+ ProbPBased	.939 (0.0%)	.779 (0.0%)	.755 (−0.5%)	.783 (1.8%)	.628 (−0.3%)	.757 (−0.3%)
+ DistBased	.946[†] (0.7%)	.772 (−0.9%)	.741 (−2.4%)	.770 (0.1%)	.635 (0.8%)	.765 (0.8%)
+ d-Latent	.941 (0.2%)	.779 (0.0%)	.733 (−3.4%)	.731 (−4.9%)	.635 (0.8%)	.752 (−0.9%)
+ dT-Stats	.939 (0.0%)	.778 (−0.1%)	.748 (−1.4%)	.751 (−2.3%)	.629 (−0.2%)	.766 (0.9%)
+ ml BaggBased	.939 (0.0%)	.785 (0.8%)	.750 (−1.2%)	.790 (2.7%)	.631 (0.2%)	.761 (0.3%)
+ ml ProbPBased	.940 (0.1%)	.774 (−0.6%)	.756 (−0.4%)	.787 (2.3%)	.639 (1.4%)	.770 (1.4%)
+ ml DistBased	.939 (0.0%)	.780 (0.1%)	.757 (−0.3%)	.777 (1.0%)	.628 (−0.3%)	.765 (0.8%)
+ ml d-Latent	.939 (0.0%)	.780 (0.1%)	.753 (−0.8%)	.788 (2.5%)	.641 (1.7%)	.761 (0.3%)
+ ml dT-Stats	.939 (0.0%)	.779 (0.0%)	.756 (−0.4%)	.766 (−0.4%)	.628 (−0.3%)	.763 (0.5%)

3.4 Application: Enhancing Classification Ensembles

Improved QPP does not automatically translate to improved retrieval [24]. Roitman et al. [27] demonstrated through simulations that a minimum correlation of $\rho > 0.35$ would be necessary for a QPP to be useful. Although this barrier has been surpassed by several QPP in the literature, their observed utility for ad-hoc retrieval has been marginal [18]. To address $Q3$, we assess the usefulness of DPP for improving text classification, by employing DPP as additional meta-features to a stacking layer, which combines the output of the eight classifiers in M.

Table 3 summarizes our results in terms of macro F1, comparing the addition of groups of DPP against the stacking strategy. We obtained significant improvements with only one strategy, *BaggBased*, which is one of our most accurate DPP. However, several other sets of features that are also accurate for the DPP task did not translate to improvements in classification. We hypothesize that having a high accuracy in the performance prediction task is not sufficient for a DPP to improve the classification ensemble, as our empirical results corroborate.

4 Conclusions

We proposed several document performance predictors (DPP) for automatic text classification. We demonstrated their effectiveness and complementarity by thorough experiments on both categorization and sentiment analysis datasets. Moreover, we showed an application for DPP in improving automatic text classification ensembles, with state-of-the-art results. As future work, we plan to investigate why predictors with high correlations on the document performance prediction task do not necessarily translate into improved text classification.

Acknowledgements. Work partially funded by project MASWeb (FAPEMIG APQ-01400-14) and by the authors' individual grants from CNPq and FAPEMIG.

References

1. Altman, N.S.: An introduction to kernel and nearest-neighbor nonparametric regression. Am. Stat. **46**(3), 175–185 (1992)
2. Bashir, S.: Combining pre-retrieval query quality predictors using genetic programming. Appl. Intell. **40**(3), 525–535 (2014)
3. Breiman, L.: Random forests. Mach. Learn. **45**(1), 5–32 (2001)
4. Campos, R., Canuto, S., Salles, T., de Sá, C.C., Gonçalves, M.A.: Stacking bagged and boosted forests for effective automated classification. In: Proceedings of SIGIR, pp. 105–114 (2017)
5. Carmel, D., Yom-Tov, E.: Estimating the query difficulty for information retrieval. Synth. Lect. Inf. Concepts Retrieval Serv. **2**(1), 1–89 (2010)
6. Chen, T., Guestrin, C.: XGBoost: a scalable tree boosting system. In: Proceedings of SIGKDD, pp. 785–794. ACM (2016)
7. Chifu, A.G., Laporte, L., Mothe, J., Ullah, M.Z.: Query performance prediction focused on summarized LETOR features. In: Proceedings of SIGIR, pp. 1177–1180 (2018)
8. Cronen-Townsend, S., Zhou, Y., Croft, W.B.: Predicting query performance. In: Proceedings of SIGIR, pp. 299–306 (2002)
9. Cummins, R., Jose, J., O'Riordan, C.: Improved query performance prediction using standard deviation. In: Proceedings of SIGIR, pp. 1089–1090 (2011)
10. Goodfellow, I., Bengio, Y., Courville, A.: Deep Learning. MIT Press, Cambridge (2016)
11. Gopal, S., Yang, Y.: Multilabel classification with meta-level features. In: Proceedings of SIGIR, pp. 315–322 (2010)
12. Hauff, C.: Predicting the effectiveness of queries and retrieval systems. Ph.D. thesis. EEMCS (2010)
13. Hauff, C., Azzopardi, L., Hiemstra, D.: The combination and evaluation of query performance prediction methods. In: Boughanem, M., Berrut, C., Mothe, J., Soule-Dupuy, C. (eds.) ECIR 2009. LNCS, vol. 5478, pp. 301–312. Springer, Heidelberg (2009). https://doi.org/10.1007/978-3-642-00958-7_28
14. Hauff, C., Hiemstra, D., de Jong, F.: A survey of pre-retrieval query performance predictors. In: Proceedings of CIKM, pp. 1419–1420 (2008)
15. He, B., Ounis, I.: Inferring query performance using pre-retrieval predictors. In: Apostolico, A., Melucci, M. (eds.) SPIRE 2004. LNCS, vol. 3246, pp. 43–54. Springer, Heidelberg (2004). https://doi.org/10.1007/978-3-540-30213-1_5
16. Hearst, M.A., Dumais, S.T., Osuna, E., Platt, J., Scholkopf, B.: Support vector machines. IEEE Intell. Syst. Appl. **13**(4), 18–28 (1998)
17. Kurland, O., Shtok, A., Carmel, D., Hummel, S.: A unified framework for post-retrieval query-performance prediction. In: Amati, G., Crestani, F. (eds.) ICTIR 2011. LNCS, vol. 6931, pp. 15–26. Springer, Heidelberg (2011). https://doi.org/10.1007/978-3-642-23318-0_4
18. Macdonald, C., Santos, R.L.T., Ounis, I.: On the usefulness of query features for learning to rank. In: Proceedings of CIKM, pp. 2559–2562 (2012)
19. Mizzaro, S., Mothe, J., Roitero, K., Ullah, M.Z.: Query performance prediction and effectiveness evaluation without relevance judgments: two sides of the same coin. In: Proceedings of SIGIR, pp. 1233–1236 (2018)
20. Mothe, J., Tanguy, L.: Linguistic features to predict query difficulty. In: Proceedings of QP Workshop at SIGIR, pp. 7–10 (2005)

21. Pang, G., Jin, H., Jiang, S.: CenKNN: a scalable and effective text classifier. Data Min. Knowl. Discov. **29**(3), 593–625 (2015)
22. Pérez-Iglesias, J., Araujo, L.: Standard deviation as a query hardness estimator. In: Chavez, E., Lonardi, S. (eds.) SPIRE 2010. LNCS, vol. 6393, pp. 207–212. Springer, Heidelberg (2010). https://doi.org/10.1007/978-3-642-16321-0_21
23. Raiber, F., Kurland, O.: Using document-quality measures to predict web-search effectiveness. In: Serdyukov, P., et al. (eds.) ECIR 2013. LNCS, vol. 7814, pp. 134–145. Springer, Heidelberg (2013). https://doi.org/10.1007/978-3-642-36973-5_12
24. Raiber, F., Kurland, O.: Query-performance prediction: setting the expectations straight. In: Proceedings of SIGIR, pp. 13–22 (2014)
25. Roitman, H.: Query performance prediction using passage information. In: Proceedings of SIGIR, pp. 893–896. ACM (2018)
26. Roitman, H., Erera, S., Weiner, B.: Robust standard deviation estimation for query performance prediction. In: Proceedings of ICTIR, pp. 245–248 (2017)
27. Roitman, H., Hummel, S., Kurland, O.: Using the cross-entropy method to re-rank search results. In: Proceedings of SIGIR, pp. 839–842 (2014)
28. Salles, T., Gonçalves, M., Rodrigues, V., Rocha, L.: BROOF: exploiting out-of-bag errors, boosting and random forests for effective automated classification. In: Proceedings of SIGIR, pp. 353–362 (2015)
29. Shtok, A., Kurland, O., Carmel, D.: Predicting query performance by query-drift estimation. In: Azzopardi, L., et al. (eds.) ICTIR 2009. LNCS, vol. 5766, pp. 305–312. Springer, Heidelberg (2009). https://doi.org/10.1007/978-3-642-04417-5_30
30. Shtok, A., Kurland, O., Carmel, D.: Using statistical decision theory and relevance models for query-performance prediction. In: Proceedings of SIGIR, pp. 259–266 (2010)
31. Tao, Y., Wu, S.: Query performance prediction by considering score magnitude and variance together. In: Proceedings of CIKM, pp. 1891–1894 (2014)
32. Zamani, H., Croft, W.B., Culpepper, J.S.: Neural query performance prediction using weak supervision from multiple signals. In: Proceedings of SIGIR, pp. 105–114 (2018)
33. Zhang, H.: The optimality of Naive Bayes. AA **1**(2), 3 (2004)
34. Zhao, Y., Scholer, F., Tsegay, Y.: Effective pre-retrieval query performance prediction using similarity and variability evidence. In: Macdonald, C., Ounis, I., Plachouras, V., Ruthven, I., White, R.W. (eds.) ECIR 2008. LNCS, vol. 4956, pp. 52–64. Springer, Heidelberg (2008). https://doi.org/10.1007/978-3-540-78646-7_8
35. Zhou, Y., Croft, W.B.: Query performance prediction in web search environments. In: Proceedings of SIGIR, pp. 543–550 (2007)

Misleading Metadata Detection on YouTube

Priyank Palod[1], Ayush Patwari[2], Sudhanshu Bahety[3], Saurabh Bagchi[2], and Pawan Goyal[1(✉)]

[1] IIT Kharagpur, Kharagpur, WB, India
priyankpalod@gmail.com, pawang.iitk@gmail.com
[2] Purdue University, West Lafayette, IN, USA
{patwaria,sbagchi}@purdue.edu
[3] Salesforce.com, San Francisco, CA, USA
sudhanshu.bahety@salesforce.com

Abstract. YouTube is the leading social media platform for sharing videos. As a result, it is plagued with misleading content that includes staged videos presented as real footages from an incident, videos with misrepresented context and videos where audio/video content is morphed. We tackle the problem of detecting such misleading videos as a supervised classification task. We develop UCNet - a deep network to detect fake videos and perform our experiments on two datasets - VAVD created by us and publicly available FVC [8]. We achieve a macro averaged F-score of 0.82 while training and testing on a 70:30 split of FVC, while the baseline model scores 0.36. We find that the proposed model generalizes well when trained on one dataset and tested on the other.

1 Introduction

The growing popularity of YouTube and associated economic opportunities for content providers has triggered the creation and promotion of fake videos and spam campaigns on this platform. There are various dimensions to this act including creating videos for political propaganda as well as choosing clickbaity or shocking title/thumbnails in order to get more views. YouTube itself has classified spam videos into many different categories including misleading metadata (metadata includes the title, description, tags, annotations, and thumbnail).

We use the following definition of "fake" videos [7]: 1. Staged videos in which actors perform scripted actions under direction, published as user generated content (UGC). 2. Videos in which the context of the depicted events is misrepresented (e.g., the claimed video location is wrong). 3. Past videos presented as UGC from breaking events. 4. Videos of which the visual or audio content has been altered. 5. Computer-generated Imagery (CGI) posing as real.

Spam detection in social media has been a widely researched topic in the academic community [10,11]. In [3], the author describes a model to detect spam in tagging systems. For video sharing platforms, most of the work has concentrated on finding spam comments [1,9].

© Springer Nature Switzerland AG 2019
L. Azzopardi et al. (Eds.): ECIR 2019, LNCS 11438, pp. 140–147, 2019.
https://doi.org/10.1007/978-3-030-15719-7_18

Recently, there have been works on creating a dataset of fake videos and computationally detecting fake videos [6,7]. As a result, there is a small but publicly available dataset of fake videos on YouTube called the Fake Video Corpus (or FVC) [8]. They had also developed a methodology to classify videos as fake or real, and reported an F-score of 79% using comment based features. After inspecting their code, we found out that the reported F-score was for the positive (fake) class only. So we reproduced their experiment to find that the macro average F-score by their method is only 36% since the classifier calls almost all videos as fake.

Through our experiments, we find that simple features extracted from metadata are not helpful in identifying fake videos. Hence we propose to use a deep neural network on comments for the task and achieve promising results. While using a 70:30 split of the FVC dataset, we find that our method achieves an F-score of 0.82 in comparison to a score of 0.36 by the baseline, and 0.73 by the feature based approach. Further, we also present a new dataset of fake videos containing 123 fake and 423 real videos called VAVD. To see the robustness of our approach, we also train UCNet on a balanced subset of VAVD, and test on FVC dataset, achieving an F-score of 0.76, better than the score obtained by the feature-based classifier trained on the same dataset. Feature-based classifiers, on the other hand, do not give robust performance while trained on our dataset, and tested on FVC.

2   Dataset Preparation

We crawled metadata and comments for more than 100,000 videos uploaded between September 2013 and October 2016 using YouTube REST data API v3. The details include video metadata (title, description, likes, dislikes, views etc.), channel details (subscriber count, views count, video count, featured channels etc.) and comments (text, likes, upload date, replies etc.).

With around 100K crawled videos and possibly very low percentage of fake content, manually annotating each and every video and searching for fake videos was infeasible. Also, random sampling from this set is not guaranteed to capture sufficient number of fake videos. Therefore, we used certain heuristics to boost the proportion of fake videos in a small sample to be annotated.

We first removed all the videos with views less than 10,000 (the average number of views in the crawled set) and with comments less than 120 (the average number of comments on a video in the crawled set). This was done to have only popular videos in the annotated dataset. A manual analysis of comments on some hand-picked spam videos gave us some comments such as "complete bullshit", "fake fake fake" etc. Then a search for more videos containing such phrases was performed on the dataset. Repeating the same process (bootstrapping the "seed phrases" as well as the set of videos) thrice gave a set of 4,284 potentially spam videos. A similar method was adopted for clustering tweets belonging to a particular rumor chain on twitter in [12] with good effect. After this, we used the ratio of dislike count: like count of the video for further filtering. Sorting the videos based on the ratio in non-ascending order and taking videos having ratio greater than 0.3 gave us a final set with 650 videos.

An online annotation task was created where a volunteer was given the link to a video and was asked to mark it as "spam" or "legitimate". An option to mark a video as "not sure" was also provided. 33 separate surveys having 20 videos per survey (one having only 10) were created and were submitted for annotation to 20 volunteering participants. This task was repeated for a second round of annotation with the same set of annotators without repetition. Statistics from the two rounds of annotation can be seen in Table 1. We see that inter-annotator agreement was not perfect, an issue which has been reported repeatedly in prior works for annotation tasks in social media [2,5]. The discrepancies in the annotations were then resolved by another graduate student volunteer and if any ambiguity still persisted in characterizing the video as spam, the video was marked as "not sure".

We call this dataset as **VAVD** (Volunteer Annotated Video Dataset) and the annotations[1] as well as our codes[2] are publicly available.

Table 1. Statistics from the two rounds of annotations

(a) Number of videos in different classes

	Round1	Round2	Final
Spam	158	130	123
Legitimate	400	422	423
Not Sure	92	98	104

(b) Annotator agreements

	Spam	Legitimate	Not Sure
Spam	70	62	26
Legitimate	54	308	38
Not Sure	6	27	59

FVC Dataset: The Fake video corpus (FVC, version 2)[3] contains 117 fake and 110 real video YouTube URLs, alongside annotations and descriptions. The dataset also contains comments explaining why a video has been marked as fake/real. Though the dataset contains annotations for 227 videos, many of them have been removed from YouTube. As a result, we could crawl only 98 fake and 72 real videos. We divide these videos into two disjoint sets, FVC70 (30), containing 70 (30)% of these videos for various experiments.

3 Experiments with Simple Features

We first tried using simple classifiers like SVMs, decision trees and random forests on VAVD and test it on FVC, which is the benchmark dataset. We hypothesized several simple features that might differentiate a fake video from a legitimate one, as described below:

---

[1] https://github.com/ucnet01/Annotations_UCNet.

[2] https://github.com/ucnet01/UCNet_Implementation.

[3] https://zenodo.org/record/1147958#.WwBS1nWWbCJ.

- has_clickbait_phrase: This feature is true if the title has a phrase commonly found in clickbaits. For e.g. *'blow your mind'*, *'here is why'*, *'shocking'*, *'exposed'*, *'caught on cam'*. We used 70 such phrases gathered manually.
- ratio_violent_words: A dictionary of several 'violent' words like *'kill'*, *'assault'*, *'hack'*, *'chop'* was used. The value of this feature is equal to the fraction of violent words in the title. We hypothesize that violent words generate fear which leads to more views for the video, hence more used in spams.
- ratio_caps: This feature is equal to the ratio of number of words in the title which are in upper case to the total number of words in the title.
- Tweet Classifier Score - Title: The Image verification corpus (IVC)[4] is a dataset containing tweets with fake/real images. We trained a multi-layer perceptron on IVC to predict the probability of a tweet (i.e., the accompanying image) being fake using only simple linguistic features on the tweet text. Now, we use the same trained network and feed it the title of a video as input. The probability of fakeness that it outputs is then taken as a feature.
- dislike_like_ratio: Ratio of number of dislikes to number of likes on the video.
- comments_fakeness: This is equal to the ratio of comments on the video which mention that the video is fake. To categorize if a comment says that the video is fake or not, we detect presence of words and regexes like *'fa+ke+'*, *'hoax'*, *'photoshopped'*, *'clickbait'*, *'bullshit'*, *'fakest'*, *'bs'*.
- comments_inappropriateness: This feature is equal to the ratio of number of comments with swear words to the total number of comments on the video.
- comments_conversation_ratio: This is the ratio of comments with at least one reply to the total number of comments on the video.

Since some of these classifiers are sensitive to correlations in the features, we first decided to remove the lesser important feature among each pair of correlated features. For this, we calculated correlations among the features on all the 100K videos and identified the pair of features with a correlation of more than 0.2. Then we generated feature importance scores using the standard random forests feature selection method and eliminated the lesser important feature from each pair. We trained our classifiers using only these remaining features. Table 2 shows the performance of some of these classifiers when trained on VAVD and tested on FVC30, as well as when trained on FVC70 and tested on FVC30 with Macro Averaged Precision (P), Recall (R) and F1 Score (F) as the metrics.

We see that although Random Forests classifier performs the best when trained and tested on FVC, its performance is very bad when trained on VAVD.

To understand the reason for such poor performance of these classifiers, we plotted the PCA of the features on FVC dataset, which is shown in Fig. 1 (left). Through the PCA, we can see that though the features may help in identifying some fake videos (the ones on the far right in the plot), for most of the videos, they fail to discriminate between the two classes.

---

[4] https://github.com/MKLab-ITI/image-verification-corpus/tree/master/mediaeval2016.

Table 2. Performance of simple classifiers tested on FVC30

(a) training dataset: VAVD

Classifier	P R F
SVM- RBF	0.74 0.60 0.49
Random Forests	0.73 0.58 0.46
Logistic Regression	0.54 0.53 0.45
Decision Tree	0.53 0.52 0.46

(b) training dataset: FVC70

Classifier	P R F
SVM- RBF	0.56 0.55 0.54
Random Forests	0.74 0.73 0.73
Logistic Regression	0.53 0.53 0.53
Decision Tree	0.73 0.67 0.67

4 UCNet: Deep Learning Approach

Our analysis during dataset preparation reveals that comments may be strong indicator of fakeness. However, not all comments may be relevant. Hence, we computed "fakeness vector" for each comment, a binary vector with each element corresponding to the presence or absence of a fakeness indicator phrase (e.g., "looks almost real"). We used 30 such fakeness indicating phrases. Now, for each comment, we passed the GoogleNews pre-trained word2vec [4] embeddings of words of the comment sequentially to the LSTM. The 300-dimensional output of the LSTM is hence referred as "comment embedding". We also took the fakeness vector and passed it through a dense layer with sigmoid activation function, to get a scalar between 0 to 1 for the comment called as the "weight" of the comment. The idea here was that the network would learn the relative importance of the phrases to finally give the weight of the comment. We then multiplied the 300-dimensional comment embeddings with the scalar weight of the comment to get "weighted comment embedding". Now we took the average of all these weighted comment embeddings to get one 300-dimensional vector representing all the comments on the video called the "unified comments embedding". The unified comments embedding was then concatenated with simple features described before and passed through 2 dense layers, first with ReLU activation and 4-dimensional output and the second with softmax to get a 2 dimensional output representing the probability of the video being real and fake, respectively. This network is called **UCNet** (Unified Comments Net)[5] and is trained using adam's optimizer with learning rate 10^{-4} and cross entropy as the loss function.

Training the network on VAVD and testing on whole FVC gives an F-score of 0.74 on both classes as shown in Table 3. Training and testing UCNet on FVC70 and FVC30 respectively gives a macro F-score of 0.82. We also reproduced the experiments that [6,7] did and found their Macro average F-score to be 0.36 on both 10-fold cross validation and on the 70:30 split.

To visualize the discriminating power of comments, we trained UCNet on VAVD. Then we gave each video of FVC as input to the network and extracted the unified comment embedding. Now we performed PCA of these unified comment embeddings to 2 dimensions and plotted it in Fig. 1 (right). We observe

---

[5] https://bit.ly/2rZ7cAT.

Table 3. Performance of UCNet tested on FVC30

(a) training dataset: VAVD

Class	P	R	F	#Videos
Real	0.64	0.88	0.74	72
Fake	0.88	0.64	0.74	98
Macro avg	0.76	0.76	0.74	170

(b) training dataset: FVC70

Class	P	R	F	#Videos
Real	0.74	0.87	0.8	23
fake	0.89	0.77	0.83	31
Macro avg	0.82	0.82	0.82	54

Table 4. Overall performance comparison of classifiers on FVC30 test set

Classifier	Training set	Precision	Recall	F-Score
UCNet	FVC70	0.82	0.82	0.82
UCNet	Class balanced subset of VAVD	0.76	0.76	0.76
Random forests	FVC70	0.74	0.73	0.73
Baseline	FVC70	0.29	0.5	0.37

that comment embeddings can discriminate among the two classes very well as compared to simple features (compare the left and right sub-figures).

Since VAVD had certain properties that overlap with our features (e.g., many fakeness indicating phrases were picked from videos in this dataset), we decided not to test our methods on VAVD as it might not be fair. Hence, we have used it only as a training corpus, and tested on an unseen FVC dataset. Although, future works may use VAVD as a benchmark dataset as well.

Finally, we present Table 4 comparing performance of different classifiers when tested on FVC30. Random forests has been reported in the table since it was the best performing simple classifier on the test set. We can see that even if trained on (a balanced subset of) VAVD, UCNet performs better than any simple classifier or the baseline.

 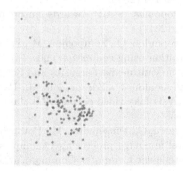

(a) Simple Features   (b) Unified Comment Embeddings

Fig. 1. PCA plots. Red dots are fake videos while blue dots are real videos from FVC. (Color figure online)

5   Conclusions

Our work presents VAVD, a new dataset for research on fake videos, and also presents UCNet, a deep learning based approach to identify fake videos with high accuracy using user comments. Future work will involve putting more emphasis on content and metadata than the comments, to be able to detect latest or 'breaking news' spam videos.

Acknowledgement. This material is based in part upon work supported by a Google Faculty Award to Saurabh. Any opinions, findings, and conclusions or recommendations expressed in this material are those of the authors and do not necessarily reflect the views of the sponsor.

References

1. Ammari, A., Dimitrova, V., Despotakis, D.: Semantically enriched machine learning approach to filter YouTube comments for socially augmented user models. In: UMAP, pp. 71–85 (2011)
2. Becker, H., Naaman, M., Gravano, L.: Learning similarity metrics for event identification in social media. In: Proceedings of the Third ACM International Conference on Web Search and Data Mining, pp. 291–300. ACM (2010)
3. Koutrika, G., Effendi, F.A., Gyöngyi, Z., Heymann, P., Garcia-Molina, H.: Combating spam in tagging systems. In: Proceedings of the 3rd International Workshop on Adversarial Information Retrieval on the Web, pp. 57–64. ACM (2007)
4. Mikolov, T., Sutskever, I., Chen, K., Corrado, G.S., Dean, J.: Distributed representations of words and phrases and their compositionality. In: Advances in Neural Information Processing Systems, pp. 3111–3119 (2013)
5. Ott, M., Choi, Y., Cardie, C., Hancock, J.T.: Finding deceptive opinion spam by any stretch of the imagination. In: Proceedings of the 49th Annual Meeting of the Association for Computational Linguistics: Human Language Technologies, vol. 1, pp. 309–319 (2011)
6. Papadopoulos, S.A.: Towards automatic detection of misinformation in social media (2017)
7. Papadopoulou, O., Zampoglou, M., Papadopoulos, S., Kompatsiaris, Y.: Web video verification using contextual cues. In: Proceedings of the 2nd International Workshop on Multimedia Forensics and Security, pp. 6–10. ACM (2017)
8. Papadopoulou, O., Zampoglou, M., Papadopoulos, S., Kompatsiaris, Y., Teyssou, D.: InVID fake video corpus v2.0, January 2018. https://doi.org/10.5281/zenodo.1147958
9. Radulescu, C., Dinsoreanu, M., Potolea, R.: Identification of spam comments using natural language processing techniques. In: 2014 IEEE International Conference on Intelligent Computer Communication and Processing, ICCP, pp. 29–35. IEEE (2014)
10. Viswanath, B., et al.: Towards detecting anomalous user behavior in online social networks. In: 23rd USENIX Security Symposium, USENIX Security 2014, pp. 223–238 (2014)

11. Wang, A.H.: Don't follow me: spam detection in Twitter. In: Proceedings of the 2010 International Conference on Security and Cryptography, SECRYPT, pp. 1–10. IEEE (2010)
12. Zhao, Z., Resnick, P., Mei, Q.: Enquiring minds: early detection of rumors in social media from enquiry posts. In: Proceedings of the 24th International Conference on World Wide Web. International WWW Conferences Steering Committee (2015)

A Test Collection for Passage Retrieval Evaluation of Spanish Health-Related Resources

Eleni Kamateri[1,2(✉)], Theodora Tsikrika[2], Spyridon Symeonidis[2],
Stefanos Vrochidis[2], Wolfgang Minker[1], and Yiannis Kompatsiaris[2]

[1] Centre for Research and Technology-Hellas, 57001 Thessaloniki, Greece
{ekamater,theodora.tsikrika,spyridons,stefanos,ikom}@iti.gr
[2] Ulm University, 89081 Ulm, Germany
{wolfgang.minker,eleni.kamateri}@uni-ulm.de

Abstract. This paper describes a new test collection for passage retrieval from health-related Web resources in Spanish. The test collection contains 10,037 health-related documents in Spanish, 37 topics representing complex information needs formulated in a total of 167 natural language questions, and manual relevance assessments of text passages, pooled from multiple systems. This test collection is the first to combine search in a language beyond English, passage retrieval, and health-related resources and topics targeting the general public.

Keywords: Test collection · Passage retrieval · Inter-rater agreement

1 Introduction

Passage retrieval aims at focusing retrieval to the most relevant parts of a document, rather than entire documents, with the goal of reducing users' information overload. It has been widely studied both in ad hoc information retrieval (e.g., [1,4,6]) and also as an intermediate step in question answering (e.g., [12]), particularly when seeking answers to non-factoid questions (e.g., [7]). Passage retrieval evaluation is particularly challenging as retrieval techniques produce answers of varying sizes, and thus requires test collections with relevance assessments at finer granularity levels, typically corresponding to short text segments [6].

To further advance the development of passage retrieval methods, given in particular the recent interest in non-factoid question answering [3], we have built a test collection for passage retrieval evaluation that addresses health-related information needs of Spanish-speaking users (available at https://mklab.iti.gr/results/spanish-passage-retrieval-dataset/). This was motivated by our need to evaluate passage retrieval from health-related Web resources in response to natural language questions expressing health-related topics of general interest; to the best of our knowledge, no such test collection is available.

L. Azzopardi et al. (Eds.): ECIR 2019, LNCS 11438, pp. 148–154, 2019.
https://doi.org/10.1007/978-3-030-15719-7_19

The evaluation task being modelled is the retrieval of short text passages from trustworthy health-related Web resources in response to a natural language question expressing a health-related information need a member of the general public might have. The user may either type their question in a conventional search interface or may interact with a retrieval system through a speech-based interface. In either case, the retrieval system aims to simulate a Spanish-speaking health expert who examines diverse health-related information in several online documents and finds the passage(s) best addressing the search topic.

For evaluating such a task, we have constructed a test collection of 10,037 health-related documents in Spanish, 37 search topics representing complex information needs, instantiated as a total of 167 natural language questions, and appropriate manual relevance assessments. The users' health-related information needs fall under three thematic categories: (i) "baby care", such as appropriate food for newborns and breastfeeding advice, (ii) "vaccination", such as the recommended vaccination schedule, and (iii) "low back pain", such as potential causes of back pain. Relevance assessments were performed by considering that a passage should correspond to the longest contiguous piece of relevant information and not split into small passages which only in combination form an appropriate answer; therefore, each relevant passage (irrespective of its size) should be able to sufficiently answer the search topic on its own. To ensure the reliability of the assessments, two different judges, together with a third "validator", were engaged. The inter-annotator agreement was also measured by kappa coefficients proposed in this work for performing pairwise comparisons at character-level.

2   Related Work

Test collections for evaluating passage retrieval have been previously built by several benchmarking activities, including TREC, INEX, and BioASQ. The TREC Hard track [1], the INEX ad hoc track [6], and the TREC Complex Answer Retrieval track [2] included tasks where the goal was to retrieve passages, instead of documents, from document collections (news articles and Wikipedia, respectively) covering a broad range of topics. The TREC Genomics track [4] and more recently the BioASQ challenge [12] focused on the biomedical domain using collections of scientific articles, rather than content targeting the general public, with the goal to answer biomedical questions, i.e., considering passage retrieval as an intermediate step to question answering. More recently, two new test collections for evaluating (answer) passage retrieval have been constructed on the basis of the large-scale TREC GOV2 [7] and ClueWeb12 [3] collections.

The above test collections mainly focus though on the English language; one exception is the recent BOLT IR test collection [10] that enables multilingual retrieval from informal discussion forum text with relevance assessments performed at the passage level. Moreover, test collections for evaluating the retrieval of (bio)medical documents in response to laypeople's health-related information needs have been constructed by CLEF eHealth [11]; these include non-English documents, but focus on document retrieval. To the best of our knowledge, our test collection is unique in its combination of passage retrieval evaluation in a language other than English, health-related documents and topics targeting the general public, and also associating each topic with multiple query formulations.

3   Document Collection

The documents in the collection were obtained from two reputable health-related websites suggested by family doctors; the first (http://enfamilia.aeped.es/) contains information about paediatric and gynaecological matters, while the second (https://medlineplus.gov/spanish/) is the Spanish-language version of Medline-Plus, offering authoritative and up-to-date information on health issues.

The two websites were crawled resulting in a total of 63,362 unique URLs (8,343 and 55,019, respectively). Then, a set of heuristic rules was applied to exclude non informative material, such as pages providing services (e.g., registration to the website). In addition, URLs referring to non-textual content were also excluded. This filtering resulted in 9,992 Web pages from the two websites. This list was further augmented by including 45 reliable health-related Web pages and PDF documents suggested by domain experts during the topic development process (see Sect. 4), reaching a total of 10,037 documents.

To extract the textual content of the webpages and remove "noise" (such as advertisements, etc.), the Default Extractor of the Boilerpipe library [8] was applied. The Apache PDFBox library was used for extracting the content of PDF documents. The cleaned "plain text" was then indexed using Apache Lucene following (i) stopword removal based on the default Lucene stopword Spanish list which was also manually expanded with some domain-specific stopwords, and (ii) stemming using the Spanish Snowball stemmer.

Overall, the document collection consists of 10,037 health-related documents in Spanish containing a total of 265,600 sentences and 2.1 billion words, with the documents' mean length being 210 words.

4   Topics

The topics were developed by four native Spanish speakers who are family doctors with health-related expertise in family and community medicine. First, they selected three broad and diverse thematic categories ("baby care", "vaccination", and "low back pain") and then developed the topics using an exploration process: each topic developer considered a topic and provided several Web resources addressing it; these were added to the document collection, if not already included.

Once the document collection was finalised, the topic developers performed preliminary searches to determine whether adequate relevant information was available in the collection. Retrieval in this case was performed using the language retrieval models described in Sect. 5 without query expansion. By studying the available relevant information, the topic developers refined this initial list of topics and formulated alternative natural language questions serving as different query instantiations for each topic, so as to cover different synonyms and ways of expressing the topic. Moreover, they wrote a textual description and guidelines of interpretation for assessing the relevance of text passages. A total of 37 topics (19, 10, and 8, respectively in the three categories) and 167 alternative query formulations were developed, with an average of 4.51 queries per topic and 11.62 terms per question. Figure 1 shows an example "baby care" topic.

```
"topic":{
  "number":"PR_ES_2",
  "category":"Baby care",
  "name":"Factores que determinan el peso y el tamaño de bebés",
  "queries":[
  {"number":"PR_ES_2_1",
   "text":"¿Cuáles son los factores que determinan el peso y el tamaño de mi bebé?"}
   ...
  {"number":"PR_ES_2_8",
   "text":"¿Influye la constitución física de los padres en el peso y la talla del bebé?"}
  ]
  "description":"This topic concerns the changes for new-born babies regarding their height and weight. We suggest to avoid highlighting the complete
         explanation because it is extremely detailed. The relevant segment might refer to the general information of the baby's adaptation after the birth."
}
```

Fig. 1. Example topic in the "baby care" category.

5   Pooling

To form a pool of relevant documents for each topic, we considered all of its alternative questions and applied four retrieval approaches: a language model with Jelinek-Mercer smoothing ($\lambda = 0.5$) and a language model with Dirichlet smoothing ($\mu = 3000$), each with and without query expansion. Query expansion was performed based on a 400-dimensional word embeddings skip-gram word2vec model [9] trained on our document collection and implemented using deepLearning4j. The most related terms identified by the model were added to the query; the number of expansion terms was empirically set to three. Then, the pool for each topic was constructed by the five top-ranked documents retrieved by each of the four retrieval methods for each alternative query resulting in pools with 19 documents per topic on average. Given that each topic is associated with multiple queries, each potentially contributing different documents to the pool, the decision was made to only consider the five top-ranked documents by the different retrieval methods so as to keep the evaluation tractable, given also the additional effort required for assessing at the finer-granularity level of passages.

6   Relevance Assessments

To ensure the reliability of the assessment, two judges were engaged; the first is one of the topic developers with background in healthcare, while the second has an ICT background and was selected as a member of the general public interested in these healthcare issues. Due to the discrepancies between these two judges (see below), the passages judged as relevant by them were then further examined by a third judge with a linguistic background, acting as a *validator* for determining the final relevance assessments.

Assessment Process. For the assessment process, we configured and adjusted the open source annotation tool developed by the BioASQ challenge [12]. This tool enables the judges to manually mark within each document the relevant text passages which are suitable answers to a topic and saves the relevance assessments information in a MongoDB database.

The judges were instructed to examine all the documents in the pool of each topic and mark the relevant passages. The annotation is performed at sentence-level marking complete sentences, based on the assumption that a full sentence is the smallest segment capable of answering such topics. For each topic, annotators were also asked to explain their decisions, so as to have a record of the rationale for their selections. The validator was asked to examine only the relevant judgments of both annotators considering the overlapping parts of two judges to correspond to correct answers and thus focusing on their discrepancies.

Table 1 provides some statistics on the assessments performed by the two judges and the validator. Compared to the second judge, the first annotated more passages and more sentences per topic. The validator noted that many of the snippets marked as relevant by the first judge contained text that did not directly answer the topic, but rather provided a more general context. The validator considered this additional text as superfluous and therefore non-relevant, and was thus more aligned to the generally shorter assessments of the second judge.

Inter-annotator Agreement. The inter-annotator agreement for all assessor pairs was first calculated by Cohen's κ at document level, i.e., a document is considered relevant if at least one of its passages is marked as relevant. The Cohen's κ values: $\kappa(Judge_1, Judge_2) = 0.52$, $\kappa(Judge_1, validator) = 0.61$, and $\kappa(Judge_2, validator) = 0.91$ indicate the moderate agreement on the document level between the two judges, and the validator and the first judge, and also the particularly strong agreement between the validator and the second judge.

To.estimate the inter-annotator agreement at a finer level of granularity, we propose the following. First, we calculate for each topic the *overlap* between the passages annotated by two assessors within each document, averaged over all documents for a topic, and then over all topics. We propose to estimate the overlap between any two passages that have at least some common parts using (i) the Largest Common Subsequence (LCS) [5] between them (as implemented by the difflib python library) normalised by the length (in characters) of the shorter of the two passages, and (ii) Cohen's κ at character-level. The proposed inter-annotation agreement at character-level takes into account not only the common relevant characters, but also the common non-relevant characters of annotated documents, since both contribute to the overlapping segments. The non-overlapping passages contribute with a zero value to these averages. We get: (i) $overlap(Judge_1, Judge_2) = 0.25$, $overlap(Judge_1, validator) = 0.32$, and $overlap(Judge_2, validator) = 0.84$, and (ii) $\kappa(Judge_1, Judge_2) = 0.28$, $\kappa(Judge_1, validator) = 0.39$, and $\kappa(Judge_2, validator) = 0.83$.

Table 1. Relevance assessments statistics

	# relevant documents			# relevant passages			# relevant sentences		
	Mean (median)	Min-max	Stdev.	Mean (median)	Min-max	Stdev.	Mean (median)	Min-max	Stdev.
Judge 1	5.57 (4)	1–32	5.72	14.62 (7)	2–127	21.35	48.22 (28)	6–243	49.31
Judge 2	5.46 (5)	1–19	3.99	7.00 (6)	1–20	4.91	24.62 (14)	1–173	32.66
Validator	5.54 (5)	1–19	4.05	7.46 (6)	1–24	5.32	29.43 (19)	1–177	35.43

Table 2. Retrieval evaluation results at each stage

Document retrieval			Paragraph retrieval			Passage retrieval		
N	Success@N	Precision@N	M	Success@M	Precision@M	Precision@1	Recall@1	F1@1
3	89.22%	42.78%	2	69.46%	34.68%	41.89%	34.26%	34.92%

In addition, we summed up the total relevant and the total non-relevant characters over all documents in the pool of each topic and calculated κ at topic-level, and then averaged κ over all topics, resulting in $\kappa(Judge_1, Judge_2) = 0.38$, $\kappa(Judge_1, validator) = 0.50$, and $\kappa(Judge_2, validator) = 0.87$. These variations for calculating inter-annotation agreement further corroborate the high consensus between the validator and the second judge, as well as the difference of opinion between the two initial judges. Moreover, the first two methods at the character level appear to be highly correlated also in terms of agreement level, while the third appears to be more lenient regarding the assessment of the agreement with the first judge.

7 Baseline Performance

While this is not an evaluation report, we would like to give an indication of a reasonable baseline performance. We perform passage retrieval in three stages. First, the union of the top-N documents retrieved by the two language models (Sect. 5) is formed. The two language models are then used for retrieving (from this document set) the union of top-M paragraphs; we set $N = 3$ and $M = 2$. Last, the passages retrieved from this set of paragraphs are ranked by the normalised sum of the two language models scores. Table 2 presents the retrieval results at each stage, averaged over the 167 questions. At least one document and one paragraph containing a relevant passage are retrieved for almost 90% and 70% of the queries, respectively. Assuming that we are only interested in retrieving the top-ranked passage, akin to non-factoid question answering, we measure precision and recall at rank 1, which are close to 42% and 34%.

8 Conclusions

We presented a new test collection for passage retrieval evaluation that contains 10,037 reputable health-related documents in Spanish, 37 topics formulated in a total of 167 natural language questions, and passage-level relevance assessments.

This test collection is expected to provide a basis for evaluating new passage retrieval methods in non-English resources, particularly in non-factoid query answering settings, such as conversational agents in the healthcare domain. In addition, we introduced new inter-annotator agreement metrics at character-level which can be used to evaluate the test collection's inter-rater reliability.

Acknowledgments. This is supported by EU H2020 KRISTINA project (645012) and GR Research-Create-Innovate REA project (T1EDK-00686).

References

1. Allan, J.: HARD track overview in TREC 2004 - high accuracy retrieval from documents. In: Proceedings of TREC 2004 (2004)
2. Dietz, L., Verma, M., Radlinski, F., Craswell, N.: TREC complex answer retrieval overview. In: Proceedings of TREC 2017 (2017)
3. Habernal, I., et al.: New collection announcement: focused retrieval over the web. In: Proceedings of ACM SIGIR 2016, pp. 701–704 (2016)
4. Hersh, W., Cohen, A., Roberts, P., Rekapalli, H.: TREC 2006 genomics track overview. In: Proceedings of TREC 2006 (2006)
5. Hirschberg, D.S.: The longest common subsequence problem. Ph.D. thesis, Princeton, NJ, USA (1975). aAI7623803
6. Kamps, J., Geva, S., Trotman, A., Woodley, A., Koolen, M.: Overview of the INEX 2008 Ad Hoc track. In: Geva, S., Kamps, J., Trotman, A. (eds.) INEX 2008. LNCS, vol. 5631, pp. 1–28. Springer, Heidelberg (2009). https://doi.org/10.1007/978-3-642-03761-0_1
7. Keikha, M., Park, J., Croft, W., Sanderson, M.: Retrieving passages and finding answers. In: Proceedings of ADCS 2014, pp. 81–84. ACM (2014)
8. Kohlschütter, C., Fankhauser, P., Nejdl, W.: Boilerplate detection using shallow text features. In: Proceedings of ACM WSDM 2010, pp. 441–450 (2010)
9. Mikolov, T., Sutskever, I., Chen, K., Corrado, G., Dean, J.: Distributed representations of words and phrases and their compositionality. Adv. Neural Inf. Process. Syst. **26**, 3111–3119 (2013)
10. Soboroff, I., Griffitt, K., Strassel, S.: The BOLT IR test collections of multilingual passage retrieval from discussion forums. In: Proceedings of ACM SIGIR 2016, pp. 713–716. ACM (2016)
11. Suominen, H., et al.: Overview of the CLEF eHealth evaluation lab 2018. In: Bellot, P., et al. (eds.) International Conference of the Cross-Language Evaluation Forum for European Languages, vol. 11018, pp. 286–301. Springer, Cham (2018). https://doi.org/10.1007/978-3-319-98932-7_26
12. Tsatsaronis, G., Balikas, G., Malakasiotis, P., Partalas, I., et al.: An overview of the BIOASQ large-scale biomedical semantic indexing and question answering competition. BMC Bioinform. **16**(1), 138 (2015)

On the Impact of Storing Query Frequency History for Search Engine Result Caching

Erman Yafay[1,2] and Ismail Sengor Altingovde[1(✉)]

[1] Middle East Technical University, Ankara, Turkey
{erman.yafay,altingovde}@ceng.metu.edu.tr
[2] ASELSAN, Ankara, Turkey

Abstract. We investigate the impact of the size of query frequency history and its compact representation in memory for search result caching.

1 Introduction

Large-scale search engines extensively employ caching of query results in main memory to improve system efficiency and scalability [3,5,8,15]. As in many other domains, frequency of past data items (or, requests) is a strong signal to decide on the items to be evicted from the query result caches of search engines, and used on its own in the well-known eviction policy Least Frequently Used (LFU) or combined with other signals, like item recency and size, in other policies [13].

In this paper, our contributions are three-fold: First, we investigate the impact of storing a large query frequency history versus just keeping the frequency of the queries that are in the cache. We show that keeping the entire history (i.e., frequency of all seen queries by the search engine) may improve the cache performance (i.e., hit ratios) for the policies that employ frequency as a signal for eviction. While similar findings have been shown for other caching applications (e.g., see [4]), as far as we know, this issue has not been explored in depth for result caching in search engines.

Secondly, we adopt a recently proposed storage scheme for exactly this purpose, i.e., storing past request frequencies in a compact manner for caching, to our application domain. The latter scheme, referred to as Tiny here, can store the query frequency history by using Simple and Counting Bloom Filters (CBFs) [4]. Our experiments reveal that the storage space for query history can be significantly reduced while cache performance still remains comparable to storing the entire query history. This is an important finding as the number of queries submitted to search engines has reached to very large numbers and storing an (almost) full history may have very demanding memory storage requirements.

As our third contribution, we investigate the performance gains when the saved memory space (using the Tiny approach) is further exploited to cache additional query results. To this end, we consider not only storing query results in HTML format (i.e., top-k results' URLs and snippets), but also adopt techniques

L. Azzopardi et al. (Eds.): ECIR 2019, LNCS 11438, pp. 155–162, 2019.
https://doi.org/10.1007/978-3-030-15719-7_20

that allow storing just the document identifiers for certain query results [9, 10]. For this latter case, we evaluate the cache performance in terms of the query processing cost, and show that using the saved space to store identifiers yields significant gains. To our best knowledge, no earlier work conducted such an exhaustive analysis for search result caching where compact schemes are employed for storing query frequency history, and our findings here shed light on the potential gains in terms of storage space, hit-rate and even query processing cost, all of which would be worthwhile in practical search systems.

In Sect. 2, we review the Tiny storage scheme as proposed in [4]. Section 3 presents our exhaustive experimental evaluation together with the conclusive summary.

2   Compact Storage for Query Frequency History

A Bloom filter (BF) allocates a vector of m bits (initially set to 0) to efficiently query the existence of an item a_i in a set $A = \{a_1, a_2 \cdots a_n\}$ of n elements. When an item a_i is inserted to A, k distinct hash functions are applied to a_i to obtain the values $h_1(a_i), \cdots, h_k(a_i)$ each within the range $[0, m]$, and bits at these positions are set to 1. When an item a_i is queried, bits at positions $h_1(a_i), \cdots, h_k(a_i)$ are read, and if any of them is 0, then it is guaranteed that $a_i \notin A$. Otherwise, we conclude that $a_i \in A$, although false positives are possible.

In order to use Bloom filters to keep track of the counts, say, to represent the frequency history of a query stream, it is adequate to allocate a vector of m counters rather than that of m bits, as proposed in [4]. We adopt the latter storage scheme, referred to as Tiny hereafter. In Tiny [4], a Minimal Increment CBF is employed to store the approximate frequency values of previous queries. The Estimate operation is used to obtain the approximate frequency of a given query, which is the minimum count among the values at the indexes computed by k different hash functions. Similarly, the Add operation computes k different hash values for a given query and increments only the minimal counters at corresponding positions. In addition to the CBF, *Tiny* contains a simple BF that is called the *doorkeeper*. The latter is intended to reduce the number or size of the counters in the CBF: As query streams are known to include a large percentage (i.e., up to 44% [3]) of singleton queries that appear only once, when a query arrives, first the doorkeeper is checked to see whether it has been seen before, and only for such queries the counters in CBF are increased.

To keep the frequency values fresh, Tiny keeps a window counter, starting from 0, that is incremented after each query. Whenever the window counter is equal to W, all of the counters in the CBF as well as the window counter are divided by 2 and the bits in the *doorkeeper* are all set to 0. Einziger et al. [4] argue that for a cache of size C, an entry should reside in the cache if it has a larger frequency than $1/C$. Therefore, for a window size of W queries, each counter can be capped to $log_2 \frac{W}{C}$ bits. Note that, since the doorkeeper can count up to 1, full counters are only necessary to count up to $\frac{W}{C} - 1$. Assuming that a doorkeeper has d bits, and the number of full counters in CBF is n (typically, $n < d$), the total number of bits required to allocate for *Tiny* is given as $d + n \times log_2 \left(\frac{W}{C} - 1 \right)$.

3   Experimental Evaluation

We seek answer to following questions: (i) Does storing the full query frequency history improve the hit-rate performance for the caching policies that exploit frequency signal (namely, LFU [13] and GDSF [1])?, (ii) Can the Tiny scheme for storing query history reduce the memory space requirements without adversely affecting the cache hit-rate?, and (iii) If the saved space is further exploited for caching, what is the gain in terms of the hit-rate and query processing cost?

Query Log and Simulation Parameters. To simulate a query stream, we use the chronologically ordered queries from the AOL query log [12] that consists of ≈ 17 million queries. We use the first 10 million queries for training. The remaining ≈ 7 million queries constitute the test set; in particular, we use the first %10 of it to warm-up the result cache and the rest is employed for evaluation.

The training set is used only to obtain past query frequencies for the caching policies that employ the frequency as a signal (i.e., LFU [13] and GDSF [1]); however, it is not directly exploited to fill the cache, as there is a separate warm-up set. For the LRU eviction policy, the training set is not used at all.

Table 1. Parameters

Parameter	Value	Parameter	Value (bytes)
Number of distinct queries (U)	≈ 6.7 M	Size of a pointer (P)	4
Ranking per posting (P_r)	200 ns	Size of an int I (long int L)	4 (8)
Decompression per posting (P_d)	100 ns	Avg. size of a query ($AvgQ$)	16.5
Snippet computation per byte (P_s)	100 ns	Avg. size of a doc (d_{avg})	16384

Table 2. Memory space cost of LRU, LFU, and GDSF variants (M^{tiny} given in text).

Policy	Memory cost	Formula (bytes)
Pure (metadata)	M^{pure}	$U \times (AvgQ + I)$
LRU	$M_{lru}(c)$	$c \times (AvgQ + 3P)$
In-memory LFU	$M_{lfu}^{inmem}(c)$	$c \times (AvgQ + 6P + I)$
Pure LFU	$M_{lfu}^{pure}(c)$	$M_{lfu}^{inmem}(c) + M^{pure}$
Tiny LFU	$M_{lfu}^{tiny}(c)$	$M_{lfu}^{inmem}(c) + M^{tiny}(c)$
In-memory GDSF-K	$M_{gdsf}^{inmem}(c)$	$c \times (AvgQ + 6P + L + I) + L$
Pure GDSF-K	$M_{gdsf}^{pure}(c)$	$M_{gdsf}^{inmem}(c) - c \times I + M^{pure}$
Tiny GDSF-K	$M_{gdsf}^{tiny}(c)$	$M_{gdsf}^{inmem}(c) - c \times I + M^{tiny}(c)$

Results for Single-Signal Caching Policies. We begin with describing our implementation of the *single-signal* cache eviction policies, namely, LRU and LFU, that exploit either query recency or frequency for the eviction, respectively.

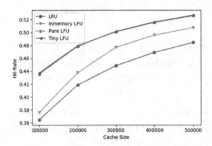
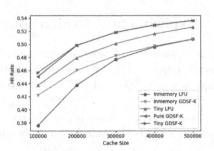

Fig. 1. Hit-rates for LRU and LFU variants (In-memory, Pure and Tiny).

Fig. 2. Hit-rates for LFU and GDSF-K variants.

Least Recently Used (LRU) policy always evicts the least recently accessed item when the cache is full. We used a doubly-linked list to keep the access order of queries and a hash-table to map the query strings to pointers to the linked-list nodes. Based on the parameters given in Table 1, we provide the worst-case space requirements for our LRU implementation in Table 2.

Least Frequently Used (LFU) policy always evicts the least frequent item when the cache is full. The variant that does not keep any history and accumulates the frequency of queries for only the cached ones is called the *In-memory LFU*. In contrary to the LRU, a list of doubly-linked lists with special frequency nodes is required (cf. [14] for details) for an efficient implementation. At the worst-case, number of frequency nodes are as many as the cached items, i.e., each frequency node contains a single cached item (and hence, we have the storage cost component $6P$ in Table 2). The second LFU variant, *Pure LFU*, keeps the full frequency history of queries in raw format and thus requires a hash table that maps query strings into frequencies (incurring the storage cost of M^{pure}) in addition to the cost of storing the aforementioned linked lists of typical LFU.

The *Tiny LFU* variant is the implementation that employs the Tiny storage scheme for query history. In our setup, we allocate $6W$ bits to doorkeeper and $3W$ full counters (where W denotes the window size and is in range $[8M, 16M]$ in our experiments) and use 4 hash functions to minimize the false positive errors [6]. Thus, we formulate the memory cost of Tiny LFU as: $M^{tiny}(c) = \left(6W + 3W \times log_2\left(\frac{W}{c} - 1\right)\right)/8$. For both Pure and Tiny LFU, during warm-up and testing, frequency statistics are updated as queries are streamed.

Figure 1 shows the hit-rates of single signal caching algorithms with respect to the cache size. Pure LFU improves the hit-ratio compared to In-memory LFU and LRU, especially for small cache sizes. This answers our first question; keeping the full history improves LFU performance. This is an important finding, because although some earlier works mentioned using a larger query history than the memory size for LFU in search engine result caching (e.g., [7]), there was no experimental evidence, as we provide here. More crucially, we see that Pure and Tiny LFU perform almost the same (i.e., the curves in Fig. 1 overlap) indicating that using the compact storage scheme (and hence, less precise frequency values) in the latter case do not reduce the cache performance.

Next, we report the ratio of the storage for metadata, M, (i.e., size of the query history stored in plain format or using the Tiny scheme) to that for the actual content of the result cache, R. For a given cache that can store C queries, the cache (content) size R is computed as $R = C \times k \times S$. We set $k = 10$ (since a typical result cache includes top-10 answers) and $S = 256$ as each query result including document's title, URL and snippet may add up to 256 bytes, as in [10]. Obviously, for a fixed cache size, the denominator R is the same for both Pure and Tiny LFU. Table 3 shows that M/R ratio varies from 0.56 to 0.13 for the smallest and largest values of C, i.e., 100K and 500K queries, respectively. In contrary, for Tiny LFU, M/R ratio is much smaller, between 0.15 and 0.05, indicating that Tiny scheme allows considerable gains in memory space. Note that, this is a finding shown for caching in other application domains [4] but not for search result caching. To our best knowledge, only [2] mentioned the possibility of using BFs for storing the query history in result caching, but their work did not provide any experimental evaluation of this idea, while we provide an exhaustive analysis in terms of the storage and cache performance metrics.

Table 3. Performance of LFU and GDSF-K variants, C: cache size, M: metadata (query history) size (in MBs), R: result cache size (in MBs), H: hit-rate.

C	R	LFU				GDSF-K					
		Pure		Tiny		In-memory		Pure		Tiny	
		H	M/R	H	M/R	H	M/R	H	M/R	H	M/R
100K	244.1	0.436	0.56	0.437	0.15	0.422	0.02	0.451	0.56	0.457	0.12
200K	488.3	0.479	0.29	0.479	0.08	0.46	0.02	0.498	0.29	0.498	0.11
300K	732.4	0.502	0.20	0.501	0.08	0.483	0.02	0.518	0.2	0.518	0.08
400K	976.6	0.516	0.15	0.516	0.06	0.497	0.02	0.529	0.15	0.529	0.07
500K	1220.7	0.527	0.13	0.526	0.05	0.508	0.02	0.536	0.13	0.536	0.05

Results for Multi-signal Caching Policy. *Greedy Dual Size Frequency (GDSF)* [1] offers a good compromise between recency and frequency, as well as the entry size and cost. We use a variant of GDSF, so called GDSF-K [11], where the frequency component is weighted by an exponent K to balance for the power-law distribution of queries in our setup. GDSF-K evicts the query result i with the minimum H_i value where $H_i = F_i^K \times \frac{C_i}{S_i} + L$. In this equation, F_i is the frequency of the query, C_i and S_i are size and cost of the query result, respectively. In our experiments, result size is set to 1, whereas we consider different alternatives for the cost (as will be discussed later). Lastly, L is the aging factor that is updated to the H_i of the cache victim whenever an item is evicted.

We again have three variants, namely In-memory (frequency information is kept only for the queries in the cache), Pure (with a raw frequency history) and Tiny. Table 2 presents the worst-case memory space usage of these variants.

While experimenting with GDSF-K variants, we first assume that all query costs C_i as 1, and measure the traditional hit-rate. Figure 2 reveals that, as before, the variants using the entire history outperform the In-memory GDSF-K, and furthermore, Pure and Tiny variants of GDSF-K perform comparably. We also see that GDSF-K formulation that combines both frequency and recency is superior to using each on its own; i.e., Pure (or, Tiny) GDSF-K outperforms the corresponding case with LFU. Finally, Tiny again provides considerable memory space gains over Pure GDSF-K, as shown in Table 3.

To investigate our third research question, i.e., assessing the value of the saved memory space, we focus on Tiny GDSF-K as the best-performing policy in the previous experiments. To this end, we first assume the saved space (i.e., the difference of memory space between Pure and Tiny variants) is also filled with query results, and measure the hit-rate. This case is denoted as "Tiny GDSF-K with Result" in Table 4, and we see that the hit-rates improve for each cache size (absolute gains being around 1.7% for the smallest cache size in comparison to the hit-rate of Tiny GDSF-K column, repeated from Table 3 for reference).

Furthermore, following the recent trend [11] that suggests using cost-aware caching policies and evaluating them again in terms of the cost savings – rather than hit rate-, we also experiment with Tiny GDSF-K using simulated query costs (i.e., in GDSF-K formula, we now set C_i using such costs). While doing so, as in [10], we consider four basic cost components: fetching the posting lists from the disk, ranking in memory, fetching top-k documents from the disk and generating snippets. Since modern search engines are known to store most of their data in memory or SSDs, here we focus on the ranking and snippet generation components. Formally, we define the former one as; $C^{rank} = |I_s| \times (P_d + P_r)$ and the latter one; $C^{snip} = 10 \times d_{avg} \times P_s$, where I_s is the shortest posting list of terms in query q (representing the conjunctive processing in the search engine, as typical) and P_d, P_r, P_s and d_{avg} are defined in Table 1 based on [10]. Posting list length of each query term is obtained from an index over the well-known Clueweb-2009 Part B collection.

In this case, our evaluation metric is the savings in query processing cost, C_s. To compute C_s, we obtain the cost of running a query stream over a system with a given type of cache divided by that of a system with no cache (where the cost of a query is simply the sum of C^{rank} and C^{snip}), and subtract the latter ratio from 1. In Table 4, we see that exploiting the saved space as a result cache further improves cost savings of Tiny GDSF-K; e.g., for a cache of 100K results, the cost savings increase from 0.698 to 0.710 (the former value, cost savings of the Tiny GDSF-K, is computed as a baseline).

Note that, recent studies in the literature propose using hybrid result caches, i.e., some part of the cache stores only doc-IDs of the query results [9,10]. In this case, if the cache-hit is for the latter part, there will be still C^{snip} cost, but much more expensive C^{rank} will be avoided. Therefore, as a final experiment, we explore what happens if the memory space saved by Tiny is reserved as a doc-ID cache, while the original cache capacity, as in previous experiments, store the query results in HTML format (i.e., top-k results' URLs and snippets). With this

Table 4. Cost saving, C_s, when saved space (by Tiny) is reserved for extending the result cache (Tiny GDSF-K w. Result) and implementing a docID cache. $\mathcal{H}$: hit-rate.

$\mathcal{C}$	Tiny GDSF-K		Tiny GDSF-K w. Result		Tiny GDSF-K w. Doc-ID
	$\mathcal{H}$	C_s	$\mathcal{H}$	C_s	C_s
$100K$	0.457	0.698	0.474	0.710	0.740
$200K$	0.498	0.724	0.504	0.728	0.744
$300K$	0.518	0.734	0.521	0.736	0.746
$400K$	0.529	0.739	0.53	0.740	0.747
$500K$	0.536	0.742	0.537	0.743	0.748

hybrid cache, we employ the Second Chance algorithm that is basically intended to keep the doc-ID results of a query for a longer time even its HTML version is evicted from the cache (please refer to [10] for details).

Our findings for the latter experiment are presented in Table 4 with column denoted as "Tiny GDSF-K with Doc-ID". We see that in terms of cost savings, there is an absolute improvement of up to 3% (note that, since the cache includes both HTML and docID results, hit-rate is meaningless, and not reported here). Furthermore, by reserving this saved space as a Doc-ID cache, the cost saving in case of a 100K cache is as good as that of a 500K result cache. These final experiments answer our third research question: the memory space saving using Tiny can yield non-negligible gains in terms of both hit-rate and query processing cost, especially for small and moderate size caches.

Concluding Summary: Our exhaustive experiments reveal that (i) Storing the entire query frequency history yields better hit-rate, (ii) *Tiny* successfully reduces memory space for the history, and (iii) The space saving is valuable, as it can be exploited to yield non-negligible gains in hit-rate and query processing cost.

Acknowledgements. This work is partially funded by The Scientific and Technological Research Council of Turkey (TÜBİTAK) under the grant no. 117E861 and Turkish Academy of Sciences Distinguished Young Scientist Award (TÜBA-GEBİP 2016).

References

1. Arlitt, M., Cherkasova, L., Dilley, J., Friedrich, R., Jin, T.: Evaluating content management techniques for web proxy caches. ACM SIGMETRICS Perform. Eval. Rev. **27**(4), 3–11 (2000)
2. Baeza-Yate, R., Junqueira, F., Plachouras, V., Witschel, H.F.: Admission policies for caches of search engine results. In: Ziviani, N., Baeza-Yates, R. (eds.) SPIRE 2007. LNCS, vol. 4726, pp. 74–85. Springer, Heidelberg (2007). https://doi.org/10.1007/978-3-540-75530-2_7
3. Cambazoglu, B.B., Baeza-Yates, R.A.: Scalability Challenges in Web Search Engines. Synthesis Lectures on Information Concepts, Retrieval, and Services. Morgan & Claypool Publishers, San Rafael (2015)

4. Einziger, G., Friedman, R.: Tinylfu: a highly efficient cache admission policy. In: 22nd Euromicro International Conference on Parallel, Distributed, and Network-Based Processing (PDP), pp. 146–153 (2014)
5. Fagni, T., Perego, R., Silvestri, F., Orlando, S.: Boosting the performance of web search engines: caching and prefetching query results by exploiting historical usage data. ACM TOIS **24**(1), 51–78 (2006)
6. Fan, L., Cao, P., Almeida, J., Broder, A.Z.: Summary cache: a scalable wide-area web cache sharing protocol. IEEE/ACM Trans. Netw. **8**(3), 281–293 (2000)
7. Gan, Q., Suel, T.: Improved techniques for result caching in web search engines. In: Proceedings of the WWW 2009, pp. 431–440 (2009)
8. Jonassen, S., Cambazoglu, B.B., Silvestri, F.: Prefetching query results and its impact on search engines. In: Proceedings of SIGIR 2012, pp. 631–640 (2012)
9. Ozcan, R., Altingovde, I.S., Cambazoglu, B.B., Junqueira, F.P.: Özgür Ulusoy: a five-level static cache architecture for web search engines. Inf. Process. Manag. **48**(5), 828–840 (2012)
10. Ozcan, R., Altingovde, I.S., Cambazoglu, B.B., Ulusoy, Ö.: Second chance: a hybrid approach for dynamic result caching and prefetching in search engines. ACM Trans. Web **8**(1), 3 (2013)
11. Ozcan, R., Altingövde, I.S., Ulusoy, O.: Cost-aware strategies for query result caching in web search engines. ACM Trans. Web **5**(2), 9:1–9:25 (2011)
12. Pass, G., Chowdhury, A., Torgeson, C.: A picture of search. In: Proceedings of the InfoScale 2006, p. 1 (2006)
13. Podlipnig, S., Böszörményi, L.: A survey of web cache replacement strategies. ACM Comput. Surv. **35**(4), 374–398 (2003)
14. Shah, K., Mitra, A., Matani, D.: An O(1) algorithm for implementing the LFU cache eviction scheme. Technical report (2010)
15. Silvestri, F.: Mining query logs: turning search usage data into knowledge. Found. Trends Inf. Retrieval **4**(1–2), 1–174 (2010). https://doi.org/10.1561/1500000013

AspeRa: Aspect-Based Rating Prediction Model

Sergey I. Nikolenko[1,4], Elena Tutubalina[1,2,4], Valentin Malykh[1,3],
Ilya Shenbin[1], and Anton Alekseev[1(✉)]

[1] Samsung-PDMI Joint AI Center, Steklov Mathematical Institute at St. Petersburg,
Saint Petersburg, Russia
anton.m.alexeyev@gmail.com
[2] Chemoinformatics and Molecular Modeling Laboratory, Kazan Federal University,
Kazan, Russia
[3] Neural Systems and Deep Learning Laboratory,
Moscow Institute of Physics and Technology, Dolgoprudny, Russia
[4] Neuromation OU, 10111 Tallinn, Estonia

Abstract. We propose a novel end-to-end Aspect-based Rating Prediction model (AspeRa) that estimates user rating based on review texts for the items and at the same time discovers coherent aspects of reviews that can be used to explain predictions or profile users. The AspeRa model uses max-margin losses for joint item and user embedding learning and a dual-headed architecture; it significantly outperforms recently proposed state-of-the-art models such as DeepCoNN, HFT, NARRE, and TransRev on two real world data sets of user reviews. With qualitative examination of the aspects and quantitative evaluation of rating prediction models based on these aspects, we show how aspect embeddings can be used in a recommender system.

Keywords: Aspect-based sentiment analysis ·
Recommender systems · Aspect-based recommendation ·
Explainable recommendation · User reviews · Neural network ·
Deep learning

1 Introduction

As the scale of online services and the Web itself grows, recommender systems increasingly attempt to utilize texts available online, either as items for recommendation or as their descriptions [1, 22, 25, 41]. One key complication is that a single text can touch upon many different features of the item; e.g., the same brief review of a laptop can assess its weight, performance, keyboard, and so on, with different results. Hence, real-world applications need to separate different aspects of reviews. This idea also has a long history [15, 26]. Many recent works in recommender systems have applied deep learning methods [10, 31, 33, 41]. In this work, we introduce novel deep learning methods for making recommendations with full-text items, aiming to learn interpretable user representations that

© Springer Nature Switzerland AG 2019
L. Azzopardi et al. (Eds.): ECIR 2019, LNCS 11438, pp. 163–171, 2019.
https://doi.org/10.1007/978-3-030-15719-7_21

reflect user preferences and at the same time help predict ratings. We propose a novel Aspect-based Rating Prediction Model (AspeRa) for aspect-based representation learning for items by encoding word-occurrence statistics into word embeddings and applying dimensionality reduction to extract the most important aspects that are used for the user-item rating estimation. We investigate how and in what settings such neural autoencoders can be applied to content-based recommendations for text items.

2   AspeRa Model

The *AspeRa* model combines the advantages of deep learning (end-to-end learning, spatial text representation) and topic modeling (interpretable topics) for text-based recommendation systems. Figure 1 shows the overall architecture of AspeRa. The model receives as input two reviews at once, treating both identically. Each review is embedded with self-attention to produce two vectors, one for author (user) features and the other for item features. These two vectors are used to predict a rating corresponding to the review. All vectors are forced to belong to the same feature space. The embedding is produced by the Neural Attention-Based Aspect Extraction Model (ABAE) [7]. As in topic modeling or clustering, with ABAE the designer can determine a finite number of topics/clusters/aspects, and the goal is to find out for every document to which extent it satisfies each topics/aspects. From a bird's eye view, ABAE is an autoencoder. The main feature of ABAE is the reconstruction loss between bag-of-words embeddings used as the sentence representation and a linear combination of aspect embeddings. A sentence embedding is additionally weighted by *self-attention*, an attention mechanism where the values are word embeddings and the key is the mean embedding of words in a sentence.

The first step in ABAE is to compute the embedding $\mathbf{z}_s \in \mathbb{R}^d$ for a sentence s; below we call it a text embedding: $\mathbf{z}_s = \sum_{i=1}^n a_i \mathbf{e}_{w_i}$, where $\mathbf{e}_{w_i}$ is a word embedding for a word w_i, $e \in \mathbb{R}^d$. As word vectors the authors use *word2vec* embeddings trained with the skip-gram model [21]. Attention weights a_i are computed as a multiplicative self-attention model: $a_i = \mathrm{softmax}(\mathbf{e}_{w_i}^\top A \mathbf{y}_s)$, where $\mathbf{y}_s$ is the average of word embeddings in a sentence, $\mathbf{y}_s = \sum_{i=1}^n \mathbf{e}_{w_i}$, and $A \in \mathbb{R}^{d \times d}$ is the learned attention model. The second step is to compute the aspect-based sentence representation $\mathbf{r}_s \in \mathbb{R}^d$ from an aspect embeddings matrix $T \in \mathbb{R}^{k \times d}$, where k is the number of aspects: $\mathbf{p}_s = \mathrm{softmax}(W\mathbf{z}_s + \mathbf{b})$, where $\mathbf{p}_s \in \mathbb{R}^k$ is the vector of probability weights over k aspect embeddings, $\mathbf{r}_s = T^\top \mathbf{p}_s$, and $W \in \mathbb{R}^{k \times d}$, $\mathbf{b} \in \mathbb{R}^k$ are the parameters of a multi-class logistic regression model. Below we call $\mathbf{r}_s$ the reconstructed embedding.

To train the model, ABAE uses the cosine distance between $\mathbf{r}_s$ and $\mathbf{z}_s$ with a contrastive max-margin objective function [39] as the reconstruction error, also adding an orthogonality penalty term that tries to make the aspect embedding matrix T to produce aspect embeddings as diverse as possible.

The proposed model's architecture includes an embedder, which provides text and reconstruction embeddings for an object similar to ABAE ("user embedding" and "item embedding" on Fig. 1). The intuition behind this separation of

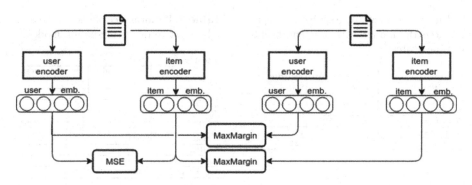

Fig. 1. Architecture of the proposed *AspeRa* model.

user and item embedding is as follows: there are some features (aspects) important in an item for a user, but the item also has other features. Hence, we want to extract user aspects from a user's reviews as well as item aspects from an item's reviews. The resulting embedding is conditioned on aspect representation of the reviews; we will see below that this model can discover interpretable topics. The model contains four embedders in total, one pair of user and item embedders for two reviews being considered at once, as shown on Fig. 1. First each review is paired with another review of the same user, grouping by users and shuffling the reviews inside a group; then with another review of the same item. Thus, the training set gives rise to only twice as many pairs as reviews available for training. The rating score for the first review in a pair is used to train the rating predictor (*MSE*); at prediction stage, only one "tower" is used.

There are two losses in *AspeRa*: MSE for rating prediction (Fig. 1) and Max-Margin loss to put user and item embeddings in the same space (Fig. 1). The MSE loss assumes that rating is predicted as the dot product of user and item embeddings for a review: $MSE = \frac{1}{N}\sum_{j=1}^{N}(\mathbf{z}_j^{u\top}\mathbf{z}_j^i - r_j)^2$, where $\mathbf{z}_j^u$ is a text embedding for the author of review j, $\mathbf{z}_j^i$ is a text embedding for the item j is about, and r_j is the true rating associated with j. Max-margin loss aims to project all user and item embeddings into the same feature (aspect) space; see Fig. 1. We use it in two ways. First, we push reconstructed and text embeddings to be closer for each user i, and pushes text embeddings for both considered items apart: $\text{MaxMargin}(i,j) = \frac{1}{N}\sum_{i,j}\max(0, 1 - \mathbf{r}_i^{u\top}\mathbf{z}_i^u + \mathbf{r}_i^{u\top}\mathbf{z}_i^i + \mathbf{r}_i^{u\top}\mathbf{z}_j^i)$, where i,j are indices of reviews, $\mathbf{r}_i^u$ is a reconstructed embedding from ABAE for user i, $\mathbf{z}_i^u$ is a text embedding for user i, $\mathbf{z}_i^i$ and $\mathbf{z}_j^i$ are text embeddings from ABAE for items i and j respectively. This loss is applied for all four possible combination of users and items, i.e., $(u_i, i_i, i_j), (u_j, i_i, i_j), (i_i, u_i, u_j), (i_j, u_i, u_j)$. Second, we keep user embeddings from two reviews of the same author close: $\text{MaxMargin}(i,j) = \frac{1}{N}\sum_{i,j}\max(0, 1 - \mathbf{z}_i^{u\top}\mathbf{z}_j^u + \mathbf{z}_i^{u\top}\mathbf{z}_i^i + \mathbf{z}_i^{u\top}\mathbf{z}_j^i)$, where i,j are indices of reviews, $\mathbf{z}_i^u$ and $\mathbf{z}_j^u$ are user embeddings from ABAE for authors of reviews i and j and $\mathbf{z}_i^i$ and $\mathbf{z}_j^i$ are text embeddings from ABAE for items i and j respectively. This second form is symmetrically applied to item and user embeddings for two reviews pf the same item from different authors.

Table 1. Two sets of AspeRa hyperparameters (for models with different initialization strategies).

Settings	AspeRa (GloVe)	AspeRa (SGNS)
Embeddings	GloVe	SGNS
Optimization alg.	Adam [12]	Adam
# aspects	11	10
Hidden layer dim.	256	64
# epochs	20	18
# words per sample	256	224

Table 2. Performance of text-based and collaborative rating prediction models.

Model	MSE	
	Instant Videos	Toys & Games
NMF	0.946	0.821
DeepCoNN	0.943	0.851
Attn+CNN	0.936	-
SVD	0.904	0.788
HFT	0.888	0.784
TransRev	0.884	0.784
NARRE	-	0.769
AspeRa (GloVe)	0.870	0.730
AspeRa (SGNS)	**0.660**	**0.571**

3 Experimental Evaluation

Datasets and Experimental Setup. We evaluated the proposed model on *Amazon Instant Videos 5-core reviews* and *Amazon Toys and Games 5-core reviews*[1] [8,19]. The first dataset consists of reviews written by users with at least five reviews on *Amazon* and/or for items with at least five reviews; it contains 37,126 reviews, 5,130 users, 1,685 items, and a total of 3,454,453 non-unique tokens. The second dataset follows 5 minimum reviews rule; it contains 167,597 reviews, 19,412 users, 11,924 items, and a total of 17,082,324 non-unique tokens. We randomly split each dataset into 10% test set and 90% training set, with 10% of the training set used as a validation set for tuning hyperparameters. Following ABAE [7], we set the aspects matrix ortho-regularization coefficient equal to 0.1. Since this model utilizes an aspect embedding matrix to approximate aspect words in the vocabulary, initialization of aspect embeddings is crucial. The work [7] used *k-means* clustering-based initialization [16,17,34], where the aspect embedding matrix is initialized with centroids of the resulting clusters of word embeddings. We compare two word embeddings for *AspeRa*: *GloVe* [27] and *word2vec* [20,21]. We adopted a *GloVe* model trained on the *Wikipedia 2014 + Gigaword 5* dataset (6B tokens, 400K words vocabulary, uncased tokens) with dimension 50. For *word2vec*, we used the training set of reviews to train a skip-gram model (SGNS) with the *gensim* library [29] with dimension 200, window size 10, and 5 negative samples; see Table 1 for details.

Rating Prediction. We evaluate the performance of *AspeRa* in comparison to state-of-the-art models: NMF [40], DeepCoNN [41], Attn+CNN [31], SVD [13], HFT [18], NARRE [4], and TransRev [5]; we introduce these models in Sect. 4. Table 2 compares the best Mean Square Error (MSE) of *AspeRa* and other models for rating prediction. Results of existing models were adopted from [5] for *Amazon Instant Videos 5-core reviews* with the ratio 80:10:10. We also used the

---

[1] http://jmcauley.ucsd.edu/data/amazon/.

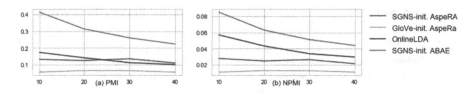

Fig. 2. Comparing *AspeRa* with GloVe (SGNS clusters), ABAE (SGNS clusters), and LDA with the same vocabulary and 10 topics on *Instant Videos*; more is better. X-axis: number of top-ranked representative words per aspect, Y-axis: topic coherence scores.

results of NARRE model [4], obtained in the same setup as [5] but with a different random seed. Note that while *AspeRa* with generic *GloVe* word embeddings still works better than any other model, adding custom word embeddings trained on the same type of texts improves the results greatly.

Topic Quality. We compared the performance of *AspeRa* with *OnlineLDA* [9] trained with the *gensim* library [29], with the same vocabulary and number of topics, and ABAE with 10 aspects and 18 epochs, initialized with the same *word2vec* vectors (SGNS) as *AspeRa* and having the same ortho-regularization coefficient as the best *AspeRa* model, evaluating the results in terms of topic coherence metrics, NPMI [2] and PMI [23,24] computed with companion software for [14]. Figure 2 shows that the quality is generally lower for larger number of representative words per aspect (horizontal axis), and that *AspeRa* achieves scores comparable to LDA and ABAE, although ABAE remains ahead. Tables 3 and 4 present several sample aspects discovered by *AspeRA*. Qualitative analysis shows that some aspects describe what could be called a *topic* (a set of words diverse by part of speech and function describing a certain domain), some encode sentiment (top words are adjectives showing attitude to certain objects discussed in the text), and some encode names (actors, directors, etc.). We also found similar patterns in the output of the basic ABAE model [7]. Thus, most aspects are clearly coherent, but there is room for improvement.

Table 3. Sample aspects from *Instant Videos* discovered by AspeRa (SGNS).

#	Aspect words
1	Communities governments incidents poverty unity hardships slaves citizens fought
2	Coppola guillermo bram kurosawa toro ridley del prolific ti festivals
3	Brisk dialogue manipulation snappy plotlines dialogues taunt camerawork muddled
4	Sock vegans peanut stifling bats buh ammonium trollstench vegetables pepsi
5	The a and to is of joe's enters that fatal

Table 4. Sample aspects from *Instant Videos* discovered by AspeRa (GloVe).

#	Aspect words
1	Protein diagnose cell genes brain membrane interacts interact oxygen spinal
2	Boost monetary raise introduce measures credit expects increase push demand
3	Towel soaked greasy towels cloth dripping tucked crisp coat buckets
4	Offbeat comic parody spoof comedic quirky cinematic campy parodies animated
5	Sheesh wham whew hurrah oops yikes c'mon shhh oooh och

4   Related Work

Classical collaborative filtering based on matrix factorization (MF) [13,40] has been extended with textual information, often in the form of topics/aspects; aspect extraction uses topic modelling [35,36,42] and phrase-based extraction [32]. Collaborative topic regression (CTR) [37] was one of the first models to combine collaborative-based and topic-based approaches to recommendation; to recommend research articles; it uses an LDA topic vector as a prior of item embeddings for MF. Hidden Factors and Hidden Topics (HTF) [18] also combines MF and LDA but with user reviews used as contextual information. A few subsequent works use MF along with deep learning approaches; e.g., Collaborative Deep Learning (CDL) [38] improves upon CTR by replacing LDA with a stacked denoising autoencoder. Unlike our approach, all these models learn in alternating rather than end-to-end manner. Recent advances in distributed word representations have made it a cornerstone of modern natural language processing [6], with neural networks recently used to learn text representations. He et al. [7] proposed an unsupervised neural attention-based aspect extraction (ABAE) approach that encodes word-occurrence statistics into word embeddings and applies an attention mechanism to remove irrelevant words, learning a set of aspect embeddings. Several recent works, including DeepCoNN [41], propose a completely different approach. DeepCoNN is an end-to-end model, both user and item embedding vectors in this model are trainable functions (convolutional neural networks) of reviews associated with a user or item respectively. Experiments on *Yelp* and *Amazon* datasets showed significant improvements over HFT. *TransNet* [3] adds a regularizer on the penultimate layer that forces the network to predict review embedding. *TransRev* [5] is based on the same idea of restoring the review embedding from user and item embeddings. *Attn+CNN* and *D-Attn* [30,31] extend *DeepCoNN* with an attention mechanism on top of text reviews; it both improves performance and allows to explain predictions by highlighting significant words. However, user and item embeddings of these models are learned in a fully supervised way, unlike the proposed model. Our model combines semi-supervised embedding learning, which makes predictions interpretable similar to HTF, with a deep architecture and end-to-end training.

5   Conclusion

We have introduced a novel approach to learning rating- and text-aware recommender systems based on ABAE, metric learning, and autoencoder-enriched learning. Our approach jointly learns interpretable user and item representations. It is expectedly harder to tune to achieve better quality, but the final model performs better at rating prediction and almost on par at aspects coherence with other state-of-the-art approaches. Our results can also be viewed as part of the research effort to analyze and interpret deep neural networks, a very important recent trend [11, 28]. We foresee the following directions for future work: (i) further improving prediction quality (especially for models that learn interpretable user representations), (ii) integrating methods that can remove "purely sentimental" aspects into interpretable models for recommendations that we have discussed above, (iii) developing visualization techniques for user profiles.

Acknowledgements. This research was done at the Samsung-PDMI Joint AI Center at PDMI RAS and was supported by Samsung Research.

References

1. Alekseev, A., Nikolenko, S.: Word embeddings for user profiling in online social networks. Computación y Sistemas **21**(2) (2017)
2. Bouma, G.: Normalized (pointwise) mutual information in collocation extraction. In: Proceedings of GSCL, pp. 31–40 (2009)
3. Catherine, R., Cohen, W.: Transnets: learning to transform for recommendation. In: Proceedings of the Eleventh ACM Conference on Recommender Systems, pp. 288–296. ACM (2017)
4. Chen, C., Zhang, M., Liu, Y., Ma, S.: Neural attentional rating regression with review-level explanations. In: Proceedings of the 2018 World Wide Web Conference, WWW 2018, pp. 1583–1592. International World Wide Web Conferences Steering Committee, Republic and Canton of Geneva, Switzerland (2018). https://doi.org/10.1145/3178876.3186070
5. García-Durán, A., Gonzalez, R., Oñoro-Rubio, D., Niepert, M., Li, H.: Transrev: modeling reviews as translations from users to items. CoRR abs/1801.10095 (2018). http://dblp.uni-trier.de/db/journals/corr/corr1801.html#abs-1801-10095
6. Goldberg, Y.: A primer on neural network models for natural language processing. CoRR abs/1510.00726 (2015). http://arxiv.org/abs/1510.00726
7. He, R., Lee, W.S., Ng, H.T., Dahlmeier, D.: An unsupervised neural attention model for aspect extraction. In: Proceedings of the 55th Annual Meeting of the Association for Computational Linguistics (Volume 1: Long Papers), vol. 1, pp. 388–397 (2017)
8. He, R., McAuley, J.: Ups and downs: modeling the visual evolution of fashion trends with one-class collaborative filtering. In: Proceedings of the 25th International Conference on World Wide Web, pp. 507–517. International World Wide Web Conferences Steering Committee (2016)
9. Hoffman, M., Bach, F.R., Blei, D.M.: Online learning for latent dirichlet allocation. In: Advances in Neural Information Processing Systems, pp. 856–864 (2010)

10. Hsieh, C.K., Yang, L., Cui, Y., Lin, T.Y., Belongie, S., Estrin, D.: Collaborative metric learning. In: Proceedings of the 26th International Conference on World Wide Web, pp. 193–201. International World Wide Web Conferences Steering Committee (2017)
11. Kádár, A., Chrupała, G., Alishahi, A.: Representation of linguistic form and function in recurrent neural networks. Comput. Linguist. **43**(4), 761–780 (2017)
12. Kingma, D.P., Ba, J.: Adam: a method for stochastic optimization. CoRR abs/1412.6980 (2014). http://arxiv.org/abs/1412.6980
13. Koren, Y., Bell, R.M., Volinsky, C.: Matrix factorization techniques for recommender systems. IEEE Comput. **42**(8), 30–37 (2009)
14. Lau, J.H., Newman, D., Baldwin, T.: Machine reading tea leaves: automatically evaluating topic coherence and topic model quality. In: Proceedings of the 14th Conference of the European Chapter of the Association for Computational Linguistics, pp. 530–539 (2014)
15. Liu, B.: Sentiment Analysis and Opinion Mining, Synthesis Lectures on Human Language Technologies, vol. 5. Morgan & Claypool Publishers, San Rafael (2012)
16. Lloyd, S.: Least squares quantization in PCM. IEEE Trans. Inf. Theory **28**(2), 129–137 (1982)
17. MacQueen, J., et al.: Some methods for classification and analysis of multivariate observations. In: Proceedings of the Fifth Berkeley Symposium on Mathematical Statistics and Probability, Oakland, CA, USA, vol. 1, pp. 281–297 (1967)
18. McAuley, J., Leskovec, J.: Hidden factors and hidden topics: understanding rating dimensions with review text. In: Proceedings of the 7th ACM Conference on Recommender Systems, pp. 165–172 (2013)
19. McAuley, J., Targett, C., Shi, Q., Van Den Hengel, A.: Image-based recommendations on styles and substitutes. In: Proceedings of the 38th International ACM SIGIR Conference on Research and Development in Information Retrieval, pp. 43–52. ACM (2015)
20. Mikolov, T., Chen, K., Corrado, G., Dean, J.: Efficient estimation of word representations in vector space. CoRR abs/1301.3781 (2013). http://arxiv.org/abs/1301.3781
21. Mikolov, T., Sutskever, I., Chen, K., Corrado, G.S., Dean, J.: Distributed representations of words and phrases and their compositionality. In: Burges, C.J.C., Bottou, L., Welling, M., Ghahramani, Z., Weinberger, K.Q. (eds.) Advances in Neural Information Processing Systems, vol. 26, pp. 3111–3119. Curran Associates, Inc. (2013). http://papers.nips.cc/paper/5021-distributed-representations-of-words-and-phrases-and-their-compositionality.pdf
22. Mitcheltree, C., Wharton, V., Saluja, A.: Using aspect extraction approaches to generate review summaries and user profiles. arXiv preprint arXiv:1804.08666 (2018)
23. Newman, D., Karimi, S., Cavedon, L.: External evaluation of topic models. In: Australasian Document Computing Symposium, 2009. Citeseer (2009)
24. Newman, D., Lau, J.H., Grieser, K., Baldwin, T.: Automatic evaluation of topic coherence. In: Human Language Technologies: The 2010 Annual Conference of the North American Chapter of the Association for Computational Linguistics, HLT 2010, pp. 100–108. Association for Computational Linguistics, Stroudsburg, PA, USA (2010). http://dl.acm.org/citation.cfm?id=1857999.1858011
25. Alekseev, A., Nikolenko, S.: User profiling in text-based recommender systems based on distributed word representations. In: Ignatov, D.I., et al. (eds.) AIST 2016. CCIS, vol. 661, pp. 196–207. Springer, Cham (2017). https://doi.org/10.1007/978-3-319-52920-2_19

26. Pang, B., Lee, L.: Opinion mining and sentiment analysis. Found. Trends Inf. Retrieval **2**(1–2), 1–135 (2008)
27. Pennington, J., Socher, R., Manning, C.: Glove: global vectors for word representation. In: Proceedings of the 2014 Conference on Empirical Methods in Natural Language Processing (EMNLP), pp. 1532–1543. Association for Computational Linguistics, Doha, Qatar (2014). http://www.aclweb.org/anthology/D14-1162
28. Radford, A., Jozefowicz, R., Sutskever, I.: Learning to generate reviews and discovering sentiment. arXiv preprint arXiv:1704.01444 (2017)
29. Rehurek, R., Sojka, P.: Software framework for topic modelling with large corpora. In: Proceedings of the LREC 2010 Workshop on New Challenges for NLP Frameworks, pp. 45–50. ELRA, Valletta, Malta, May 2010. http://is.muni.cz/publication/884893/en
30. Seo, S., Huang, J., Yang, H., Liu, Y.: Interpretable convolutional neural networks with dual local and global attention for review rating prediction. In: Proceedings of the Eleventh ACM Conference on Recommender Systems, pp. 297–305. ACM (2017)
31. Seo, S., Huang, J., Yang, H., Liu, Y.: Representation learning of users and items for review rating prediction using attention-based convolutional neural network. In: 3rd International Workshop on Machine Learning Methods for Recommender Systems (MLRec) (SDM 2017) (2017)
32. Solovyev, V., Ivanov, V.: Dictionary-based problem phrase extraction from user reviews. In: Sojka, P., Horák, A., Kopeček, I., Pala, K. (eds.) TSD 2014. LNCS (LNAI), vol. 8655, pp. 225–232. Springer, Cham (2014). https://doi.org/10.1007/978-3-319-10816-2_28
33. Srivastava, A., Sutton, C.: Autoencoding variational inference for topic models. arXiv preprint arXiv:1703.01488 (2017)
34. Steinhaus, H.: Sur la division des corp materiels en parties. Bull. Acad. Polon. Sci **1**(804), 801 (1956)
35. Tutubalina, E., Nikolenko, S.: Inferring sentiment-based priors in topic models. In: Lagunas, O.P., Alcántara, O.H., Figueroa, G.A. (eds.) MICAI 2015. LNCS (LNAI), vol. 9414, pp. 92–104. Springer, Cham (2015). https://doi.org/10.1007/978-3-319-27101-9_7
36. Tutubalina, E., Nikolenko, S.: Constructing aspect-based sentiment lexicons with topic modeling. In: Ignatov, D.I., et al. (eds.) AIST 2016. CCIS, vol. 661, pp. 208–220. Springer, Cham (2017). https://doi.org/10.1007/978-3-319-52920-2_20
37. Wang, C., Blei, D.M.: Collaborative topic modeling for recommending scientific articles. In: Proceedings of the 17th ACM SIGKDD International Conference on Knowledge Discovery and Data Mining, pp. 448–456. ACM (2011)
38. Wang, H., Wang, N., Yeung, D.Y.: Collaborative deep learning for recommender systems. In: Proceedings of the 21th ACM SIGKDD International Conference on Knowledge Discovery and Data Mining, pp. 1235–1244. ACM (2015)
39. Weston, J., Bengio, S., Usunier, N.: WSABIE: scaling up to large vocabulary image annotation. IJCAI **11**, 2764–2770 (2011)
40. Zhang, S., Wang, W., Ford, J., Makedon, F.: Learning from incomplete ratings using non-negative matrix factorization. In: Proceedings of the 2006 SIAM International Conference on Data Mining, pp. 549–553. SIAM (2006)
41. Zheng, L., Noroozi, V., Yu, P.S.: Joint deep modeling of users and items using reviews for recommendation. In: Proceedings of the Tenth ACM International Conference on Web Search and Data Mining, pp. 425–434. ACM (2017)
42. Zhu, X., Blei, D., Lafferty, J.: TagLDA: bringing document structure knowledge into topic models. Tech. report, UWisc Technical Report TR-1533 (2006). http://ftp.cs.wisc.edu/pub/techreports/2006/TR1553.pdf

Heterogeneous Edge Embedding
for Friend Recommendation

Janu Verma[1]([✉]), Srishti Gupta[1], Debdoot Mukherjee[1],
and Tanmoy Chakraborty[2]

[1] Hike Messenger, New Delhi, India
{janu,srishti,debdoot}@hike.in
[2] IIIT Delhi, New Delhi, India
tanmoy@iiitd.ac.in

Abstract. We propose a friend recommendation system (an application of link prediction) using edge embedding on social networks. Most real world social networks are multi-graphs, where different kinds of relationships (*e.g.*, chat, friendship) are possible between a pair of users. Existing network embedding techniques do not leverage signals from different edge types and thus perform inadequately on link prediction in such networks. We propose a method to mine network representation that effectively exploits edge heterogeneity in multi-graphs. We evaluate our model on a real-world, active social network where this system is deployed for friend recommendation for millions of users. Our method outperforms various state-of-the-art baselines on Hike's social network in terms of accuracy metrics as well as user satisfaction.

1 Introduction

Users need to find relevant friends in order to engage on any social network. Social platforms like Facebook, LinkedIn, Twitter facilitate friend discovery via *Friend Recommendation* [1,6]. A good recommendation system strengthens the network by aiding creation of new social connections between existing users. It also helps in retention of new users by helping them find friends as they join the platform. Hence, efficacy of friend recommendation method is of utmost importance to drive growth and engagement on the platform.

The problem of friend recommendation fits into the classical link prediction problem [12]. Given a snapshot of a social network at time t, can we accurately predict the edges that will be added to the network during the interval from time t to a given future time t'? In a nutshell, can the current state of the network be used to predict future links? Traditional methods for link prediction were based on measures for analyzing the "proximity" of nodes in a network. Specifically, if the neighborhoods of two nodes have a large overlap, existing methods will indicate that they are highly likely to share a link. Common neighbors, Jaccard coefficient, Adamic-Adar [12], preferential attachment [16] are different measures

---

J. Verma and S. Gupta—equal contribution.

© Springer Nature Switzerland AG 2019
L. Azzopardi et al. (Eds.): ECIR 2019, LNCS 11438, pp. 172–179, 2019.
https://doi.org/10.1007/978-3-030-15719-7_22

that have been devised to assess overlap between node neighborhoods. Supervised models [6] have also been trained with these features for link prediction.

Recently, there has been a lot of work in the development of methods for learning continuous representations [4,7–10,14,15] of nodes that can effectively preserve their network neighborhoods. Such representations have proven to be more effective than hand-engineered features in encoding structural information for nodes in classification problems setup for link prediction [7,14].

In this work, we present and evaluate a friend recommendation system for a real-world social network. We propose a framework for graph representation learning on heterogeneous networks with multiple edge types for link prediction. The system has two components - network embedding for a large heterogeneous network, and training a friend recommendation model on a large set of known friend and non-friend pairs by leveraging the learned embedding.

We split a heterogeneous network with multiple edge types into homogeneous components and obtain edge embedding for each component. Our friend recommendation system contains a multi-tower neural network which takes the homogeneous embeddings as inputs and combines them to obtain a unified edge embedding for the link prediction problem.

Our contribution are as follows:

- We demonstrate the efficacy of network embedding for link prediction on a large real-world network.
- We provide a formulation of friend recommendation problem as link prediction in an edge heterogeneous network.
- We propose methods to obtain unified edge embedding by combining segregated embedding from homogeneous components.
- We present a multi-tower neural network architecture for learning unified edge embedding for the link prediction problem.
- We evaluated the method by comparing it with the state-of-the-art approaches offline and also by deploying the system on an active platform.

2 Network Embedding for Friend Recommendation

We consider friend recommendation as a binary classification problem where a pair of users will be classified as *friends* or *not-friends*. Given a collection of user-pairs (i.e., edges), we build a model that can learn to predict new edges.

2.1 Network Embedding

Network embedding has shown great success in various social network applications, e.g., link prediction, node clustering, multi-label classification of nodes, etc. The idea is to learn a node-centric function that can map nodes into a low dimensional vector space by preserving the structural information about their neighborhoods. The node embedding of two nodes can be combined to form a representation of the edge connecting them. In case of link prediction, such an edge embedding can be given to a classifier to predict whether the edge is likely to exist or not. Two popular node embedding methods are as follows: (i) **Deep-Walk** [14] learns node embedding in a homogeneous network. Unbiased, uniform, fixed number of random walks of pre-decided length are generated starting at each node in the network to produce 'sentences' of nodes, similar to sentences of words in a natural language. The Skip-gram algorithm devised by Mikolov et al. [13] is used to obtain node embedding from the random walks, which are expected to capture the contextual properties of the network nodes – the nodes that occur in same context have similar vector embedding. (ii) **Node2vec** [7] is another method for homogeneous node embedding. A biased random walk to navigate the neighborhood of a node can be parameterized to make a transition from breadth-first search (BFS) to depth-first search (DFS). However, a proper parameterization is critical for good performance and this requires heavy tuning.

2.2   Extending Network Embedding for Heterogeneous Multi-graph

Models like DeepWalk [14] and Node2vec [7] are restricted to homogeneous networks. However, real-world social networks are heterogeneous in nature – nodes are of multiple types such as users, posts, etc., and edges can be drawn based on different relationships such as friendship, like, comment, follow, etc. Existing network embedding methods designed for homogeneous networks may not be directly applicable to heterogeneous networks. Recently, metapath2vec [5], an embedding technique for heterogeneous networks was proposed, which defines *metapath* (a sequence of node types) to restrict the random walks. sHowever, it is not obvious how to define such a metapath as we often lack an intuition for the paths and metapaths, and the length of the metapath. Moreover for a heterogeneous network with different edge types and multiple edges between two nodes (heterogeneous multi-graph), there is no intuitive way to define a meta-path of edges types. Chang et al. [3] propose a deep architecture to learn node embedding in a multi-modal network with image and text nodes. However, it doesn't generalize to a multi-graph where different types of edges exist between a pair of nodes. Next, we study straightforward extensions of DeepWalk for a heterogeneous multi-graph.

Equal Probability of Edges: Similar to DeepWalk, we generate unbiased random walks by assigning equal probability of walking through any edge between two nodes. This increases the probability of the random walk going to a node which has multiple edges from the current node. For instance, if two users A and B have each other listed as a contact, they chat and are also friends, then the random walker would be thrice as likely to traverse from A to B as compared to a setting where they shared a single edge. This technique suffers from the following limitation. For any node in the Hike network, there are far more contact edges than friends and even fewer chat connections. Thus the random walk has higher chances of going via contact edges, and in some cases completely avoid chat edges. This method will be referred as **HeteroDeepWalk**, henceforth.

Equal Probability of Edge Types: Here is a simple way to resolve the problem of HeteroDeepwalk, i.e., random walks being biased by the dominant edge type. The random walk is generated in two steps: (i) an edge type is chosen randomly from all possible edge types, (ii) an edge is randomly chosen from all edges of the selected edge type. This amounts to biasing the random walks uniformly with equal weights for each edge type. This method will be referred as **UniformBiasDeepWalk**, henceforth. In reality, different edge types contribute differently to the random walks and would have unequal weights. We don't have an intuitive way to obtain these weights.

3   Proposed Solution: Heterogenous Edge Embedding

HeteroDeepWalk and UniformBiasDeepWalk are two straightforward extensions of DeepWalk to deal with multi-graphs. However, the main drawback of these methods is that there is no obvious way to figure out how to bias the random walks for each edge-type; there is no reason to think that edge-types have equal importance. We propose a method to automatically estimate the weights of each edge-type in the embedding and obtain an edge embedding comprising of the contributions from various edge-types. Our method has four steps:

1. Split the multi-graph into homogeneous sub-graphs each with one edge type.
2. Obtain node embedding from each of these subnetworks.
3. Obtain edge embedding from node embedding for each of these subnetworks.
4. Train a unified, heterogeneous edge embedding for link prediction

For example, we can split a social network into friend subnetwork, contact subnetwork and chat subnetwork, where users are connected via friendship, contact list and chatting, respectively. Each of these networks is homogeneous and can be embedded to a low dimensional space using DeepWalk or Node2vec. Each node in the original heterogeneous network thereby has an embedding in three different spaces, e.g., for a node v, we have vector representations – v^{friend}, $v^{contact}$, and v^{chat}. For simplicity, we assume the dimension of 3 spaces is equal.

Given these node representations, we can combine them in various ways to obtain the embedding of the edge $\langle u, v \rangle$ connecting two nodes u and v. If u^x is the node embedding of u for the homogeneous x-subnetwork (where x = contact/chat/friend), then a homogeneous edge embedding $e^x_{\langle u,v \rangle}$, can be computed by taking average, Hadamard product or concatenation of u^x and v^x.

Next, we combine the segregated edge representations, $e^x_{\langle u,v \rangle}$, for different edges of type x between a node pair, $\langle u, v \rangle$, to obtain a unified, heterogeneous edge embedding, $E_{\langle u,v \rangle}$. We discuss two methods based on neural network and logistic regression for doing this.

Neural Network: Figure 1 (Left) shows the architecture of a neural network to train a heterogeneous edge embedding for link prediction. It takes the edge vectors from different sub-networks (*e.g.*, contact, friend and chat) as inputs.

Each edge vector is passed through a hidden layer with 256 units that use RELU activation. The outputs are then concatenated and fed into another hidden layer that creates a 256 dimensional unified edge embedding. This unified embedding is passed into a Sigmoid layer to predict the class label - *link* or *no-link*.

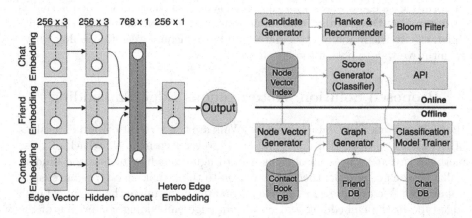

Fig. 1. (Left) Proposed neural network architecture (described in Sect. 3). (Right) System diagram (described in Sect. 4).

Logistic Regression: We can learn a unified edge embedding as a linear combination of different dimensions present in the homogeneous edge representations, $e_{\langle u,v \rangle}^x$, for the set of edge types, $\{x\}$. The weights can be obtained by training a logistic regression model for link prediction that uses the edge vectors as features.

4   System Description

Delivering friend recommendations in a massive online social network within strict Service Level Agreements (SLAs) poses significant engineering challenges. Traditional recommendations, which utilized features to assess overlap in network neighborhood (*e.g.*, common neighbors) couldn't scale to deliver recommendations beyond 1 or 2 hops in the user's neighborhood. The reason being online computation of such features between a pair of nodes is quite expensive. Again, pre-computing such features is not an option for fast evolving social networks as they become stale soon. Since node embedding effectively captures network neighborhood for a node, it provides an elegant solution to this problem of finding recommendations that go beyond 1 or 2 hops. We discuss how this is achieved as we describe our system below.

Our system, as seen in Fig. 1 (Right), has two parts: Offline and Online. The offline set up recomputes the social graph daily, stores node embedding for each user in a scalable similarity search index [11], and trains a new recommendation model based on heterogeneous edge embedding as described in Sect. 3. In the online system, friend recommendations for a user are created in two steps. First, we look up the user's node embedding and perform a nearest neighbour search to fetch recommendation candidates. Second, we score these candidates using our neural network based model. A bloom filter check ensures that recommendations are not repeated for a user. Top k recommendations generated in this manner are served in different recommendation widgets on the Hike app.

5   Experimental Results

5.1   Dataset Description

The data for this work is taken from a subgraph of Hike network which contains users from a selected set of closely connected districts in one state. We divide the graph into two parts: pre-July containing the network structure on 30th June 2018, and edges added in the month of July. We employ the pre-July network comprising of 3.3 million nodes and 32 million edges for training network embedding. Another 10 million node-pairs (5 million edges and 5 million non-edges), which were not the part of the pre-July network, are used for training the friend recommendation model. Finally, the trained model was evaluated on a set of 1 million node-pairs, which also contains friendships made in July 2018 as positive examples. For training the embedding, we used a heterogeneous network with three edge types: 'contact' if the users are in each other's contact book, 'friend' if the users are friends, and 'chat' if the users have chatted at least once. This setting allows for multiple edges between two nodes, e.g., if they are present in each other's contact list, friends and have chatted in July, there will be three edges between them.

Our network is much larger than what has been previously studied in friend recommendation paradigm using network embedding [7,14]. The edge heterogeneity is also unique to our network which is a multi-network with multiple edge-types. Another confounding factor in our data is that 46% of node-pairs in our validation data share no mutual friend – there are many isolated and loosely connected nodes. These challenges compound the difficulty of the friend recommendation problem.

5.2   Comparative Evaluation

We compare different methods of computing edge embedding from node embedding for both node2vec and DeepWalk style random walks. Also, we evaluate the logistic regression model and the neural network model described in Sect. 3. The evaluation based on Area Under the ROC Curve (AUC) is performed on the link prediction problem cast as a classification of node pairs into 'link' or 'no-link'. To learn node embedding, we generate 10 random walks of length 30 emanating from each node. The context window size for the Skip-gram model [2] is taken to be 10, and we optimize the negative sampling loss with a learning rate of 0.01 to obtain embedding in 128-dimensional space. The results are

Table 1. AUC of diff. edge embedding

Node embedding	Edge combiner	LogReg	NeuralNet
DeepWalk	Average	0.79	0.81
	Hadamard	0.78	0.82
	Concatenate	0.80	0.81
Node2Vec	Average	0.79	0.82
	Hadamard	0.78	0.81
	Concatenate	0.80	**0.84**

Table 2. Offline and real-world evaluation

Model	AUC	P @ 5	% increase in CTR
DeepWalk	0.70	0.91	0
Node2Vec	0.72	0.90	2.29
HeteroDeepWalk	0.71	0.83	−0.67
UniformBiasDeepWalk	0.72	0.92	1.26
Node2Vec-NeuralNet	**0.84**	**0.95**	**7.61**

presented in Table 1. We find that the best performing edge embedding arises as a result of: (i) using 'concatenate' on node2vec node-embeddings, (ii) unifying homogeneous edge vectors with the trained multi-tower neural network (Fig. 1 (Left)). We refer to this variant as **Node2Vec-NeuralNet**.

Further, we evaluate Node2Vec-NeuralNet, against the following baselines: (i) DeepWalk: learning node embedding using DeepWalk algorithm on the homogeneous friendship subnetwork, (ii) Node2vec: node embedding using node2vec on homogeneous friendship subgraph, (iii) HeteroDeepWalk, and (iv) UniformBiasDeepWalk. The embeddings obtained using the above methods are fed to a logistic regression model for classification of node-pairs into 'friend' and 'nonfriend'. We employ three metrics for evaluation - AUC, Precision at 5 (P @ 5), and Real world Click-Through Rate (CTR) for top 1 recommendation.

The results are shown in Table 2. Node2Vec-NeuralNet is significantly more accurate compared to other methods for predicting 'friend' edges in our test dataset of 1 million node-pairs. Node2Vec-NeuralNet beats the DeepWalk baseline by 20% in AUC and 4.4% in Precision @ 5.

We conducted controlled experiments on the Hike app to test the relevance of the top friend recommendation generated by each of the five model variants in the real world. We used a sample set of 10 K users for each variant. The candidate set used to find recommendations for a user comprised of 1 hop neighbours of the user on the Hike network. In Table 2, the last column compares the click through rates (CTR) for each variant to the DeepWalk baseline. Again, Node2Vec-NeuralNet shows a relative improvement of 7.61% over DeepWalk.

6   Conclusion

In this work, we developed and compared methods to learn edge embedding in a heterogeneous multi-graph. We showed the efficacy of such an edge embedding for deriving friend recommendation on Hike's network with a large scale offline evaluation as well as real world user experiments. Friend recommendation based on this method is deployed and currently running live at Hike.

References

1. Backstrom, L., Leskovec, J.: Supervised random walks: predicting and recommending links in social networks. In: Proceedings of the Fourth ACM International Conference on Web Search and Data Mining, pp. 635–644. ACM (2011)
2. Bojanowski, P., Grave, E., Joulin, A., Mikolov, T.: Enriching word vectors with subword information. Trans. Assoc. Comput. Linguist. **5**, 135–146 (2017)
3. Chang, S., Han, W., Tang, J., Qi, G.J., Aggarwal, C.C., Huang, T.S.: Heterogeneous network embedding via deep architectures. In: Proceedings of the 21th ACM SIGKDD International Conference on Knowledge Discovery and Data Mining, pp. 119–128. ACM (2015)
4. Cui, P., Wang, X., Pei, J., Zhu, W.: A survey on network embedding. IEEE Trans. Knowl. Data Eng. (2018)
5. Dong, Y., Chawla, N.V., Swami, A.: Metapath2vec: scalable representation learning for heterogeneous networks. In: Proceedings of the 23rd ACM SIGKDD International Conference on Knowledge Discovery and Data Mining, KDD 2017, pp. 135–144. ACM, New York (2017)
6. Dong, Y., et al.: Link prediction and recommendation across heterogeneous social networks. In: 2012 IEEE 12th International Conference on Data Mining, pp. 181–190. IEEE (2012)
7. Grover, A., Leskovec, J.: node2vec: scalable feature learning for networks. CoRR abs/1607.00653 (2016). http://arxiv.org/abs/1607.00653
8. Hamilton, W.L., Ying, R., Leskovec, J.: Representation learning on graphs: methods and applications. CoRR abs/1709.05584 (2017). http://arxiv.org/abs/1709.05584
9. Hong, S., Chakraborty, T., Ahn, S., Husari, G., Park, N.: SENA: preserving social structure for network embedding. In: Proceedings of the 28th ACM Conference on Hypertext and Social Media, pp. 235–244. ACM (2017)
10. Hong, S., Park, N., Chakraborty, T., Kang, H., Kwon, S.: PAGE: answering graph pattern queries via knowledge graph embedding. In: Chin, F.Y.L., Chen, C.L.P., Khan, L., Lee, K., Zhang, L.-J. (eds.) BIGDATA 2018. LNCS, vol. 10968, pp. 87–99. Springer, Cham (2018). https://doi.org/10.1007/978-3-319-94301-5_7
11. Johnson, J., Douze, M., Jégou, H.: Billion-scale similarity search with GPUs. arXiv preprint arXiv:1702.08734 (2017)
12. Liben-Nowell, D., Kleinberg, J.: The link-prediction problem for social networks. J. Am. Soc. Inf. Sci. Technol. **58**(7), 1019–1031 (2007)
13. Mikolov, T., Chen, K., Corrado, G., Dean, J.: Efficient estimation of word representations in vector space. CoRR abs/1301.3781 (2013). http://arxiv.org/abs/1301.3781
14. Perozzi, B., Al-Rfou, R., Skiena, S.: DeepWalk: online learning of social representations. In: Proceedings of the 20th ACM SIGKDD International Conference on Knowledge Discovery and Data Mining, KDD 2014, pp. 701–710. ACM, New York (2014)
15. Tang, J., Qu, M., Wang, M., Zhang, M., Yan, J., Mei, Q.: LINE: large-scale information network embedding. In: Proceedings of the 24th International Conference on World Wide Web, pp. 1067–1077. International World Wide Web Conferences Steering Committee (2015)
16. Zeng, S.: Link prediction based on local information considering preferential attachment. Phys. A: Stat. Mech. Appl. **443**, 537–542 (2016)

Public Sphere 2.0: Targeted Commenting in Online News Media

Ankan Mullick[1]($\boxtimes$), Sayan Ghosh[2]($\boxtimes$), Ritam Dutt[2]($\boxtimes$), Avijit Ghosh[2]($\boxtimes$), and Abhijnan Chakraborty[3]($\boxtimes$)

[1] Microsoft, Hyderabad, India
ankan.mullick@microsoft.com
[2] Indian Institute of Technology Kharagpur, Kharagpur, India
{sgdgp,ritam,avijitg22}@iitkgp.ac.in
[3] Max Planck Institute for Software Systems, Saarbrücken, Germany
achakrab@mpi-sws.org

Abstract. With the increase in online news consumption, to maximize advertisement revenue, news media websites try to attract and retain their readers on their sites. One of the most effective tools for reader engagement is commenting, where news readers post their views as comments against the news articles. Traditionally, it has been assumed that the comments are mostly made against the full article. In this work, we show that present commenting landscape is far from this assumption. Because the readers lack the time to go over an entire article, most of the comments are relevant to only particular sections of an article. In this paper, we build a system which can automatically classify comments against relevant sections of an article. To implement that, we develop a deep neural network based mechanism to find comments relevant to any section and a paragraph wise commenting interface to showcase them. We believe that such a data driven commenting system can help news websites to further increase reader engagement.

1 Introduction

Recent years have witnessed a paradigm shift in the way people consume news. Online news media has become more popular than the traditional newsprint, especially to younger news readers[1]. To further engage them, in addition to presenting news, online news platforms also allow readers to comment and share their points of view on the matter reported in stories. Irrespective of concerns about quality of the comments, especially their language and tone, comments are considered to be the most effective tool to increase reader engagements [1].

Several prior works in media and communication studies have highlighted the importance of discussions in the evolution of a democratic society. In a seminal work, Habermas established the notion of 'Public Sphere' where public opinion gets formed via *rational-critical debates* [2]. Ruiz *et al.* [3] argued that online news media provide a new manifestation of the public sphere – *Public Sphere 2.0*, where commenting acts as the facilitator of public debates.

---

[1] http://news.bbc.co.uk/2/hi/business/8542430.stm.

L. Azzopardi et al. (Eds.): ECIR 2019, LNCS 11438, pp. 180–187, 2019.
https://doi.org/10.1007/978-3-030-15719-7_23

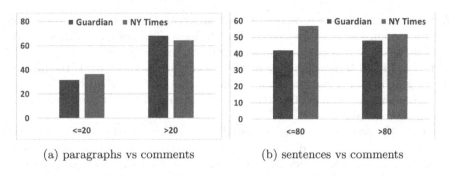

(a) paragraphs vs comments   (b) sentences vs comments

Fig. 1. Comment count varies with paragraph and sentence count.

However, the myriad plethora of news websites today has resulted in a gradual decline of the attention span of an user to a particular news story. In an earlier work, Nielson [4] has noted that the readers predominately read online web pages in an F-shaped pattern i.e., two horizontal stripes in the top of the page followed by a vertical stripe along the page. This implies that the attention span of users wanes as they go through an article and most of their attention is focused on the initial paragraphs. In this context, it is important to understand whether the commenting options in news websites today can felicitate discussions on the news stories and play the role of public sphere 2.0.

To investigate this issue, we gather articles and corresponding comments from two popular news websites – The Guardian (`theguardian.com`) and The New York Times (`nytimes.com`). We observe that a large number of comments are made targeting particular sections of an article, rather than the entire article itself. Yet, most news media websites allow their readers to comment only on the full article. In this paper, we propose to revamp the commenting UI by automatically placing the most relevant comments against each section of an article. For this, we develop a neural network based mechanism to map comments to particular paragraphs. Extensive evaluations show that our proposed methodology outperforms state-of-the-art baselines. Finally, we build a system which allows a reader to check for comments made against any section of an article and comment on the same. We believe that such system can help news websites in increasing reader engagement further.

2   Dataset and Motivation

In recent years, news media sites have seen huge increase in user engagement through commenting, liking, sharing etc. However, users do not spend similar time over the entire news article. Nielsen [4] observed that, for news articles, users mostly focus on initial paragraphs or few sentences of a paragraph to consume the summary of an article, possibly due to limited time to read the whole story.

To investigate how this influences the commenting behavior, we gathered news articles from two popular news websites - 'The Guardian' and 'The New

York Times'. In total, we collected 1, 352 Guardian and 1, 020 NYTimes news articles encompassing various topics like Business, Technology, Politics, Sports and Editorials and all comments made against these articles[2].

Figure 1a and b show how the number of comments varies w.r.t. the number of paragraphs and sentences in an article (Y-axis is % distribution). Figure 1a points out that more than 60% comments are posted to the articles having more than 20 paragraphs.

Figure 1b shows how comment distribution varies for 80 sentence threshold (∼20 paragraphs) for two online news papers. Overall, we see that having more paragraphs in an article increases the number of comments posted against it. Thus, we can conclude that comment-paragraph relation is important.

Table 1. Distribution of different labels for two datasets.

Relevance label	% in The Guardian	% in NY Times
1	31.05	40.11
2	19.09	10.23
3	17.77	17.50
4	19.08	12.29
5	13.01	19.87

From the collected articles, we randomly selected 50 articles from each media site for manual annotation, where two annotators were asked to give one of five possible relevance scores for a comment to a paragraph. The relevance scores are 1 (strongly irrelevant), 2 (weakly irrelevant), 3 (neutral), 4 (weakly relevant) and 5 (strongly relevant), where the relevance is judged by the presence and absence of common words or a common thought between the paragraph and the comment text. Both annotators provided a relevance score for each paragraph-comment pairs in all 100 articles. Inter-annotator agreement (Cohen κ) was 0.71. A particular relevance score to a comment-paragraph pair was granted when both the annotators agreed.

We observed that around 42.7% of the comments (in total) were relevant to the whole article as those were not mapped to a particular paragraph. We consider a comment to be related to the entire article if the comment has a relevance score ≥ 4 for at least 3 paragraphs or has a relevance score of ≤ 2 for all the paragraphs of the article.

However, approximately half of the comments (48.9% and 48.8%) of the Guardian and NYTimes articles are centered towards 2–3 particular paragraphs as opposed to the entire article. Similar to [4], we also observe that the mean relevance of a comment decreases along the article's length. This exemplifies that more relevant comments are related to the beginning paragraphs of an article and such a trend holds true for both Guardian and NYT articles.

---

[2] https://tinyurl.com/paragraph2comment.

Thus it is an interesting problem to find out how comments are related to individual paragraphs rather than the whole article. To automatically find out this association, we created the gold standard annotated datasets of 1834 and 1114 comments for 'The Guardian' and 'New York Times' respectively. The detailed statistics of the different annotated labels are provided in Table 1. Using this data (after class balancing using the SMOTE [5] algorithm), we design an automated approach as explained next.

Table 2. Performance of different models on the two datasets.

Model	The Guardian						New York Times					
	Precision			Recall			Precision			Recall		
	Macro	Micro	Weighted	Macro	Micro	Weighted	Macro	Micro	Weighted	Macro	Micro	Weighted
NB	46.2	42.6	61.6	42.6	42.5	42.6	33.9	35.2	60.9	40.9	35.2	35.2
DT	42.9	49.6	53.5	35.6	50.9	50.9	36.9	59.2	52.7	31.2	59.1	59.2
RF	37.9	44.7	45.1	24.4	44.6	44.7	17.5	57.6	37.8	20.2	57.3	57.6
K-NN	48.5	63.5	61.3	48.1	63.4	63.5	37.6	58.9	55.9	34.8	61.2	61.2
R-SVM	49.9	63.1	60.9	45.4	63.0	63.1	39.4	61.9	51.9	27.3	61.9	60.3
AdaBoost	38.3	49.3	48.2	35.7	49.3	49.2	29.3	56.1	48.1	28.5	55.1	54.6
LR	41.2	54.0	51.3	38.8	54.1	54.0	34.1	60.6	50.7	25.8	60.7	60.1
LSTM	64.1	74.4	74.5	63.6	74.5	73.3	56.6	76.8	76.1	57.8	76.9	76.8
GRU	**64.2**	**75.3**	**75.9**	**63.7**	**75.3**	**75.4**	**64.8**	**79.1**	**78.4**	**64.3**	**79.3**	**79.1**

3 Linking Comments to Paragraphs

In this paper, we propose an approach to correctly identify paragraph-comment pairs and encourage users to comment towards the paragraphs, instead of only commenting on the whole article. Our proposed framework is based on deep neural networks. We have used two different neural network models - Long Short-Term Memory (LSTM) and Gated Recurrent Unit (GRU) where inputs are paragraph and comment vectors. We have used the pre-trained 300 dimension Google News Vectors for each word and in case a pre-trained embedding for a word is not found we take it to 0 (in 300 dimension space). In order to calculate the vector for the entire paragraph and comment, we take the average of all word vectors corresponding to each word in the paragraph and comment respectively. Deep neural network models - (i) LSTM and (ii) GRU were applied on top of the paragraph and comment vectors to get a 150 dimension vector for both paragraph and comment[3]. Thereafter these two vectors were merged and on top of it a fully connected layer with 5 units (for five classes) and soft-max activation is applied to get the probability for each class. The proposed model is shown in Fig. 2a. No explicit feature extraction, using POS Tagger or LIWC was required for these models.

---

[3] After experimenting with different dimensions, results (in terms of precision, recall) were best for 150 dimension.

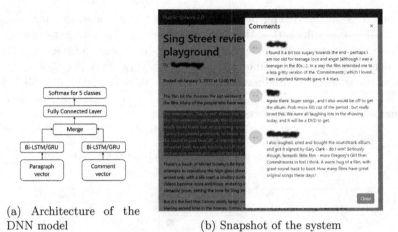

(a) Architecture of the DNN model

(b) Snapshot of the system

Fig. 2. Our proposed system.

3.1 Baselines

Other than neural network models, we have experimented with various traditional machine learning models - Naive Bayes (NB), Decision Tree (DT), Random Forest (RF), K-Nearest Neighbors (K-NN), RBF Support Vector Machine (R-SVM), Logistic Regression (LR) and Adaboost. We have extracted different features for these models, which can be grouped into three different categories.

POS Tag and Dependency Features: Stanford Part-Of-Speech Tagger [6] and Stanford dependency parser [7] were used to get different Parts-Of-Speech based features. Total 45 features were extracted.

LIWC Features: Total 63 psycholinguistic features were extracted using the LIWC tool [8].

Others: Uni-gram, bi-gram, tri-gram features for paragraphs and comments.

After generating the feature matrix, dimensions were reduced using Latent Semantic Indexing (LSA) before feeding into the traditional ML-classifiers.

3.2 Evaluation

After feature extraction of the annotated datasets, various ML-classifiers were used to calculate 10-fold cross validation tests. For the deep learning model, we have trained for 5 epochs for each step in the 10-fold cross validation. Results are shown in terms of Macro, Micro and Weighted averaged precision and recall for 'The Guardian' and 'New York Times' datasets[4]. Table 2 shows that LSTM

---

[4] For ML-classifiers, we have computed precision and recall for different combination of (i) POS Tag and Dependency, (ii) LIWC and (iii) Others features but due to space constraint only the best results were shown.

and GRU models outperform ML-classifier models in terms of all metrics and GRU model performs the best. Figure 2b shows the snapshot of our model where top k (here k = 3) relevant comments are highlighted when the cursor is placed around the second paragraph of a particular story.

To check the effectiveness of our system, we showed to 20 volunteers the same 10 Guardian news stories on the original website and through our system. At the end, the volunteers were asked to rate the interface better for commenting against the articles. 17 out of 20 volunteers gave higher rating to our system interface, and the main reason they cited is the ability to see old comments and post new comments against different portions of the articles.

4   Related Works

Here, we briefly survey the prior works on commenting in online news media.

Comment Ranking: Hsu et al. [9] developed a regression model for identifying and ranking comments within a Social Web community based on the community's expressed preferences. Dalal et al. [10] built Hodge decomposition based rank aggregation technique to rank online comments on the social web.

Comment Recommendation: Bansal et al. [11] proposed 'Collaborative Correspondence Topic Models' to recommend comment-worthy blogs or news stories to a particular user (i.e., where she would be interested to leave comments on them), where user feature profile is generated by content analysis. Shmueli et al. [12] combined content-based approach with a collaborative-filtering approach (utilizing users' co-commenting patterns) for personalized recommendation of stories to users for discussing through comments. Agarwal et al. [13] focused on personalized user preference based ranking of the comments in an article.

Comment Analysis: Liu [14] ranked interest based news sections and articles by using a passage retrieval algorithm. Stroud et al. [15] analyzed demographics, attitudes and behaviors of user population who comment on different sections. Similar analysis has also been done by Chakraborty et al. [16,17] for social media posts. Mullick et al. [18,19] classified online comments into opinion and fact and respective subcategories. Mullick et al. [20] developed opinion-detection algorithm for news articles. Almgren et al. [21] compared commenting, sharing, tweeting and measured user participation in them. Chakraborty et al. [22,23] utilized these different popularity signals for online news recommendations. Mullick et al. [24] experimented topic drift event and characteristics in online comments.

Our present work is complementary to these earlier works, where our focus is to explore paragraph oriented commenting pattern and build a model to show relevant comments to a paragraph for felicitating more commenting.

5   Conclusion

To play the role of the Public Sphere, online news websites need to encourage readers to comment on their articles. In this paper, we argued for a revamp of the traditional commenting interface, and for enabling commenting on selective sections of an article. We developed a deep neural network approach to link comments to particular section. We showed that Gated Recurrent Unit (GRU) model provides best results in terms of macro and micro level precision and recall. Then, we built a basic user interface to increase user engagement in online comment sections. There are few issues to be resolved in our framework - for example, the scenario where a comment belongs to multiple paragraphs, how can a viewer select two non-consecutive paragraphs to read the respective comments and showing scores for comments. Our immediate future step is to develop an end-to-end system after resolving the issues in the model to show a user top K relevant comments (further divided into different sentiment expressed in the comments), while scrolling down the paragraphs. We believe such data driven selective commenting systems can bring more specific and targeted reader engagement for online publishing houses.

References

1. Park, D., Sachar, S., Diakopoulos, N., Elmqvist, N.: Supporting comment moderators in identifying high quality online news comments. In: Proceedings of the 2016 CHI Conference on Human Factors in Computing Systems, pp. 1114–1125. ACM (2016)
2. Habermas, J.: Moral Consciousness and Communicative Action. MIT press, Cambridge (1990)
3. Ruiz, C., Domingo, D., Micó, J.L., Díaz-Noci, J., Meso, K., Masip, P.: Public sphere 2.0? The democratic qualities of citizen debates in online newspapers. Int. J. Press/Politics **16**(4), 463–487 (2011)
4. Nielsen, J.: Usability 101: Introduction to usability (2003)
5. Chawla, N.V., Bowyer, K.W., Hall, L.O., Kegelmeyer, W.P.: SMOTE: synthetic minority over-sampling technique. J. Artif. Intell. Res. **16**, 321–357 (2002)
6. Manning, C., Surdeanu, M., Bauer, J., Finkel, J., Bethard, S., McClosky, D.: The stanford CoreNLP natural language processing toolkit. In: Proceedings of 52nd Annual Meeting of the Association for Computational Linguistics: System Demonstrations, pp. 55–60 (2014)
7. De Marneffe, M.C., Manning, C.D.: Stanford typed dependencies manual. Technical report, Technical report, Stanford University (2008)
8. Tausczik, Y.R., Pennebaker, J.W.: The psychological meaning of words: LIWC and computerized text analysis methods. J. Lang. Soc. Psychol. **29**(1), 24–54 (2010)
9. Hsu, C.F., Khabiri, E., Caverlee, J.: Ranking comments on the social web. In: International Conference on Computational Science and Engineering, CSE 2009, vol. 4, pp. 90–97. IEEE (2009)
10. Dalal, O., Sengemedu, S.H., Sanyal, S.: Multi-objective ranking of comments on web. In: Proceedings of the 21st International Conference on World Wide Web, pp. 419–428. ACM (2012)

11. Bansal, T., Das, M., Bhattacharyya, C.: Content driven user profiling for comment-worthy recommendations of news and blog articles. In: Proceedings of the 9th ACM Conference on Recommender Systems, pp. 195–202. ACM (2015)

12. Shmueli, E., Kagian, A., Koren, Y., Lempel, R.: Care to comment?: recommendations for commenting on news stories. In: Proceedings of the 21st International Conference on World Wide Web, pp. 429–438. ACM (2012)

13. Agarwal, D., Chen, B.C., Pang, B.: Personalized recommendation of user comments via factor models. In: Proceedings of the Conference on Empirical Methods in Natural Language Processing, pp. 571–582. Association for Computational Linguistics (2011)

14. Liu, X.: Comment centric news analysis for ranking. Proc. Am. Soc. Inf. Sci. Technol. **46**(1), 1–8 (2009)

15. Stroud, N.J., Van Duyn, E., Peacock, C.: News commenters and news comment readers. Engaging News Project (2016)

16. Chakraborty, A., Sarkar, R., Mrigen, A., Ganguly, N.: Tabloids in the era of social media? Understanding the production and consumption of clickbaits in Twitter. arXiv preprint arXiv:1709.02957 (2017)

17. Chakraborty, A., Messias, J., Benevenuto, F., Ghosh, S., Ganguly, N., Gummadi, K.P.: Who makes trends? Understanding demographic biases in crowdsourced recommendations. arXiv preprint arXiv:1704.00139 (2017)

18. Mullick, A., Maheshwari, S., Goyal, P., Ganguly, N., et al.: A generic opinion-fact classifier with application in understanding opinionatedness in various news section. In: Proceedings of the 26th International Conference on World Wide Web Companion, International World Wide Web Conferences Steering Committee, pp. 827–828 (2017)

19. Mullick, A., Ghosh D.S., Maheswari, S., Sahoo, S., Maity, S.K., Goyal, P., et al.: Identifying opinion and fact subcategories from the social web. In: Proceedings of the 2018 ACM Conference on Supporting Groupwork, pp. 145–149. ACM (2018)

20. Mullick, A., Goyal, P., Ganguly, N.: A graphical framework to detect and categorize diverse opinions from online news. In: Proceedings of the Workshop on Computational Modeling of People's Opinions, Personality, and Emotions in Social Media (PEOPLES), pp. 40–49 (2016)

21. Almgren, S.M., Olsson, T.: Commenting, sharing and tweeting news. Nordicom Rev. **37**(2), 67–81 (2016)

22. Chakraborty, A., Ghosh, S., Ganguly, N., Gummadi, K.P.: Optimizing the recency-relevancy trade-off in online news recommendations. In: Proceedings of the 26th International Conference on World Wide Web, International World Wide Web Conferences Steering Committee, pp. 837–846 (2017)

23. Chakraborty, A., Patro, G.K., Ganguly, N., Gummadi, K.P., Loiseau, P.: Equality of voice: towards fair representation in crowdsourced top-k recommendations. In: ACM FAT* (2019)

24. Mullick, A., et al.: Drift in online social media. In: IEEE IEMCON, pp. 302–307, November 2018

An Extended CLEF eHealth Test Collection for Cross-Lingual Information Retrieval in the Medical Domain

Shadi Saleh[(✉)] and Pavel Pecina

Institute of Formal and Applied Linguistics, Faculty of Mathematics and Physics,
Charles University, Prague, Czech Republic
{saleh,pecina}@ufal.mff.cuni.cz

Abstract. We present a test collection for medical cross-lingual information retrieval. It is built on resources used by the CLEF eHealth Evaluation Lab 2013–2015 in the patient-centered information retrieval tasks and improves applicability and reusability of the official data. The document set is identical to the official one used for the task in 2015 and contains about one million English medical webpages. The query set contains 166 items used during the three years of the campaign as test queries, now available in eight languages. The extended test collection provides additional relevance judgements which almost doubled the amount of the officially assessed query-document pairs. This paper describes the content of the extended collection, details of query translation and relevance assessment, and state-of-the-art results obtained on this collection.

Keywords: Cross-lingual Information Retrieval · eHealth · Benchmarking

1 Introduction

Cross-lingual Information Retrieval (CLIR) allows users to search for documents using queries in a language different from the language of the documents. Evaluation of CLIR system is difficult mainly due to limited availability of appropriate benchmarks and their reusability. In this paper, we present an extended version of the test collection used in the CLEF eHealth Evaluation Lab in 2013–2015 [6,7,16] for the patient-centered information retrieval task. This benchmark (available via the LINDAT/CLARIN repository)[1] contains about one million documents (medical webpages in English), 166 queries (generated in English and translated to other languages), and relevance assessments based on pooling the officially submitted results. Our main contribution is providing complete manual translations of the queries into seven languages (Czech, French, German, Hungarian, Polish, Spanish, Swedish) and extending the relevance judgements by assessing highly ranked documents in additional cross-lingual experiments.

---

[1] http://hdl.handle.net/11234/1-2925.

© Springer Nature Switzerland AG 2019
L. Azzopardi et al. (Eds.): ECIR 2019, LNCS 11438, pp. 188–195, 2019.
https://doi.org/10.1007/978-3-030-15719-7_24

We also include machine translation of the queries into English, propose a new training/test data split and report state-of-the-art CLIR results on this benchmark.

2 Related Work

CLIR has been studied since the 1990's, and several benchmarks have been produced within various evaluation challenges. A brief overview of the major ones follows. **TREC** (Text REtrieval Conference) is an annual event organized by NIST[2]: In 1997, TREC-6 [22] was the first TREC event accommodating a CLIR track. The document collection included three sets of English, French and German documents taken from news agencies. 25 test topics in the same languages were created based on the interest of the participated assessors who performed binary relevance assessment for these queries. The TREC-7 CLIR track used the same document collection as in TREC-6 plus a set of documents and topics (28) in Italian [20]. The TREC-8 CLIR track used the same document collection as in TREC-7 with new set of 28 queries in the same four languages [21]. TREC-9 ran a CLIR track with document collection aggregated from Chinese news agencies and 25 queries in English and Chinese [4]. In the TREC-10 CLIR track, an Arabic newswire document collection was used with a set of 25 topics created by assessors in Arabic and English and afterwards translated into French [5]. In TREC-11 [14], the same Arabic document collection as in TREC-10 was used with newly 25 created English topics then translated into Arabic. **NTCIR** (NII Testbeds and Community for Information access Research) is a project of NII[3]. The first NTCIR workshop (NTCIR-1) was held on 1999 and aimed to improve linguistic research of Asian languages [9]. NTCIR-1 released test collection which included scientific documents in Japanese and English, plus 83 Japanese topics with graded relevance assessment. NTCIR-2 worked with a collection of academic conference papers in Japanese and English and 49 topics in both languages. NTCIR-3 used a document collection of news in Chinese, Japanese and English with 50 topics in Chinese and 30 topics in Japanese and their translations into Chinese, Korean, Japanese and English. The same dataset was used in NTCIR-4 CLIR. The NTCIR-5 CLIR test collection included documents from news agencies in Chinese, Japanese, Korean and English and 50 search topics in all these languages with graded relevance assessment. NTCIR-6 exploited a document collection of newspaper articles. It reused the collection from NTCIR-5, 4 and 3 CLIR tasks and included 50 topics in Chinese, Japanese, Korean and English and additional documents from newspaper articles in Chinese, Japanese and Korean with graded relevance assessment too. NTCIR-7 ACLIA included CLIR as a subtask which included news articles in Chinese, Japanese and Korean, with 100 topics in Japanese and 100 topics in Chinese and 300 English topics and 3-level relevance assessment. NTCIR-8 ACLIA also launched CLIR subtask with documents in Chinese and Japanese with 300 topics in English. **FIRE** (Forum

---

[2] http://trec.nist.gov.
[3] http://ntcir.nii.ac.jp.

for Information Retrieval Evaluation) [13] has been running since 2008 and aims to support research in multilingual information access for Asian languages. In FIRE 2008, a document collection of news articles in English, Hindi and Marathi was used with 50 queries in the same languages. In FIRE 2010, the 2008 document collection was enriched with new documents in Bengali. A set of 50 topics is manually translated into English, Gujarati, Marathi, Tamil and Telugu. FIRE 2011 used the same collection as in 2010, the queries were refined and interactive search was used to improve the relevance assessment. **CLEF** (Cross-Language Evaluation Forum)[4] has organised multiple tasks of multilingual information access. The Ad-hoc track was organised from 2000 to 2009. The document collections in 2000–2007 were collected from news agencies in several European languages and topics were generated in multiple languages to allow CLIR evaluation. In 2008 and 2009, the document collection was created in cooperation with the European Library [2]. The CLEF CL-SR (Cross-Language Speech Retrieval) task was organized annually in 2003–2007 and focused on searching in spoken English news archives using queries in five languages (Czech, English, French, German and Spanish)[17]. **CLEF ShARe/eHealth**[5] has been organized since 2013 aiming at improving access to the medical and health-related documents by laypeople and medical experts in monolingual and cross-lingual settings. In ShARe/CLEF eHealth 2013 Task 3 [6], the English queries were generated by clinical documentation reporters and nurses based on real discharge summaries to mimic the realistic patients' queries. Five queries were used for development purposes and 50 queries for testing. The document collection contained about one million English pages crawled from medical websites. No CLIR task was organized that year. In ShARe/CLEF eHealth 2014 Task 3 [7], the queries were generated in the same fashion as in the previous year. In addition to the monolingual task, a CLIR task was introduced. Five development and 50 test queries were generated in English and then manually translated into Czech, German and French to simulate cross-lingual setting. The document collection was the same as in 2013. In CLEF eHealth 2015 Task 2 [16], the query creation aimed to implement self-diagnosing case. Non-expert student volunteers were shown images of symptoms of specific conditions and asked to create three different queries (in English) for each symptom. 66 queries were then randomly selected and used for testing (plus 5 queries for development). The queries were manually translated into Arabic, Czech, French, German, Farsi and Portuguese. The 2015 collection was a subset of the 2014's collection (a few websites were removed). In CLEF eHealth 2016 Task 3 [10], a new document collection was introduced (ClueWeb12 B13[6]). The collection contained web documents from both medical and non-medical domain in an attempt to give more realistic representation when users look-up information from the web (generic collection). An initial query pool was created from online posts that contain questions about health conditions. Then for each query, six query variations were created by three medical experts

---

[4] http://www.clef-initiative.eu/.
[5] https://sites.google.com/site/clefehealth/.
[6] http://lemurproject.org/clueweb12/specs.php.

and three people without medical knowledge resulting into the final set of 300 queries representing 50 topics. The queries were translated (by medical experts) into Czech, French, German, Hungarian, Polish, Spanish and Swedish to allow CLIR experiments. CLEF eHealth 2017 IR Task used the same collection and queries as in 2016. However, an additional assessment was performed [15]. CLEF eHealth 2018 Consumer Health Search Task released a document collection created using CommonCrawl platform [8] containing more than five million documents from more than thousand websites. 50 queries were provided in English in the monolingual task (IRTask 1 Ad-hoc search). In IRTask 4 (Multilingual Ad-hoc Search) the same English queries were provided in French, German and Czech.

Table 1. Examples of test queries.

Id	Year	Title
qtest2013.38	2013	*MI and hereditary*
qtest2013.41	2013	*right macular hemorrhage*
qtest2014.1	2014	*Coronary artery disease*
qtest2014.6	2014	*Aortic stenosis*
clef2015.test.1	2015	*many red marks on legs after traveling from US*
clef2015.test.57	2015	*infant labored breathing and tight wheezing cough*

3   Test Collection

The presented test collection is based on the CLEF eHealth resources used in 2013–2015. We adopt the document collection, the original English queries, their translations to other languages (where available), and the relevance assessments. The set of **documents** is identical to the one used in the CLEF eHealth 2015 Task 2: User-Centred Health Information Retrieval [16]. It includes a total of $1,104,298$ web pages in HTML that are automatically crawled from various medical English websites (e.g. Genetics Home Reference, ClinicalTrial.gov, Diagnosia). The average length of a document is 911 words.

The **queries** include all test queries from the IR tasks in 2013 (50 queries), 2014 (50 queries), and 2015 (66 queries). The nature of the queries varies from year to year (see Sect. 2 and Table 1). We mixed them to get more representative and balanced query set, and then split this set into a subset of 100 queries for training (33 queries from 2013 test set, 32 from 2014 and 35 from 2015) and 66 queries for testing (17 queries from 2013 test set, 18 queries from 2014 and 31 from 2015). The two sets are stratified in terms of distribution of the year of origin, number of relevant/not-relevant documents, and the query length (number of words). The query ID tags in the package preserve the original IDs which allows mapping the queries to their original year. All the queries are available in a total of 8 languages (the original English plus human translations into Czech, French, German, Hungarian, Polish, Spanish, and Swedish) to allow monolingual (queries in English) and cross-lingual retrieval (queries in the other languages). The query translations come from two sources: official translations provided by the CLEF eHealth organisers (in 2015 and 2014, queries were officially released in Czech, French and German, while in 2013, only English queries were available) and newly created translations (when the official translations were not existing). The new translations were conducted by medical experts fluent in English and the target language. They followed the same instructions as the official translators [7,19] (i.e. preserve syntax where possible and translate term-by-term otherwise).

In addition to the human translation of the queries from English into the target languages, we included queries machine-translated back to English to allow CLIR experiments without having access to a machine translation system. We employed the phrase-based SMT system that is adapted to translate medical-domain queries described in [3]. For each input query, the system generates a list of 1000 ranked translation hypotheses (*n-best-list*) including internal system information and scores for each one of them (e.g., alignment between source and target language, scores of language model, translation model, reordering model and word penalty).

Table 2. Relevance assessment statistics.

	2013	2014	2015	Extension	Total
Relevant	1,174	3,209	2,515	2,517	9,415
Irrelevant	3,676	3,591	9,576	11,851	28,694

The query-document relevance assessment in the presented test collection was substantially improved. The original assessment of 23,741 query-document pairs (6898 relevant, 16843 irrelevant) was enriched by additional 14,368 judgements (2,571 relevant, 11,851 irrelevant) obtained by domain experts instructed the same way as the official CLEF eHealth assessors. Table 2 shows statistics of the assessment information in the 2013–2015 CLEF eHealth IR tasks and our contribution to the assessment information in the test set. The newly assessed query-document pairs were selected by pooling results of various experiments. These experiments were conducted after the end of CLEF 2015 IR Task, using the queries and the original assessment from 2013–2015 IR tasks. The major pooling experiment is described in [18] – it is the state-of-the-art result obtained using this collection. This approach exploits multiple hypothesis translations (for an input query) produced by the MT system [3] which are reranked using a supervised machine-learning method trained to directly optimise the retrieval quality. The document pool contained unjudged documents from the top 10 retrieved documents for each query translation. Although the assessors were different from the official ones, we attempted to mimic the official assessment procedure to the maximum possible extent. The assessors used the same software (Relevation) [11], the same topic descriptions, and the same instructions. Each topic was assessed by a single assessor by randomly splitting the topics among the assessors, and the pooled documents were judged using three grades (irrelevant, somewhat relevant, highly relevant). To get binary assessment (relevant, irrelevant) from the graded assessment, we followed the CLEF eHealth organisers' approach where somewhat relevant documents are considered to be relevant too. To confirm the assessment quality we performed two dual assessment experiments and measured agreement between: (i) the two new assessors and (ii) the new assessors vs. the official assessment. In both experiments, we randomly selected 2 relevant and 2 irrelevant documents for each topic and asked for additional (independent) relevance judgement. The first experiment (i) showed 86% agreement rate, the second experiment (ii) showed 79% agreement rate (measured as accuracy of binarized relevance), which is generally considered to be sufficient [1]. The dual assessment is included separately in the package. The new relevance assessments are very important for reusability of the presented test collection. Test collections without exhaustive relevance assessment tend to underestimate the evaluation scores (by treating unjudged documents as irrelevant) and enriching the assessment helps to reduce this problem. A major effect also comes from the query-document pairs assessed as not relevant. These are useful for methods employing supervised learning e.g. learning to rank [12] and supervised query expansion [23]. The extended relevance assessment also helped in training and evaluation of the hypotheses reranking model in [18] which predicts the optimal query translation out of 15-best translation hypotheses generated by an SMT system, which lead to the best results achieved using this collection (see Table 3).

Table 3. State-of-the-art results (in terms of common retrieval evaluation measures in %) obtained using the extended CLEF eHealth test collection. See [18] for details.

Language	English	Czech	French	German	Hungarian	Polish	Spanish	Swedish
P@10	50.30	48.03	51.67	46.21	48.48	43.18	50.15	41.36
NDCG@10	55.26	49.51	53.27	47.21	49.88	44.01	52.57	43.16
BPREF	39.94	37.59	37.33	36.46	38.00	38.90	34.61	33.44
MAP	28.31	24.02	25.66	23.09	25.28	21.76	25.11	21.29

4   Conclusion

We presented an extended version of the CLEF eHealth test collection for cross-lingual information retrieval in the medical domain based on the collection used in the CLEF eHealth Evaluation Lab 2013–2015 IR tasks. The extended collection improves reusability of the officially provided resources and allows investigating supervised learning approaches (in both cross-lingual and monolingual IR) on the proposed test set. The test set contains English queries and their manual translations into seven languages to allow cross-lingual retrieval, additional relevance assessment, and a new training/test split of the query set. We also added various data for experimenting with machine translation of queries. The data package containing the official and newly added data is publicly available.

Acknowledgments. The language resources presented in this paper are distributed by the LINDAT/CLARIN project of the Ministry of Education of the Czech Republic. This work was supported by the Czech Science Foundation (grant n. P103/12/G084).

References

1. Cohen, J.: A coefficient of agreement for nominal scales. Educ. Psychol. Measure. **20**(1), 37–46 (1960)
2. Di Nunzio, G.M., Ferro, N., Mandl, T., Peters, C.: CLEF 2007: ad hoc track overview. In: Peters, C., et al. (eds.) CLEF 2007. LNCS, vol. 5152, pp. 13–32. Springer, Heidelberg (2008). https://doi.org/10.1007/978-3-540-85760-0_2
3. Dušek, O., Hajič, J., Hlaváčová, J., Novák, M., Pecina, P., Rosa, R., et al.: Machine translation of medical texts in the Khresmoi project. In: Proceedings of the Ninth Workshop on Statistical Machine Translation, pp. 221–228. ACL, Baltimore (2014)
4. Gey, F.C., Chen, A.: TREC-9 cross-language information retrieval (English-Chinese) overview. In: Proceedings of the Ninth Text REtrieval Conference (TREC-9), pp. 15–23. NIST, Gaithersburg (2000)
5. Gey, F.C., Oard, D.W.: The TREC-2001 cross-language information retrieval track: searching Arabic using English, French or Arabic queries. In: The Tenth Text REtrieval Conference (TREC 2001), pp. 16–26. NIST, Gaithersburg (2001)
6. Goeuriot, L., et al.: ShARe/CLEF eHealth evaluation lab 2013, task 3: information retrieval to address patients' questions when reading clinical reports. CLEF 2013 Online Working Notes **8138**, pp. 1–16 (2013)
7. Goeuriot, L., et al.: ShARe/CLEF eHealth evaluation lab 2014, task 3: user-centred health information retrieval. In: CLEF Online Working Notes. CEUR Workshop Proceedings, vol. 1180, pp. 43–61. CEUR-WS, Sheffield (2014). http://ceur-ws.org/Vol-1180/. ISSN: 1613-0073
8. Suominen, H., et al.: Overview of the CLEF 2018 consumer health search task. In: CLEF 2018 Evaluation Labs and Workshop: Online Working Notes, pp. 1–15. CEUR-WS, Avignon (2018)
9. Kando, N.: NTCIR Workshop: Japanese- and Chinese-English cross-lingual information retrieval and multi-grade relevance judgments. In: Peters, C. (ed.) CLEF 2000. LNCS, vol. 2069, pp. 24–35. Springer, Heidelberg (2001). https://doi.org/10.1007/3-540-44645-1_3
10. Kelly, L., Goeuriot, L., Suominen, H., Névéol, A., Palotti, J., Zuccon, G.: Overview of the CLEF eHealth evaluation lab 2016. In: Fuhr, N., et al. (eds.) CLEF 2016. LNCS, vol. 9822, pp. 255–266. Springer, Cham (2016). https://doi.org/10.1007/978-3-319-44564-9_24

11. Koopman, B., Zuccon, G.: Relevation!: an open source system for information retrieval relevance assessment. In: Proceedings of the 37th International ACM SIGIR Conference on Research and Development in Information Retrieval, pp. 1243–1244. ACM, Gold Coast (2014)
12. Liu, T.Y., Xu, J., Qin, T., Xiong, W., Li, H.: LETOR: benchmark dataset for research on learning to rank for information retrieval. In: Proceedings of SIGIR 2007 Workshop on Learning to Rank for Information Retrieval, pp. 3–10. ACM, New York (2007)
13. Majumder, P., Pal, D., Bandyopadhyay, A., Mitra, M.: Overview of FIRE 2010. In: Majumder, P., Mitra, M., Bhattacharyya, P., Subramaniam, L.V., Contractor, D., Rosso, P. (eds.) FIRE 2010-2011. LNCS, vol. 7536, pp. 252–257. Springer, Heidelberg (2013). https://doi.org/10.1007/978-3-642-40087-2_24
14. Oard, D.W., Gey, F.C.: The TREC 2002 Arabic/English CLIR track. In: The Eleventh Text Retrieval Conference (TREC 2002), pp. 1–15. NIST, Gaithersburg (2002)
15. Palotti, J., Zuccon, G., Jimmy, P.P., Lupu, M., Goeuriot, L., Kelly, L., Hanbury, A.: CLEF 2017 task overview: the IR task at the eHealth evaluation lab. In: Working Notes of Conference and Labs of the Evaluation (CLEF) Forum. CEUR Workshop Proceedings, pp. 1–10. CEUR-WS, Dublin (2017)
16. Palotti, J.R.M., et al.: CLEF eHealth evaluation lab 2015, task 2: retrieving information about medical symptoms. In: Working Notes of CLEF 2015 - Conference and Labs of the Evaluation Forum. CEUR Workshop Proceedings, vol. 1391, pp. 1–22. CEUR-WS, Toulouse (2015)
17. Pecina, P., Hoffmannová, P., Jones, G.J.F., Zhang, Y., Oard, D.W.: Overview of the CLEF-2007 cross-language speech retrieval track. In: Peters, C., et al. (eds.) CLEF 2007. LNCS, vol. 5152, pp. 674–686. Springer, Heidelberg (2008). https://doi.org/10.1007/978-3-540-85760-0_86
18. Saleh, S., Pecina, P.: Reranking hypotheses of machine-translated queries for cross-lingual information retrieval. In: Fuhr, N., et al. (eds.) CLEF 2016. LNCS, vol. 9822, pp. 54–66. Springer, Cham (2016). https://doi.org/10.1007/978-3-319-44564-9_5
19. Urešová, Z., Hajič, J., Pecina, P., Dušek, O.: Multilingual test sets for machine translation of search queries for cross-lingual information retrieval in the medical domain. In: Proceedings of LREC 2014, pp. 3244–3247. ERLA, Reykjavik (2014)
20. Voorhees, E.M., Harman, D.: Overview of the seventh text retrieval conference TREC-7. In: Proceedings of the Seventh Text REtrieval Conference (TREC-7), pp. 1–24. NIST, Gaithersburg (1998)
21. Voorhees, E.M., Harman, D.: Overview of the eighth text retrieval conference (TREC-8). In: Proceedings of the Eighth Text REtrieval Conference (TREC-8), pp. 1–24. NIST, Gaithersburg (2000)
22. Voorhees, E.M., Harman, D.: Overview of the sixth text retrieval conference (TREC-6). Inf. Process. Manage. **36**, 3–35 (2000)
23. Zhang, Z., Wang, Q., Si, L., Gao, J.: Learning for efficient supervised query expansion via two-stage feature selection. In: Proceedings of the 39th International ACM SIGIR Conference on Research and Development in Information Retrieval, SIGIR 2016, pp. 265–274. ACM, New York (2016)

An Axiomatic Study of Query Terms Order in Ad-Hoc Retrieval

Ayyoob Imani[1][(✉)], Amir Vakili[1], Ali Montazer[2], and Azadeh Shakery[1]

[1] University of Tehran, Tehran, Iran
{ayyoub.imani,shakery,a_vakili}@ut.ac.ir
[2] University of Massachusetts Amherst, Amherst, USA
montazer@umass.edu

Abstract. Classic retrieval methods use simple bag-of-word representations for queries and documents. This representation fails to capture the full semantic richness of queries and documents. More recent retrieval models have tried to overcome this deficiency by using approaches such as incorporating dependencies between query terms, using bi-gram representations of documents, proximity heuristics, and passage retrieval. While some of these previous works have implicitly accounted for term order, to the best of our knowledge, term order has not been the primary focus of any research. In this paper, we will show that documents that have two query terms in the same order as in the query have a higher probability of being relevant than documents that have two query terms in the reverse order. Using the axiomatic framework for information retrieval, we introduce a constraint that retrieval models must adhere to in order to effectively utilize term order dependency among query terms. We modify two existing robust retrieval models based on this constraint. Our empirical evaluation using both TREC newswire and web corpora demonstrates that the modified retrieval models significantly outperform their original counterparts.

Keywords: Query term order · Axiomatic analysis · SDM · PLM

1 Introduction

Classic information retrieval systems such as BM25 [11] use a very simple bag of word representation. These models have proven to be effective and offer a compromise between efficiency and good results. However, query terms have associations that are not considered when using a bag of word representation. Recent research has shown that taking these associations into consideration can effectively improve retrieval performance [1,2,6,7,10,12].

Some previous models capture dependencies between terms using information like proximity and co-occurrences of terms in documents [7,12]. While these previous works on term associations totally neglect the role of term order, methods that use n-grams [1,2,8,10], implicitly consider the order of terms for adjacent

© Springer Nature Switzerland AG 2019
L. Azzopardi et al. (Eds.): ECIR 2019, LNCS 11438, pp. 196–202, 2019.
https://doi.org/10.1007/978-3-030-15719-7_25

terms to some extent but don't consider the order of terms that are separated by a few other terms. These methods suffer from data sparsity and using a bigger n-gram to cover this small distance causes even more sparsity. Our proposed method aims to consider term order dependency not only between two adjacent terms but also for more distant terms inside a specified window size thus alleviating the data sparsity problem.

In this paper, we hypothesize that if a query contains term pairs whose semantics change if the order of terms is reversed, documents where these terms appear in the same order as they did in the query are more likely to be relevant. To verify the hypothesis, we conduct data exploratory analysis on various TREC collections. We use the axiomatic framework proposed by Fang [4] to model this hypothesis as a formal constraint and modify SDM [8] and PLM to satisfy this constraint [7].

2   Methodology

Axiomatic analysis provides an approach for developing retrieval models based on formalized constraints and has received much attention in the information retrieval community [3–5,9]. In this section, we explain the intuition behind our term order constraint before formally defining it. Then, we propose our modifications to the PLM and SDM retrieval models.

2.1   Term Order and Document Relevance

In this section we investigate whether our intuition regarding the effect of term order is correct. To achieve this, we compute $p(Rel|\text{ordered match})$ and $p(Rel|\text{reversed match})$ and test whether $p(Rel|\text{ordered match})$ is significantly higher than $p(Rel|\text{reversed match})$.

For all queries we find term pairs $q_1 q_2$ where q_1 appears before q_2 in a window of size 5 within a single document. Having relevance judgments for the queries, we define the following probabilities:

$$p(Rel|\text{ordered match}) = p(Rel|q_1 q_2) = \frac{Rdf(q_1 q_2)}{Rdf(q_1 q_2) + Rdf(q_2 q_1)}$$

$$p(Rel|\text{reversed match}) = p(Rel|q_2 q_1) = \frac{Rdf(q_2 q_1)}{Rdf(q_1 q_2) + Rdf(q_2 q_1)}$$

where $Rdf(q_1 q_2)$ is the relevant document frequency of the two terms q_1 and q_2 appearing in order inside a window of size 5. The results suggest the probability of relevance for a document having the terms in the same order as query is more than a document that has them in reverse order. For all four datasets, the difference is statistically significant using the two tailed paired t-test computed at a 95% confidence level.

2.2   Definition of the Query Term Order Constraint

This constraint is defined to capture term order in documents and queries which is lost in existing retrieval models. If the semantics of a pair of terms in a query differs when their order is reversed, this constraint will ensure that a document with these terms in the correct order will have a higher relevance score than a document which has them in the reverse order.

Formally, let $D = \langle w_1, \cdots, w_m \rangle$ be a document where w_i is the term at position i and $Q = \langle q_1, q_2 \rangle$ be a query such that $sem(q_1\ q_2) \neq sem(q_2\ q_1)$. Where $sem(q_1\ q_2)$ denotes the semantic meaning of the phrase "$q_1\ q_2$". The above equation indicates that the semantic meaning of the phrase "$q_1\ q_2$" is not the same as "$q_2\ q_1$".

We then define $D_1 = D\|\langle q_1 q_2\rangle = \langle d_1, \cdots, d_m, q_1, q_2 \rangle$ and $D_2 = D\|\langle q_2, q_1 \rangle = \langle d_1, \cdots, d_m, q_2, q_1 \rangle$. We expect that $S(D_2, Q) \leq S(D_1, Q)$ where $S(D, Q)$ denotes the relevance score of document D with respect to query Q. Based on this constraint, we want the retrieval function to give a higher score to a document which has the two query terms in the same order as the query.

2.3   Modification of Existing Retrieval Methods

To the best of our knowledge, no existing retrieval model satisfies the proposed constraint. In this section, we select SDM [8] and PLM [7] as examples of n-gram retrieval and robust passage retrieval models respectively, and then modify them so as to satisfy the proposed term order constraint.

The proposed constraint imposes a stipulation that we should only consider term order for terms whose order is semantically important so we need to define a function that captures whether the order of two query terms is important or not. For this purpose we define:

$$sem(w, w') = \left| 1/2 - \frac{Df(w, w')}{Df(w, w') + Df(w', w)} \right| \tag{1}$$

where $Df(w, w')$ is the frequency of documents that contain terms w and w' in this order in a window of specific size. This function ranges from 0 to $1/2$. When the difference between document frequency of ww' and $w'w$ is large, we can conclude that different orders of these two terms are pointing to different concepts and the sem function will evaluate to a value close to $1/2$. But when the difference between document frequencies for the different orders of these two terms is not high we do not have enough evidence to decide with certainty whether the different orders are pointing to different concepts and the sem function evaluates to 0. We call this simple and computationally efficient function, which gives satisfactory results, semantic importance of term order (SITO).

Modification of SDM. The Sequential dependency model is a retrieval function that incorporates both term bigrams and term proximity. The score of a document D with respect to query Q is calculated as:

$$P(D|Q) = \lambda_T \sum_{q \in Q} f_T(q, D) + \lambda_O \sum_{q_i, q_{i+1} \in Q} f_O(q_i, q_{i+1}, D) + \lambda_U \sum_{q_i, q_{i+1} \in Q} f_U(q_i, q_{i+1}, D) \quad (2)$$

where f_T, f_O, and f_U are functions dictating the importance of unigram frequency, ordered bigram frequency, and unordered term co-occurrence frequency within a window, and λ_T, λ_O and λ_U are hyper-parameters.

In order for SDM to satisfy the proposed constraint, it should take into account the order of all term pairs within a window, not just bi-grams. Therefore we add a component to SDM which calculates ordered term co-occurance. The function should also take into account the semantic importance of word order (SITO) when rewarding terms appearing in order. The modified SDM function is as follows:

$$P(D|Q) = \lambda_T \sum_{q \in Q} f_T(q, D) + \lambda_O \sum_{q_i, q_{i+1} \in Q} f_O(q_i, q_{i+1}, D) g(q_i, q_{i+1}) +$$

$$\lambda_U \sum_{q_i, q_{i+1} \in Q} f_U(q_i, q_{i+1}, D) h(q_i, q_{i+1}) + \sum_{\substack{q_i, q_j \in Q, \\ i+1 < j}} \lambda_{OW} f_{OW}(q_i, q_j, D) g(q_i, q_j)$$

where $g(w_1, w_2) = \frac{3}{4} + sem(w_1, w_2)$ and $h(w_1, w_2) = \frac{5}{4} - sem(w_1, w_2)$. We define them as such since sem has a range of $[0, 1/2]$ and larger values indicate term order is semantically important. $g(\cdot, \cdot)$ increases or decreases the weight based on whether term order is semantically important or not. $h(\cdot, \cdot)$ does the opposite. λ_{OW} is the weight we would like to give to the ordered co-occurrence component. f_{OW} is defined as

$$f_{OW}(q_i, q_j, D) = \log \left[\frac{tf_{\#owN(q_i, q_j, D)} + \mu \frac{cf_{\#owN(q_i, q_j)}}{|C|}}{|D| + \mu} \right] \quad \begin{array}{l} \text{weight of ordered (span=N)} \\ \text{window "}q_i\text{", "}q_j\text{" in } D \end{array}$$

Modification of the PLM Model. Before we introduce our modification to PLM, we provide a short overview of the model. Let $D = (w_1, w_2, \cdots, w_N)$ be a document of size N where w_i shows the ith term of the document. Let $c(w, j)$ be the count of term w at position i in document D (if w occurs at position i, it is 1, otherwise 0) and $k(i, j)$ be the propagated count to position i from a term at position j. PLM defines the total propagated count of term w at position i from the occurrences of w in all the positions as:

$$c'(w, i) = \sum_{j=1}^{N} c(w, j) k(i, j)$$

Based on this term propagation, PLM has a frequency vector $\langle c'(w_1, i), c'(w_2, i), \cdots, c'(w_N, i) \rangle$ at position i forming a virtual document D'_i. PLM then computes the score of document D'_i using KL-divergence retrieval model. Finally, PLM calculates the overall score of D based on the scores of these virtual documents.

In order for PLM to satisfy our word order constraint, we need to reward documents in which matched query terms appear in order with some other query terms in the document. Therefore, if a term in position i appears in order with another query term, we increase the score the document receives from this term. To achieve this we multiply $c(w, i)$ with a weight that captures the semantic importance of term order.

$$c'(w, i, D, Q) = \sum_{j=1}^{N} c(w, j)k(i, j)\text{weight}(w_j, D, Q)$$

If the term at position j is not in order with any other query terms around position j, this weight will be 1, but if another term appears in order with this query term around position j, the weight will be increased proportionally to the SITO of these two terms (Eq. 1). We define the weight function as follows:

$$\text{weight}(w_j, D, Q) = 1 + \sum_{w' \in Q} \lambda \cdot sem(w_j, w') \cdot I(w_j, w', D, Q)$$

where $I(w_j, w', D, Q)$ is true if w_j and w' appear together in the same order they appear in the query within a specified window size around position j and λ is a free parameter.

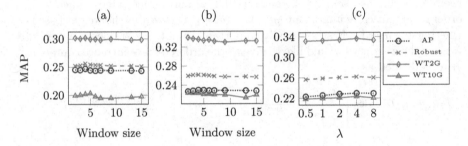

Fig. 1. Figures depict the effects of window size on MAP for SDM-M, window size on MAP for PLM-M, and λ on MAP for PLM-M respectively

3 Experiments

In this section, we evaluate our proposed modified versions of SDM and PLM. We used four standard TREC collections in our experiment: AP88-89, Robust, WT2G, and WT10G. The first two collections are news collections, and the last two are web collections with more noisy documents. We take the titles of topics as queries. We stem the documents and queries using the Porter stemmer. The experiments on PLM and SDM were carried out on the Lemur toolkit and the Galago toolkit respectively[1].

We use mean average precision (MAP) of the top 1000 ranked documents as our evaluation metric. Statistical significance testing is performed using two-tailed paired t-test at a 95% confidence level.

---

[1] http://www.lemurproject.org/.

Table 1. Comparison of the modified retrieval methods with the baselines.

	AP		Robust		WT2g		WT10g	
	MAP	P@10	MAP	P@10	MAP	P@10	MAP	P@10
SDM	0.2358	0.3483	0.2472	0.4149	0.2902	0.4300	0.1951	0.2369
SDM-M	0.2446*	0.3839*	0.2562*	0.4197*	0.3021*	0.4260	0.2034*	0.2500*
PLM	0.2198	0.3483	0.2538	0.4305	0.3287	0.4520	0.2073	0.2640
PLM-M	0.2299*	0.3678*	0.2619*	0.4378	0.3364*	0.4520	0.2236*	0.2560

3.1   Evaluation of Modified Methods

We compare each modified method with its unmodified counterpart as the baseline. The results are summarized in Table 1. Modified methods result in a statistically significant improvement for all four datasets. The modifications have a greater effect on the WT2G and WT10g datasets. This is most likely due to the fact that AP88-89 and Robust are homogeneous collections but WT2G and WT10G are heterogeneous and therefore noisier. As reported previously in [7] term dependency information is more helpful on noisy datasets.

We conducted the experiment with window sizes between 2 and 15. Figure 1a and b shows the sensitivity of MAP to the window size parameter for SDM-M and PLM-M. The best results are achieved at a window size of around 4. This is expected as term order between distant terms is meaningless and small windows sizes fail to detect semantic importance between all terms and therefore lose some information.

Figure 1c shows the sensitivity of the modified PLM method to parameter λ. Increasing this parameter to large numbers increases document scores by an unreasonable amount and if we choose a very small value for this parameter, changes to document score will be ineffective. The best choice for all four datasets is to set this parameter to 4. To see whether different values of λ may affect window size, for each window size we further compared the results of different values of λ and observed that the effect of λ on MAP is unaffected by window size, and the best choice for λ for any window size is still 4.

4   Conclusions

In this paper we used the axiomatic framework to propose a query term order constraint for ad-hoc retrieval which states that if the order of two query terms is semantically important, a document that has these two terms in the same order as the query should get a higher score compared to a document that has them in the reverse order. Furthermore, we proposed modifications to two well-known and robust information retrieval methods SDM and PLM so as to satisfy the proposed constraint. Experimental results show the proposed modifications cause a significant improvement over the baselines and a window of size 4 is the best choice for considering term order dependency. A future research direction is to search for a better SITO function and a more integrated way to make state-of-the-art retrieval methods satisfy the proposed constraint.

References

1. Bendersky, M., Croft, W.B.: Modeling higher-order term dependencies in information retrieval using query hypergraphs. In: Proceedings of the 35th ACM SIGIR Conference, pp. 941–950. ACM (2012)
2. Bendersky, M., Metzler, D., Croft, W.B.: Learning concept importance using a weighted dependence model. In: Proceedings of the Third ACM WSDM Conference, pp. 31–40 (2010)
3. Cummins, R., O'Riordan, C.: An axiomatic comparison of learned term-weighting schemes in information retrieval: clarifications and extensions. Artif. Intell. Rev. **28**(1), 51–68 (2007)
4. Fang, H., Tao, T., Zhai, C.: A formal study of information retrieval heuristics. In: Proceedings of the 27th Annual ACM SIGIR Conference, pp. 49–56. ACM (2004)
5. Fang, H., Zhai, C.: Semantic term matching in axiomatic approaches to information retrieval. In: Proceedings of the 29th Annual International ACM SIGIR Conference on Research and Development in Information Retrieval, pp. 115–122. ACM (2006)
6. Huston, S., Croft, W.B.: A comparison of retrieval models using term dependencies. In: Proceedings of the 23rd ACM CIKM, pp. 111–120. ACM (2014)
7. Lv, Y., Zhai, C.: Positional language models for information retrieval. In: Proceedings of the 32nd ACM SIGIR Conference, pp. 299–306. ACM (2009)
8. Metzler, D., Croft, W.B.: A Markov random field model for term dependencies. In: Proceedings of the 28th Annual International ACM SIGIR Conference on Research and Development in Information Retrieval, pp. 472–479. ACM (2005)
9. Montazeralghaem, A., Zamani, H., Shakery, A.: Axiomatic analysis for improving the log-logistic feedback model. In: Proceedings of the 39th ACM SIGIR Conference, pp. 765–768. SIGIR 2016. ACM (2016)
10. Peng, J., Macdonald, C., He, B., Plachouras, V., Ounis, I.: Incorporating term dependency in the DFR framework. In: Proceedings of the 30th ACM SIGIR Conference, pp. 843–844. ACM (2007)
11. Robertson, S.E., Walker, S.: Some simple effective approximations to the 2-Poisson model for probabilistic weighted retrieval. In: Croft, B.W., van Rijsbergen, C.J. (eds.) SIGIR 1994, pp. 232–241. Springer, New York (1994). https://doi.org/10.1007/978-1-4471-2099-5_24
12. Yu, C.T., Buckley, C., Lam, K., Salton, G.: A generalized term dependence model in information retrieval. Technical report, Cornell University (1983)

Deep Neural Networks for Query Expansion Using Word Embeddings

Ayyoob Imani[1(✉)], Amir Vakili[1], Ali Montazer[2], and Azadeh Shakery[1]

[1] University of Tehran, Tehran, Iran
{ayyoub.imani,shakery,a_vakili}@ut.ac.ir
[2] University of Massachusetts Amherst, Amherst, USA
montazer@umass.edu

Abstract. Query expansion is a method for alleviating the vocabulary mismatch problem present in information retrieval tasks. Previous works have shown that terms selected for query expansion by traditional pseudo-relevance feedback methods such as mixture model are not always helpful to the retrieval process. In this paper, we show that this is also true for more recently proposed embedding-based query expansion methods. We then introduce an artificial neural network classifier, which uses term word embeddings as input, to predict the usefulness of query expansion terms. Experiments on four TREC newswire and web collections show that using terms selected by the classifier for expansion significantly improves retrieval performance compared to competitive baselines. The results are also shown to be more robust than the baselines.

Keywords: Query expansion · Word embeddings · Siamese network

1 Introduction

Query expansion is a method for alleviating the vocabulary mismatch problem present in information retrieval tasks. This is a fundamental problem where users and authors often use different terms describing the same concepts. In this paper, we aim to differentiate terms helpful to query expansion through the use of an artificial neural network classifier.

Various methods for selecting expansion words exist that leverage various data sources and employ very different principles and techniques [3]. Though, not all of the terms suggested by these methods are actually helpful. For instance, [2] showed that many of the terms extracted using the mixture model method have a negative effect on the retrieval process. [2] divided terms into three categories: good, bad and neutral depending on whether their addition to the query improved, diminished or had no effect on retrieval performance. [2] then used a feature vector to train a classifier for separating good expansion terms from bad ones. This feature vector included the following: terms distribution in the feedback documents and terms distribution in the whole collection, co-occurrences of

© Springer Nature Switzerland AG 2019
L. Azzopardi et al. (Eds.): ECIR 2019, LNCS 11438, pp. 203–210, 2019.
https://doi.org/10.1007/978-3-030-15719-7_26

the expansion term with the original query terms, and proximity of the expansion terms to the query terms.

Another approach for improving the selection of query expansion terms is considering the semantic similarity of a candidate term with the query terms [1,11,14,21,22]. [14] proposed a semantic similarity constraint for PRF methods and showed that adhering to it improves retrieval performance. [11] proposed to expand queries with terms semantically related to query terms. To this end, they trained word embeddings on document corpora using the Word2Vec Continuous bag of words approach [13]. These vectors are both semantic and syntactical representations of their corresponding words. [11] then uses these word embeddings to expand queries either by adding terms closest to the centroid of the query word embedding vectors (referred to in following sections as AWE) or selecting terms closest to individual query terms in the word embedding space.

Another approach is embedding based query expansion models [21] where two models were proposed, the first of which assumes that query terms are independent of each other (referred to in following sections as EQE1), and the second assumes that the semantic similarity between two given terms is independent of the query. In another related work, the local training of word embeddings on retrieved documents was used to improve query expansion effectiveness [6]. Using supervised training and word embeddings to learn term weights to be used in retrieval models such as BM25 has also been proposed [24].

In this paper, We propose an artificial neural network (ANN) model for selecting suitable expansion terms. We use a siamese neural network architecture inspired by [10] in order to lessen the impact of limited training data. Siamese network architectures have been gaining popularity in recent years in the information retrieval community [7,9,17,19,20] and have achieved impressive performance in various tasks. Using pre-trained word embedding of terms, this network learns whether a term is semantically suitable for expanding a query. Our neural network approach intends to go beyond simple vector similarity and learns the latent features present within word embeddings responsible for term effectiveness or ineffectiveness when used for query expansion. In short, the main advantage of the proposed method is that it no longer requires manual feature design and uses a data-driven approach for selecting expansion terms.

We evaluate the effectiveness of our approach on four TREC collections. We compare results with traditional approaches and more recent methods. Results show incorporating terms identified as suitable expansion terms by our ANN model into the retrieval process significantly improves retrieval performance. We also show that our proposed method is more robust compared to the baselines. The rest of this paper is organized as follows: In Sect. 2 we discuss [2]'s method for labeling expansion terms in greater detail, Sect. 3 we introduce our classifier model and explain its integration with the retrieval process, Sects. 4 and 5, we present the experiments and results.

Table 1. Query expansion term statistics for used collections

	Embedding-based				Pseudo relevance docs				QLM (MAP)	QLM +MM (MAP)
	Good (%)	Neutral (%)	Bad (%)	Oracle (MAP)	Good (%)	Neutral (%)	Bad (%)	Oracle (MAP)		
AP	3.8	55.5	40.5	0.2981	16.2	53.6	30.1	0.3983	0.2206	0.2749
Robust	5.6	62.4	31.9	0.3122	21.9	55.4	22.6	0.4021	0.2176	0.2658
WT2g	6.2	37.8	55.8	0.3442	15.7	61.3	22.9	0.4383	0.2404	0.2593
WT10g	3.4	75.6	20.9	0.2410	14.1	63.6	22.2	0.2871	0.1837	0.1902

2   Good, Bad, and Neutral Expansion Terms

In order to identify terms helpful to query expansion, we follow [2] and divide candidate terms into three classes: good, bad and neutral. For a particular query, a good/bad term will increase/decrease retrieval performance and neutral terms have no effect. For examining the potential impact of selecting only good terms for query expansion, we perform query expansion using pseudo-relevance and embedding based methods and calculate the ratio of good/bad/neutral terms selected. We then calculate the MAP of a hypothetical oracle retrieval model that only uses good terms for expansion. Embedding based selects top 1000 closest words to the average word embedding of the query and pseudo-relevance uses terms in the top 10 pseudo-relevant documents. The results are shown in Table 1. The small percentage of good terms returned by embedding models may be indicative of their relatively weaker performance compared to PRF methods.

For training our classifier, we require a set of queries and candidate expansion terms labelled as Good/Bad/Neutral. Therefore, we take each query of a corpus and first average the embedding vectors of its query terms; this approach has been proposed in [13] and used in other works such as [11,21]. Then, we use cosine-similarity to find the top 1000 terms that are closest to the averaged vector. Finally, to identify the class of a term, we use the method proposed in [2]. Briefly, we add the expansion to the query and perform retrieval using query likelihood method; if the mean average precision increases/decreases, the term is a good/bad expansion term. If the change is not tangible, the term is neutral.

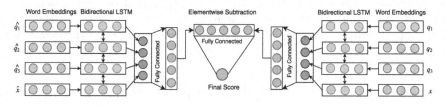

Fig. 1. Architecture of the proposed siamese network. The architecture consists of two identical models projecting two separate inputs (query and expansion term pair) into a common embedding space and then comparing the two projections to get a final similarity score. This score tells us whether the candidate expansion terms of the two classes belong to the same class (good, bad or neutral).

3 Expansion Term Classification

In this section, we first present the model used for the classification task, then we explain how the classifier results are integrated into the retrieval process.

3.1 Problem Formulation

Suppose we have a dataset $\mathcal{D} = \{(q_i, x_i, l_i)\}_{i=1}^N$ where $q_i = \{q_{i,1}, \cdots, q_{i,k}\}$ represents a query and x_i is a candidate expansion term. $l_i \in \{\text{Good}, \text{Neutral}, \text{Bad}\}$ is the label denoting the expansion terms effectiveness for the query. Our goal is to learn a model $g(\cdot, \cdot)$, using $\mathcal{D}$ as training data, which will classify any query and candidate expansion term pair (q, x), as either good, neutral or bad.

3.2 Model Overview

We propose the deep expansion classifier (DEC). The architecture is depicted in Fig. 1. A major roadblock when using artificial neural network approaches in information retrieval tasks is the lack of training data. Various methods such as using weak supervision [5] have been put forward for tackling this challenge. In this paper, we use the learning technique proposed in [10]. The *siamese network* was proposed to overcome the lack of training data by learning whether two samples are of the same class or not rather than directly predicting which class a sample belongs to. We compare each sample to a random group of previously classified samples and use majority voting for classification.

As we have few queries for our corpora, if we naively train a neural network as a vanilla cross-entropy-loss softmax classifier, it will severely overfit. Such a network will not be able to distinguish suitable expansion terms for a query it has never seen before. The siamese architecture learns what features make a suitable expansion term by learning to differentiate between good and bad samples.

3.3 Modeling the Query and Candidate Expansion Term Relation

Given a query q and a candidate expansion term x, the model maps the terms to their embeddings. A bidirectional long short term memory [8] (BiLSTM) network is then used to construct new embeddings for the sequence. In essence, the BiLSTM receives query term word embeddings and one expansion term as input and outputs contextualized word embeddings for each of them. The outputs of the BiLSTM layer are fed to a fully connected layer. This final layer represents the relation between query terms and the candidate expansion term.

3.4 Expansion Term Classification and Term Re-weighting

The representations of query-expansion term pairs are compared in order to determine whether two pairs belong to the same class or not. This is achieved by an element-wise subtraction of the two representations and the result being fed into a fully connected layer which outputs a final score.

To determine whether a candidate expansion term is good for a query, we feed the pair along with a set of query-expansion term pairs whose classes are known. We calculate the probability of a candidate expansion term being good as $P(l = \text{Good}|(q, x)) = \frac{N_g + N_{nb}}{N}$ where N is the total number of pairs we compare to and $N_g + N_{nb}$ is the number of times the pair being tested was determined to be in the same class a good pair or a different class than a bad pair.

Finally, we re-weight the expansion term weights obtained using the AWE method by multiplying them by $P(l = \text{Good}|(q, x))$. The AWE method weights the candidate expansion terms based on the cosine similarity of the expansion term embedding and the query term embeddings centroid. So the final weight for an expansion term will be: $(1 + \alpha \cdot p(l = \text{Good})) \cdot \delta(\bar{q}, x)$ where δ is the cosine similarity function and α is a hyper-parameter.

4   Experiment

For evaluating the proposed method we use four standard TREC collections: AP (Associated Press 1988–1989), Robust (TREC Robust Track 2004), WT2G and WT10G (TREC Web track 1999, 2000–2001). The first two collections contain news articles and the second two are general web crawls. We used the title of topics as queries. The words are stopped using the inquiry stop word list and no stemming was used. We used pre-trained word embeddings with a dimension of 200 extracted using the GloVe [16] method on a 6 billion token collection.

The parameters were updated with the Adam algorithm using a learning rate of 0.001 and mini-batch size of 32. Dimensions for the BiLSTM hidden vector and final representation vector are 200 and 400 respectively. We use k-fold cross validation and average the evaluation metrics.

4.1   Comparison Approaches

For evaluation, we only compare to methods that select expansion term candidates based on only word embeddings (no initial retrieval required). As we use general purpose word embeddings, we also do not compare to methods that train embeddings specifically for query expansion such as [18,23]. We consider three baselines: (1) the standard query likelihood model using maximum likelihood estimation, (2) AWE where expansion terms closest to the centroid of query term embeddings are selected [11], and (3) EQE1 where expansion terms are scored by their multiplicative similarity to the query terms [21].

4.2   Evaluation Metrics

We use mean average precision (MAP) and precision for the top 10 retrieved documents (P@10) as our evaluation metrics. Statistical significance tests are performed using two-tailed paired t-test at a 95% confidence level. For evaluating robustness of the proposed method we use robustness index (RI) [4] which is defined as $\frac{N_+ - N_-}{|Q|}$ where $|Q|$ is the number of queries and N_+/N_- are the number of queries which have improved/diminished compared to the baseline.

Table 2. Evaluation results on the four datasets. The superscripts 1/2/3 denote that the MAP improvements over QLM/AWE/EQE1 are statistically significant. The highest value in each column is marked in bold.

	AP			Robust			WT2g			WT10g		
	MAP	P@10	RI	MAP	P@10	RI	MAP	P@10	RI	MAP	P@10	RI
QLM	0.2206	0.3432	-	0.2176	0.3856	-	0.2404	0.4167	-	0.1837	0.2420	-
AWE	0.2312^1	0.3392	0.12	0.2230^1	0.3899	0.10	0.2456^1	0.4169	0.08	0.1849	0.2399	0.10
EQE1	0.2344^{12}	**0.3442**	0.29	0.2278^{12}	0.4016	0.25	0.2463^1	0.4188	**0.17**	0.1867	0.2432	0.18
DEC	$\mathbf{0.2403^{123}}$	0.3434	**0.31**	$\mathbf{0.2358^{123}}$	**0.4057**	0.31	$\mathbf{0.2489^1}$	**0.4213**	0.16	$\mathbf{0.1891^{12}}$	**0.2434**	**0.20**

5　Results and Discussion

Table 2 presents the performance of the baselines and our proposed method. These methods expand queries with semantically related terms. The results show that the query expansion classifier DEC outperforms all baselines in terms of MAP. For all three embeddings-based methods, the performance gains are more pronounced in the two newswire collections. This may be due to the fact that these collections are more homogeneous than the web corpora. Web corpora will generally be noisier which in turn will affect classifier performance. Another reason could be due to the fact that our word-embeddings (GloVe) were pre-trained on Wikipedia and newswire articles. This would make them more suitable for use in newswire collections. Using word embeddings pre-trained on common crawl data may yield better performance in web corpora.

To summarize, the proposed method outperforms other state-of-the-art methods utilizing word embeddings for query expansion. The results are also more robust than previous approaches. This indicates that for selecting candidate expansion terms, simple similarity functions such as cosine similarity can be improved upon using ANN classifiers.

6　Conclusion

In this paper, we proposed a neural network architecture for classifying terms based on their effectiveness in query expansion. The neural network uses only pre-trained word embeddings and no manual feature selection or initial retrieval is necessary. We evaluated the proposed methods using four TREC collections. The results showed that the proposed method significantly outperforms other word embedding based approaches and traditional pseudo-relevance feedback. The method is also shown to be more robust compared to the baselines. For future work, one possible method is integrating topic vectors trained using methods such as Latent Dirichlet Allocation into the classification process. Another is using word embeddings trained using methods other than Word2Vec and GloVe (such as Lda2Vec [15] or Paragraph2Vec [12]) or training on domain-specific corpora.

References

1. AlMasri, M., Berrut, C., Chevallet, J.-P.: A comparison of deep learning based query expansion with pseudo-relevance feedback and mutual information. In: Ferro, N., et al. (eds.) ECIR 2016. LNCS, vol. 9626, pp. 709–715. Springer, Cham (2016). https://doi.org/10.1007/978-3-319-30671-1_57

2. Cao, G., Nie, J.Y., Gao, J., Robertson, S.: Selecting good expansion terms for pseudo-relevance feedback. In: Proceedings of the 31st Annual International ACM SIGIR Conference on Research and Development in Information Retrieval, pp. 243–250. ACM (2008)

3. Carpineto, C., Romano, G.: A survey of automatic query expansion in information retrieval. ACM Comput. Surv. (CSUR) 44(1), 1 (2012)

4. Collins-Thompson, K.: Reducing the risk of query expansion via robust constrained optimization. In: Proceedings of the 18th ACM Conference on Information and Knowledge Management, pp. 837–846. ACM (2009)

5. Dehghani, M., Zamani, H., Severyn, A., Kamps, J., Croft, W.B.: Neural ranking models with weak supervision. In: Proceedings of the 40th International ACM SIGIR Conference on Research and Development in Information Retrieval, pp. 65–74. ACM (2017)

6. Diaz, F., Mitra, B., Craswell, N.: Query expansion with locally-trained word embeddings. arXiv preprint arXiv:1605.07891 (2016)

7. He, H., Lin, J.: Pairwise word interaction modeling with deep neural networks for semantic similarity measurement. In: Proceedings of the 2016 Conference of the North American Chapter of the Association for Computational Linguistics: Human Language Technologies, pp. 937–948 (2016)

8. Hochreiter, S., Schmidhuber, J.: Long short-term memory. Neural Comput. 9(8), 1735–1780 (1997)

9. Huang, P.S., He, X., Gao, J., Deng, L., Acero, A., Heck, L.: Learning deep structured semantic models for web search using clickthrough data. In: Proceedings of the 22nd ACM International Conference on Conference on Information & Knowledge Management, pp. 2333–2338. ACM (2013)

10. Koch, G., Zemel, R., Salakhutdinov, R.: Siamese neural networks for one-shot image recognition. In: ICML Deep Learning Workshop, vol. 2 (2015)

11. Kuzi, S., Shtok, A., Kurland, O.: Query expansion using word embeddings. In: Proceedings of the 25th ACM International on Conference on Information and Knowledge Management, pp. 1929–1932. ACM (2016)

12. Le, Q., Mikolov, T.: Distributed representations of sentences and documents. In: International Conference on Machine Learning, pp. 1188–1196 (2014)

13. Mikolov, T., Chen, K., Corrado, G., Dean, J.: Efficient estimation of word representations in vector space. arXiv preprint arXiv:1301.3781 (2013)

14. Montazeralghaem, A., Zamani, H., Shakery, A.: Axiomatic analysis for improving the log-logistic feedback model. In: Proceedings of the 39th International ACM SIGIR Conference on Research and Development in Information Retrieval, pp. 765–768. ACM (2016)

15. Moody, C.E.: Mixing Dirichlet topic models and word embeddings to make lda2vec. arXiv preprint arXiv:1605.02019 (2016)

16. Pennington, J., Socher, R., Manning, C.: Glove: global vectors for word representation. In: Proceedings of the 2014 Conference on Empirical Methods in Natural Language Processing (EMNLP), pp. 1532–1543 (2014)

17. Severyn, A., Moschitti, A.: Learning to rank short text pairs with convolutional deep neural networks. In: Proceedings of the 38th International ACM SIGIR Conference on Research and Development in Information retrieval, pp. 373–382. ACM (2015)
18. Sordoni, A., Bengio, Y., Nie, J.Y.: Learning concept embeddings for query expansion by quantum entropy minimization. In: AAAI, vol. 14, pp. 1586–1592 (2014)
19. Wang, S., Jiang, J.: A compare-aggregate model for matching text sequences. arXiv preprint arXiv:1611.01747 (2016)
20. Yang, L., Zamani, H., Zhang, Y., Guo, J., Croft, W.B.: Neural matching models for question retrieval and next question prediction in conversation. arXiv preprint arXiv:1707.05409 (2017)
21. Zamani, H., Croft, W.B.: Embedding-based query language models. In: Proceedings of the 2016 ACM International Conference on the Theory of Information Retrieval, pp. 147–156. ACM (2016)
22. Zamani, H., Croft, W.B.: Estimating embedding vectors for queries. In: Proceedings of the 2016 ACM International Conference on the Theory of Information Retrieval, pp. 123–132. ACM (2016)
23. Zamani, H., Croft, W.B.: Relevance-based word embedding. In: Proceedings of the 40th International ACM SIGIR Conference on Research and Development in Information Retrieval, pp. 505–514. ACM (2017)
24. Zheng, G., Callan, J.: Learning to reweight terms with distributed representations. In: Proceedings of the 38th International ACM SIGIR Conference on Research and Development in Information Retrieval, pp. 575–584. ACM (2015)

Demonstration Papers

Online Evaluations for Everyone: Mr. DLib's Living Lab for Scholarly Recommendations

Joeran Beel[1,2(✉)], Andrew Collins[1], Oliver Kopp[3], Linus W. Dietz[4], and Petr Knoth[5]

[1] School of Computer Science and Statistics, ADAPT Centre,
Trinity College Dublin, Dublin, Ireland
{beelj, ancollin}@tcd.ie

[2] Digital Content and Media Sciences Division, National Institute of Informatics,
Tokyo, Japan

[3] IPVS, University of Stuttgart, Stuttgart, Germany
kopp@informatik.uni-stuttgart.de

[4] Department of Informatics, Technical University of Munich,
Garching, Germany
linus.dietz@tum.de

[5] Knowledge Media Institute, The Open University, London, UK
petr.knoth@open.ac.uk

Abstract. We introduce the first 'living lab' for scholarly recommender systems. This lab allows recommender-system researchers to conduct online evaluations of their novel algorithms for scholarly recommendations, i.e., recommendations for research papers, citations, conferences, research grants, etc. Recommendations are delivered through the living lab's API to platforms such as reference management software and digital libraries. The living lab is built on top of the recommender-system as-a-service Mr. DLib. Current partners are the reference management software JabRef and the CORE research team. We present the architecture of Mr. DLib's living lab as well as usage statistics on the first sixteen months of operating it. During this time, 1,826,643 recommendations were delivered with an average click-through rate of 0.21%.

Keywords: Recommender system evaluation · Living lab · Online evaluation

1 Introduction

'Living labs' for recommender systems enable researchers to evaluate their recommendation algorithms with real users in realistic scenarios. Such living labs – sometimes also called 'Evaluations-as-a-Service' [1–3] – are usually built on top of production recommender systems in real-world platforms such as news websites [4]. Via an API, external researchers can 'plug-in' their experimental recommender systems

---

This publication has emanated from research conducted with the financial support of Science Foundation Ireland (SFI) under Grant Number 13/RC/2106. We are further grateful for the support received by Samuel Pearce and Siddharth Dinesh.

© Springer Nature Switzerland AG 2019
L. Azzopardi et al. (Eds.): ECIR 2019, LNCS 11438, pp. 213–219, 2019.
https://doi.org/10.1007/978-3-030-15719-7_27

to the living lab. When recommendations for users of the platform are needed, the living lab sends a request to the researcher's experimental recommender system. This system then returns a list of recommendations that are displayed to the user. The user's actions (clicks, downloads, purchases, etc.) are logged and can be used to evaluate the recommendation algorithms' effectiveness.

Living labs are available in information retrieval and for many recommender-system domains, particularly news [4–6], and they attracted dedicated workshops [7]. There is also work on living labs in the context of search and browsing behavior in digital libraries [8]. However, to the best of our knowledge, there are no living labs for scholarly recommendations, i.e., recommendations for research articles [9, 10], citations [11, 12], conferences [13, 14], reviewers [15, 16], quotes [17], research grants, or collaborators [18]. Consequently, researchers in the field of scholarly recommender systems predominately rely on offline evaluations, which tend to be poor predictors of how algorithms will perform in a production recommender system [19, 20].

In this paper, we present the first living lab for scholarly recommendations, built on top of *Mr. DLib*, a scholarly recommendations-as-a-service provider [21, 22]. Mr. DLib's main feature is to provide third parties such as digital libraries with recommendations for their users. This way, digital libraries do not need to maintain their own recommender system, which would usually be costly and require advanced skills in machine learning and recommender systems. So far, Mr. DLib relied only on its own recommender system to generate recommendations [21, 22]. The system was not open to external researchers. The newly added living lab opens Mr. Lib and provides an environment for any researcher in the field of scholarly recommendations to evaluate novel recommendation algorithms with real users in addition to, or instead of, conducting offline evaluations.

2   Mr. DLib's Scholarly Living Lab

Mr. DLib's living lab is open for two types of partners. First, platform operators, who want to provide their users with scholarly recommendations. Second, research partners, who want to evaluate their novel scholarly recommendation algorithms with real users. The current platform partner of Mr. DLib is the reference-management software JabRef [23, 24]. The current research partner of Mr. DLib is CORE [25–27]. Mr. DLib acts as an intermediate between these partners. Mr. DLib also operates its own internal recommendation engine, which applies content-based filtering with terms, key-phrases, and word embeddings as well as stereotype and most-popular recommendations [22, 28]. Thus, Mr. DLib's internal recommendation engine establishes a baseline for research partners to compare their novel algorithms against.

The workflow of Mr. DLib's living lab is illustrated in Fig. 1: (1) A JabRef user selects a source article in the list, and then selects the "Related Articles" tab; JabRef sends a request to Mr. DLib's API. The request comprises of the selected article's title. Mr. DLib's API accepts the request, and its A/B engine randomly forwards the request either to (2a) Mr. DLib's internal recommender system or (2b) to CORE's recommender system. CORE or Mr. DLib's internal recommender system creates a list of recommendations and (3) returns them to JabRef, which displays them to the user.

(4) When a user clicks a recommendation, a notification it sent to Mr. DLib for evaluation purposes.

While, currently, Mr. DLib only has one research and one platform partner, there will potentially be numerous such partners in the future. Mr. DLib's living lab is open to any research partner whose experimental recommender system recommends scholarly items; is available through a REST API; accepts a string as input (typically a source article's title); and returns a list of related-articles including URLs to web pages on which the recommended articles can be downloaded, preferably open access. Also, recommendations must be returned within less than 2 s.

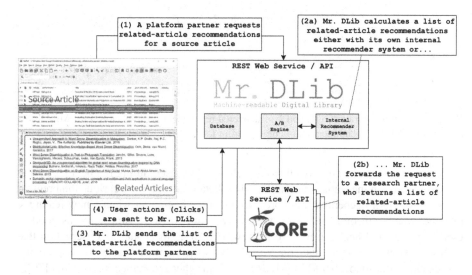

Fig. 1. Illustration of the recommendation process

All data on Mr. DLib's recommendations is available publicly [29]. This data can be used to replicate our calculations and perform additional analyses. JabRef's client software, including the recommender system, can be downloaded at http://jabref.org. Source code of the API is available on http://mr-dlib.org.

3 Usage Statistics

Mr. DLib started its general recommendation service in 2016 [21] and its living lab in June 2017. The living lab was integrated first in a beta version of JabRef. During the beta phase (until September 2017), JabRef sent around 4,200 requests per month to Mr. DLib (Fig. 2). For each request, Mr. DLib returned typically 6 recommendations (25 k recommendations in total), whereas between 20% to 30% of the recommendations were generated by CORE, and the remaining by Mr. DLib's internal recommendation engine. Click-through rate (**CTR**) on the recommendations decreased from 0.76% in June to 0.34% in September (Fig. 2). After the beta phase, i.e., from October 2017 on,

216   J. Beel et al.

the number of delivered recommendations increased to around 150 k per month, again with 20% to 30% of the recommendations generated by CORE. The overall click-through rate decreased to around 0.18% but remained stable until today.

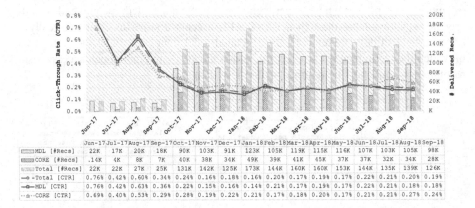

	Jun-17	Jul-17	Aug-17	Sep-17	Oct-17	Nov-17	Dec-17	Jan-18	Feb-18	Mar-18	Apr-18	May-18	Jun-18	Jul-18	Aug-18	Sep-18
MDL [#Recs]	22K	17K	20K	18K	90K	103K	91K	123K	105K	119K	115K	116K	107K	103K	105K	98K
CORE [#Recs]	.14K	4K	8K	7K	40K	38K	34K	49K	39K	41K	45K	37K	37K	32K	34K	28K
Total [#Recs]	22K	22K	27K	25K	131K	142K	125K	173K	144K	160K	160K	153K	144K	135K	139K	126K
—o —Total [CTR]	0.76%	0.42%	0.60%	0.34%	0.24%	0.16%	0.18%	0.16%	0.20%	0.17%	0.19%	0.17%	0.22%	0.21%	0.20%	0.19%
—□—MDL [CTR]	0.76%	0.42%	0.63%	0.36%	0.22%	0.15%	0.16%	0.14%	0.21%	0.17%	0.19%	0.17%	0.22%	0.21%	0.18%	0.18%
··△··CORE [CTR]	0.69%	0.40%	0.53%	0.29%	0.28%	0.19%	0.22%	0.21%	0.17%	0.18%	0.20%	0.17%	0.21%	0.21%	0.27%	0.24%

Fig. 2. Click-through rate (CTR) and # of delivered recommendation in JabRef for Mr. DLib's (MDL) and CORE's recommendation engine and in total.

We can only speculate why click-through rate decreased during the beta phase and decreased again in the stable version. Possibly, beta users are more curious than regular users. Maybe users generally are more curious in the beginning when a new feature is released. Maybe, recommendations worsen over time, or were simply not as good as users expected and hence users lose interest. However, we made the observation that CTR decreases over time also on Mr. DLib's other partner platforms that do not participate in the living lab [22, 28], as well as in other recommender systems [30].

Interestingly, click-through rates for both CORE and Mr. DLib's internal recommendation engine are almost identical over the entire data collection period. Both systems mostly use Apache Lucene for their recommendation engine, yet there are notable differences in the algorithms and document corpora. We will not elaborate further on the implementations but refer the interested reader to [22, 27, 28]. The interesting point here is that two separately implemented recommender systems perform almost identically. It is also interesting that the click-through rate in the reference management software JabRef (0.18%) is quite similar to the click-through rate in the social-science repository Sowiport [28, 31–33], although the two platforms differ notably.

4   Future Work

In the long-run, we hope to provide a platform to the information retrieval, digital library, and recommender systems community that helps conducting more reproducible and robust research in real-world scenarios [34, 35]. To achieve this, we plan to add more partners on both sides – platform partners who provide access to real users, and research partners who evaluate their novel algorithms via the living lab. We also aim for personalized recommendations in addition to the current focus on related-article recommendations. We will also enable the recommendation of other scholarly items such as research grants, or research collaborators. We also plan to develop a more automatic process for the integration of partners, with standard protocols and data formats, and pre-implemented clients, to ease the process. Another major challenge in the future will be to select the best algorithms for each platform partner [36].

References

1. Hanbury, A., et al.: Evaluation-as-a-service: overview and outlook. arXiv preprint arXiv: 1512.07454 (2015)
2. Hopfgartner, F., et al.: Report on the evaluation-as-a-service (EaaS) expert workshop. In: ACM SIGIR Forum, pp. 57–65. ACM (2015)
3. Hopfgartner, F., et al.: Evaluation-as-a-service for the computational sciences: overview and outlook. J. Data Inf. Qual. (JDIQ) **10**, 15 (2018)
4. Brodt, T., Hopfgartner, F.: Shedding light on a living lab: the CLEF NEWSREEL open recommendation platform. In: Proceedings of the 5th Information Interaction in Context Symposium, pp. 223–226. ACM (2014)
5. Hopfgartner, F., et al.: Benchmarking news recommendations: the clef newsreel use case. In: ACM SIGIR Forum, pp. 129–136. ACM (2016)
6. Kille, B., et al.: Overview of NewsREEL'16: multi-dimensional evaluation of real-time stream-recommendation algorithms. In: Fuhr, N., et al. (eds.) CLEF 2016. LNCS, vol. 9822, pp. 311–331. Springer, Cham (2016). https://doi.org/10.1007/978-3-319-44564-9_27
7. Balog, K., Elsweiler, D., Kanoulas, E., Kelly, L., Smucker, M.D.: Report on the CIKM workshop on living labs for information retrieval evaluation. In: ACM SIGIR Forum, pp. 21–28. ACM (2014)
8. Carevic, Z., Schüller, S., Mayr, P., Fuhr, N.: Contextualised browsing in a digital library's living lab. In: Proceedings of the 18th ACM/IEEE on Joint Conference on Digital Libraries, Fort Worth, Texas, USA, pp. 89–98. ACM (2018)
9. Li, S., Brusilovsky, P., Su, S., Cheng, X.: Conference paper recommendation for academic conferences. IEEE Access **6**, 17153–17164 (2018)
10. Vargas, S., Hristakeva, M., Jack, K.: Mendeley: recommendations for researchers. In: Proceedings of the 10th ACM Conference on Recommender Systems, Boston, Massachusetts, USA, pp. 365–365. ACM (2016)
11. Färber, M., Thiemann, A., Jatowt, A.: CITEWERTs: a system combining cite-worthiness with citation recommendation. In: Pasi, G., Piwowarski, B., Azzopardi, L., Hanbury, A. (eds.) ECIR 2018. LNCS, vol. 10772, pp. 815–819. Springer, Cham (2018). https://doi.org/10.1007/978-3-319-76941-7_82
12. Jia, H., Saule, E.: Graph Embedding for Citation Recommendation, *arXiv preprint* arXiv: 1812.03835 (2018)
13. Beierle, F., Tan, J., Grunert, K.: Analyzing social relations for recommending academic conferences. In: Proceedings of the 8th ACM International Workshop on Hot Topics in Planet-scale mObile computing and online Social neTworking, pp. 37–42. ACM (2016)
14. Yu, S., Liu, J., Yang, Z., Chen, Z., Jiang, H., Tolba, A., Xia, F.: PAVE: personalized academic venue recommendation exploiting co-publication networks. J. Netw. Comput. Appl. **104**, 38–47 (2018)

15. Kou, N.M., Mamoulis, N., Li, Y., Li, Y., Gong, Z., et al.: A topic-based reviewer assignment system. Proc. VLDB Endowment **8**, 1852–1855 (2015)
16. Lian, J.W., Mattei, N., Noble, R., Walsh, T.: The conference paper assignment problem: using order weighted averages to assign indivisible goods. In: Thirty-Second AAAI Conference on Artificial Intelligence (2018)
17. Tan, J., Wan, X., Liu, H., Xiao, J.: QuoteRec: toward quote recommendation for writing. ACM Trans. Inf. Syst. (TOIS) **36**, 34 (2018)
18. Kong, X., Jiang, H., Wang, W., Bekele, T.M., Xu, Z., Wang, M.: Exploring dynamic research interest and academic influence for scientific collaborator recommendation. Scientometrics **113**, 369–385 (2017)
19. Moreira, G.S.P., de Souza, G.A., da Cunha, A.M.: Comparing offline and online recommender system evaluations on long-tail distributions. In: Proceedings of the ACM Recommender Systems Conference RecSys (2015)
20. Rossetti, M., Stella, F., Zanker, M.: Contrasting offline and online results when evaluating recommendation algorithms. In: Proceedings of the 10th ACM Conference on Recommender Systems, Boston, Massachusetts, USA, pp. 31–34. ACM (2016)
21. Beel, J., Aizawa, A., Breitinger, C., Gipp, B.: Mr. DLib: recommendations-as-a-service (RaaS) for academia. In: Proceedings of the 17th ACM/IEEE Joint Conference on Digital Libraries, Toronto, Ontario, Canada, pp. 313–314. IEEE Press (2017)
22. Beel, J., Collins, A., Aizawa, A.: The architecture of Mr. DLib's scientific recommender-system API. In: Proceedings of the 26th Irish Conference on Artificial Intelligence and Cognitive Science (AICS), CEUR-WS, pp. 78–89 (2018)
23. Feyer, S., Siebert, S., Gipp, B., Aizawa, A., Beel, J.: Integration of the scientific recommender system Mr. DLib into the reference manager JabRef. In: Jose, Joemon M., et al. (eds.) ECIR 2017. LNCS, vol. 10193, pp. 770–774. Springer, Cham (2017). https://doi.org/10.1007/978-3-319-56608-5_80
24. Kopp, O., Breitenbuecher, U., Mueller, T.: CloudRef - towards collaborative reference management in the cloud. In: Proceedings of the 10th Central European Workshop on Services and their Composition (2018)
25. Hristakeva, M., et al.: Building recommender systems for scholarly information. In: Proceedings of the 1st Workshop on Scholarly Web Mining, pp. 25–32. ACM (2017)
26. Knoth, P., et al.: Towards effective research recommender systems for repositories. In: Proceedings of the Open Repositories Conference (2017)
27. Pontika, N., Anastasiou, L., Charalampous, A., Cancellieri, M., Pearce, S., Knoth, P.: CORE recommender: a plug in suggesting open access content (2017). http://hdl.handle.net/1842/23359
28. Beel, J., Dinesh, S., Mayr, P., Carevic, Z., Raghvendra, J.: Stereotype and most-popular recommendations in the digital library sowiport. In: Proceedings of the 15th International Symposium of Information Science (ISI), pp. 96–108 (2017)
29. Beel, J., Smyth, B., Collins, A.: RARD II: The 2nd Related-Article Recommendation Dataset. arXiv:1807.06918 [cs.IR] (2018)
30. Beel, J., Langer, S., Gipp, B., Nuernberger, A.: The architecture and datasets of Docear's research paper recommender system. D-Lib Mag. **20** (2014)
31. Hienert, D., Sawitzki, F., Mayr, P.: Digital library research in action-supporting information retrieval in sowiport. D-Lib Mag. **21** (2015)
32. Mayr, P.: Sowiport User Search Sessions Data Set (SUSS). GESIS Datorium (2016)
33. Stempfhuber, M., Schaer, P., Shen, W.: Enhancing visibility: integrating grey literature in the SOWIPORT information cycle. In: International Conference on Grey Literature, pp. 23–29 (2008)

34. Beel, J., Breitinger, C., Langer, S., Lommatzsch, A., Gipp, B.: Towards reproducibility in recommender-systems research. User Model. User-Adap. Inter. (UMUAI) **26**, 69–101 (2016)
35. Ferro, N., Fuhr, N., Rauber, A.: Introduction to the special issue on reproducibility in information retrieval: tools and infrastructures. J. Data Inf. Qual. (JDIQ) **10**, 14 (2018)
36. Collins, A., Tkaczyk, D., Beel, J.: A novel approach to recommendation algorithm selection using meta-learning. In: Proceedings of the 26th Irish Conference on Artificial Intelligence and Cognitive Science (AICS). CEUR-WS, pp. 210–219 (2018)

StyleExplorer: A Toolkit for Textual Writing Style Visualization

Michael Tschuggnall$^{(\boxtimes)}$, Thibault Gerrier, and Günther Specht

Department of Computer Science, Universität Innsbruck, Innsbruck, Austria
{michael.tschuggnall,guenther.specht}@uibk.ac.at,
thibault.gerrier@student.uibk.ac.at

Abstract. The analysis of textual writing styles is a well-studied problem with ongoing and active research in fields like authorship attribution, author profiling, text segmentation or plagiarism detection. While many features have been proposed and shown to be effective to characterize authors or document types in terms of high-dimensional feature vectors, an intuitive, human-friendly view on the computed data is often lacking. For example, machine learning algorithms are able to attribute previously unseen documents to a set of known authors by utilizing those features, but a visualization of the most discriminating features is usually not provided. To this end, we present StyleExplorer, a freely available web tool that is able to extract textual features from documents and to visualize them in multiple variants. Besides analyzing single documents intrinsically, it is also possible to visually compare multiple documents in single views with respect to selected metrics, making it a valuable analysis tool for various tasks in natural language processing as well as for areas in the humanities that work and analyze textual data.

Keywords: Text mining · Visualization · Natural language processing

1 Introduction

With the advent of freely accessible online collections and social media platforms, large amounts of text have emerged on the Internet which can be easily captured and processed. From a scientific perspective, many approaches have been proposed in various research fields that deal with the analysis of such texts, including authorship attribution [9] (who is the author of a given document?), author profiling [8] (what information about the author can be extracted from a text, e.g., gender or age?), text segmentation (how can a document automatically be divided by, e.g., topic [3]?) or plagiarism detection [6]. Except for the latter, where only *intrinsic* algorithms [11] operate solely on the document to reveal potential plagiarism, all approaches rely on an analysis of the writing style. For example, in the pioneer study of Mosteller and Wallace [7], three authors have been attributed to nearly 100 political essays by (manually) computing

© Springer Nature Switzerland AG 2019
L. Azzopardi et al. (Eds.): ECIR 2019, LNCS 11438, pp. 220–224, 2019.
https://doi.org/10.1007/978-3-030-15719-7_28

simple statistics of frequencies of common words like articles (e.g., "a", "the") or prepositions (e.g., "in", "of").

To capture the writing style of authors various so-called stylometric features [15] have been proposed, including lexical metrics like frequencies of (n-grams of) characters or words (e.g., [10]), syntactic metrics that capture grammatically related indicators (e.g., [13]), metrics related to semantics (e.g., [1]) and structural information (e.g., [14]), or metrics capturing other information like occurrences of spelling/grammar errors (e.g., [5]). While the effectiveness of each feature (type) has been evaluated thoroughly for the individual tasks a human-friendly view to the data is often lacking in existing approaches or simply not intended within the scope of the respective research. For example, by computing frequencies of nouns, verbs and adjectives used in a text, a three-dimensional vector is produced which can be utilized by machine learning algorithms to classify authors, but which is hard to interpret for humans. This problem becomes even worse when taking high-dimensional feature types like character n-grams into account, often resulting in several thousand dimensions.

To bridge this gap, we propose *StyleExplorer*, a web-based tool that visualizes the writing style of documents. By incorporating a rich amount of commonly used features, it provides the means to analyze documents for specific metrics, and also to compare multiple documents within single views. Consequently, it is a useful facility in many research fields working with textual data like authorship attribution, where the significant differences between authors can be made visible. Moreover, it may be utilized for linguistic analyses of various kinds, e.g., in the field of forensic linguistics [2].

2   StyleExplorer

The proposed *StyleExplorer* tool is designed as a free web-based application[1], targeting scientists as well as interested persons who want to visualize and compare statistics and metrics of text documents. From a technical perspective, it is based on the Meteor framework[2] using a React[3] frontend communicating with a Java library [4] in the backend, which computes the features and stores results in a MongoDB. In the following, a short overview about the workflow, the available features as well as the provided visualization types is given.

2.1   Workflow

StyleExplorer is generally free to use, but nevertheless requires a user account to be able to connect documents to users. After uploading the documents of interest, the computation of stylometric features (see Sect. 2.2) can be triggered, which is done asynchronously on the server. Thereby, all selected features can either be extracted for each paragraph or for individual user-defined units (e.g.,

---

[1] Available at https://dbis-styleexplorer.uibk.ac.at, [Review login: ecir2019/ecir2019].
[2] https://www.meteor.com, visited October 2018.
[3] https://reactjs.org, visited October 2018.

whole chapters). Once the document is processed, all computed features can be visually inspected using several visualization methods (see Sect. 2.3)[4]. Thus it is possible to compare features for each paragraph/unit and to see how they evolve within the document. Additionally, all documents of a user as well as other public documents from other users can be compared to each other, allowing to visually approach tasks like authorship attribution or genre detection, for example.

2.2 Available Text Features

The tool is capable of computing several stylometric features including lexical, syntactic and error [5] features. Specifically, the 40 metrics depicted in Table 1 are currently supported, whereby features marked with * yield a set of values per processing unit (e.g., character n-grams).

Table 1. Available stylometric features

Type	#	Features
Lexical	14	Avg word length, avg words/sentence, avg syllables/sentences, hapax dis-/legomena; frequencies of characters, special characters, words, character n-grams*, word n-grams*; type-token ratio, compression rate, compression time
Readability	8	Avg word frequency class, index of diversity, Honore's measure, Sichel's measure, Brunet's measure, Flesch-Kincaid reading ease, Gunning Fog index, Yule's K measure
Syntactic	6	Frequencies of POS tags*, POS tag n-grams*, function words; avg function words/sentence, punctuation-word ratio
Error	12	Frequencies of matching error rules*, error suggestions*, sentence whitespace, comma whitespace, double punctuation, uppercase sentence start, multiple whitespace, unpaired brackets, word repetitions, compound errors, contraction spelling, morphologic american spelling

2.3 Visualization

StyleExplorer offers several visualization options[5] that can be applied to either a single document or multiple documents for comparison. An excerpt of visualizations is depicted in Fig. 1, which are computed for a sample document. For features which produce a single value per paragraph, the default option is a line chart (see Fig. 1a), but also box plots are available (Fig. 2b). To be able to compare documents of different lengths (paragraphs) within a single chart, the x-axis

---

[4] And also be downloaded in JSON format for individual further postprocessing.

[5] Utilizing Highcharts, https://www.highcharts.com, visited October 2018.

represents the text position in percent with respect to the whole document. For multi-valued features like n-grams or frequencies of words, *StyleExplorer* uses bar charts by default (see an example in Fig. 1b). Alternatively, word clouds and pie charts can be utilized (see Fig. 1c and d). As it is not feasible to display all resulting values per document, only the top n features with respect to their frequency are displayed, whereby n can be defined by the user. Finally, Fig. 2 shows two examples of comparing multiple documents within single charts.

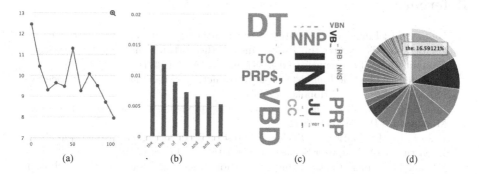

(a)   (b)   (c)   (d)

Fig. 1. Visualization examples showing the Dale-Chall readability formula (a), the most frequent function words (b), a word cloud of POS-tags (c) and a pie chart of most frequent character 3-grams (d) of a sample document.

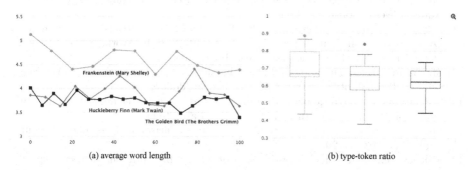

(a) average word length   (b) type-token ratio

Fig. 2. Example of comparing three documents in terms of average word length and type-token ratio.

3   Conclusion and Future Work

In this paper we presented *StyleExplorer*, a freely available web-tool for visual textual analysis. By uploading arbitrary documents it is capable of extracting several common features used in stylometric analyses like authorship attribution or author profiling. Due to individual corresponding visualization variants it is possible to inspect a document with respect to selected features, and also to compare different documents against each other.

With respect to future work, several optimizations are possible. From the visualization perspective, new types of graphs as well as combinations thereof would be beneficial. Further, as the core of *StyleExplorer* relies on the extraction of features from text documents, new features (types) that emerge in text analysis research[6] should be added continuously. Finally, language-dependent features should be enhanced to support more languages than English.

References

1. Gamon, M.: Linguistic correlates of style: authorship classification with deep linguistic analysis features. In: Proceedings of the 20th International Conference on Computational Linguistics (COLING), p. 611. ACL (2004)
2. Gibbons, J.: Forensic Linguistics: An Introduction to Language in the Justice System. Wiley-Blackwell, Hoboken (2003)
3. Misra, H., et al.: Text segmentation: a topic modeling perspective. Inf. Process. Manage. **47**(4), 528–544 (2011)
4. Huber, B.: Evaluation of Style Features of Text Documents, Bachelor thesis. Department of Computer Science, Universität Innsbruck (2016)
5. Koppel, M., Schler, J.: Exploiting stylistic idiosyncrasies for authorship attribution. In: Proceedings of the 18th International Joint Conference on AI, vol. 69, pp. 72–80 (2003)
6. Potthast, M., et al.: Overview of the 5th international competition on plagiarism detection. In: Notebook Papers of the 9th PAN Evaluation Lab (2013)
7. Mosteller, F., Wallace, D.: Inference and Disputed Authorship: The Federalist. Addison-Wesley, Boston (1964)
8. Rangel, F., Rosso, P., Verhoeven, B., Daelemans, W., Potthast, M., Stein, B.: Overview of the 4th author profiling task at PAN 2016. In: Working Notes Papers of the CLEF 2016 Evaluation Labs, vol. 1609 (2016)
9. Stamatatos, E.: A survey of modern authorship attribution methods. J. Am. Soc. Inf. Sci. Technol. **60**(3), 538–556 (2009). https://doi.org/10.1002/asi.v60:3
10. Stamatatos, E.: Intrinsic plagiarism detection using character n-gram profiles. In: Notebook Papers of the 5th PAN Evaluation Lab (2011)
11. Stein, B., Lipka, N., Prettenhofer, P.: Intrinsic plagiarism analysis. Lang. Resour. Eval. **45**(1), 63–82 (2011)
12. Mikolov, T., et al.: Efficient estimation of word representations in vector space. arXiv preprint arXiv:1301.3781 (2013)
13. Tschuggnall, M., Specht, G.: Using grammar-profiles to intrinsically expose plagiarism in text documents. In: Métais, E., Meziane, F., Saraee, M., Sugumaran, V., Vadera, S. (eds.) NLDB 2013. LNCS, vol. 7934, pp. 297–302. Springer, Heidelberg (2013). https://doi.org/10.1007/978-3-642-38824-8_28
14. Zheng, R., Li, J., Chen, H., Huang, Z.: A framework for authorship identification of online messages: writing-style features and classification techniques. J. Am. Soc. Inf. Sci. Technol. **57**(3), 378–393 (2006)
15. Eissen, S.M., Stein, B.: Intrinsic plagiarism detection. In: Lalmas, M., MacFarlane, A., Rüger, S., Tombros, A., Tsikrika, T., Yavlinsky, A. (eds.) ECIR 2006. LNCS, vol. 3936, pp. 565–569. Springer, Heidelberg (2006). https://doi.org/10.1007/11735106_66

---

[6] E.g., recent and popular techniques like word2vec [12].

MedSpecSearch: Medical Specialty Search

Mehmet Uluç Şahin[1], Eren Balatkan[1], Cihan Eran[1], Engin Zeydan[2], and Reyyan Yeniterzi[1(✉)]

[1] Özyeğin University, Istanbul, Turkey
reyyan.yeniterzi@ozyegin.edu.tr
[2] Türk Telekom Labs, Istanbul, Turkey

Abstract. MedSpecSearch (www.medspecsearch.com) is a search engine for helping users to find the relevant medical specialty for a doctor visit based on users' description of symptoms. This system is useful for users who are not sure of which medical specialty they should consult to. Furthermore, the API of the search engine can be used as part of the online doctor appointment and medical consultation sites to route the patient or question to the right medical specialty. The system returns the top three relevant specialties when the estimated confidence score is high. Otherwise, it asks users to input more data.

Keywords: Medical text classification · Word embeddings · Confidence estimation · Data collection

1 Introduction

A recent survey [1] on physician appointment wait times performed over 15 major metropolitan cities in the United States revealed that the time to schedule an appointment has reached an average of 24 days. In addition to the longer wait times, the cost of doctor appointments are getting more expensive worldwide [3,4]. Given these circumstances, getting appointment from a doctor on the right medical specialty is becoming more crucial for patients in order to not delay the diagnosis and treatment process any further and spend more. This becomes a more significant problem in places where patients are allowed to schedule their doctor appointments from any medical department without any guidance. For such conditions, the proposed MedSpecSearch system aims to help patients on finding the right medical specialty to visit, by providing a publicly available and user friendly medical specialty search engine. Given the user's description of the medical case like patient information (gender, age etc.) and observed symptoms, the MedSpecSearch will return the top 3 medical specialties that can be relevant to the medical case only if the estimated confidence of the returned results is good enough. In case the user's description of the medical condition is not enough for a confident prediction, the system asks for more information from the user. MedSpecSearch comes with an Application Programming Interface (API) which

This work has been funded by Türk Telekom R&D Center.

L. Azzopardi et al. (Eds.): ECIR 2019, LNCS 11438, pp. 225–229, 2019.
https://doi.org/10.1007/978-3-030-15719-7_29

can be used by online physician appointment sites to direct the patient to the right medical specialty for appointment, or by online medical question answer sites to route user's question to the right medical specialist.

2   System Description

The Association of American Medical Colleges (AAMC) defined around 30 general and 100 sub-medical specialties [8]. For instance, *Internal Medicine* is categorized as a general specialty, while *Hematology*, *Rheumatology* and *Pulmonary Disease* etc. are categorized as its sub-specialties. The proposed MedSpecSearch system initially focuses on the general specialties and for a given patient description aims to identify the relevant medical specialty within the AAMC categories. The system pipeline is shown in Fig. 1 and described in the following sections.

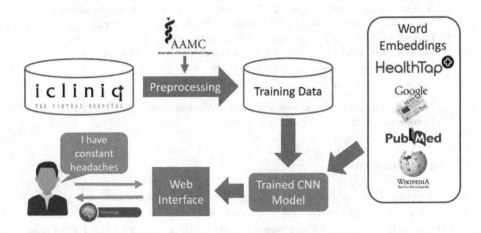

Fig. 1. MedSpecSearch pipeline

2.1   Data

There are some publicly available de-identified health record datasets (such as the MIMIC [5]) which contain a preliminary, free text diagnosis for the patient on hospital admission. These informative diagnoses are usually generated by the admitting clinicians after listening the complaints and medical histories of patients. These notes are very useful for routing patients to the right specialty within the hospital. However, they contain too much medical terminology which general public do not use or even know that existed. Therefore, they are not useful for our task which use only the context generated by patients without any medical expert assistance. As a result, this proposed system explores patient generated content collected from online medical platforms.

Data is collected from two online medical consultation platforms; iCliniq[1] and HealthTap[2]. In iCliniq platform, patients provide details about their conditions and complaints, such as their gender, age, previous critical health history if relevant, previous and current medications, and their symptoms, in their questions. Later on, these questions are categorized into a specialty and replied by corresponding medical specialists. We use initial contexts of questions provided by patients as our training input data. The iCliniq has around 90 categories. These categories are not directly used but instead matched to the predefined specialty categories by using the AAMC hierarchy. For instance, questions under *Endocrinology* are categorized under *Internal Medicine* category. In order to check the accuracy of iCliniq category labels, 73 iCliniq questions are randomly chosen and given to two medical doctors for labeling. The inter-rater agreement is calculated with Cohen's [2] kappa statistics for both doctors and doctor-iCliniq pairs. Agreement score of doctors' among themselves is 0.77 which is substantial as being within the 0.6–0.8 range [7]. The agreement scores between iCliniq and doctors are 0.62 and 0.67 which are also substantial. These agreement scores indicate that iCliniq labels are in good quality for supervised learning, hence more than 7K iCliniq questions are retrieved, pre-processed and used for training.

HealthTap is a similar medical consultation platform with more questions. Around 1.6 million questions are retrieved from HealthTap, but unfortunately HealthTap uses more diverse set of labels to categorize their questions. It has 230 main categories and around 5400 subcategories. Some example main categories are *abdominal pain*, *ankle* or *blood* which cannot be easily mapped to the AAMC categories. Therefore, HealthTap data is not used in supervised learning but instead explored in an unsupervised manner by training word embeddings.

2.2   The Model

Convolutional Neural Networks (CNN) has achieved competitive performance on many NLP tasks, especially on text classification tasks due to their capability of capturing useful n-grams. Our classification model follows Kim's [6] CNN architecture; however, leverages several pre-trained word embeddings. In order to analyze the individual effects of these embeddings, we use a single channel model as only one embedding is input to the model at once. In addition to our trained Word2Vec model of HealthTap, additional publicly available pretrained word embeddings like Word2Vec [9] embeddings trained with Google News[3] and GloVe [10] embeddings trained on Wikipedia 2014 + Gigaword 5[4] are used. Furthermore, the Word2Vec embedding[5] which is trained by Pyysalo et al. [11] on 22 Million PubMed records, 672K PubMed Central Open-Access texts and a recent Wikipedia dump is also used. A 0.8/0.2 split of data to test the performance of these word embeddings returned very similar accuracies: PubMed W2V

---

[1] https://www.icliniq.com.
[2] https://www.healthtap.com.
[3] https://code.google.com/archive/p/word2vec/.
[4] http://nlp.stanford.edu/data/glove.6B.zip.
[5] http://bio.nlplab.org/.

72.7%, HealthTap W2V 72.6%, Google News W2V 74.1% and Glove 76.1%. In the advanced options part of the search engine, users can play around with these embeddings by selecting any one them.

In hospitals if doctors cannot be sure of a diagnosis, they ask for more tests. Even though MedSpecSearch is not a diagnostic system, it uses a similar idea (asking for more user input) when predicting the medical specialty. Being aware of the cost of wrong predictions, the system uses a confidence threshold to decide either to return the ranked list of specialties or not. If the estimated confidence is below the threshold, the system asks user to input more context for describing the medical situation. The default confidence threshold value is 90% but it can be set to any value between 10%–90% in the advanced options part.

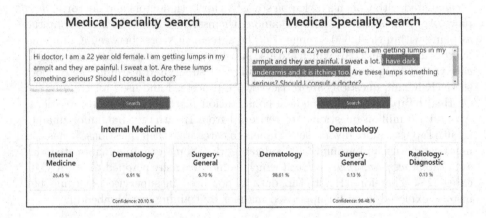

Fig. 2. MedSpecSearch front end with two example queries

An example is provided in Fig. 2, where the system's confidence threshold is specifically kept low to show the results of the provided user input. This is an example question from iCliniq. The question on the right is the original question with label *Dermatology* predicted correctly by MedSpecSearch with confidence 98.48%. On the left is the same question with one sentence removed. With this incomplete question the system cannot make a confident prediction, therefore asking for more information from the user is the right call. Using an arbitrarily chosen confidence threshold like 90%, significantly reduces the amount of misclassifications of the system and increases the general prediction accuracy to 88.1%. In our model, we have used the modification proposed by Sensoy et al. [12] to calculate these confidence estimates. All these functionalities are available as a web service (www.medspecsearch.com) and API. The API and the trained HealthTap word embeddings are available online[6].

As a future work, we will extend the existing system for other languages as well. Furthermore, we will improve the interpretability of the returned results by

---

[6] https://github.com/OzU-NLP/MedSpecSearch.

highlighting the terms in user input in terms of their significance based on the predicted specialty. Additionally, we will explore ways of getting useful additional information from the user when the confidence value is below the threshold. In such a case, the system should identify several key symptoms to be asked to the user for better prediction.

References

1. Survey of Physician Appointment Wait Times. https://www.merritthawkins.com/news-and-insights/thought-leadership/survey/survey-of-physician-appointment-wait-times/. Accessed 25 Sept 2018
2. Cohen, J.: A coefficient of agreement for nominal scales. Educ. Psychol. Meas. **20**(1), 37–46 (1960)
3. Doctor visits may cost at least three times as much next year. https://www.thestar.com.my/news/nation/2018/08/27/doctor-visits-may-cost-at-least-three-times-as-much-next-year/. Accessed 25 Sept 2018
4. Doctors' visits are about to get more expensive in France. https://www.thelocal.fr/20170915/some-doctors-visits-are-about-to-get-more-expensive-in-france/. Accessed 25 Sept 2018
5. Johnson, A., et al.: MIMIC-III, a freely accessible critical care database. Sci. Data **3** (2016)
6. Kim, Y.: Convolutional neural networks for sentence classification. ArXiv e-prints (2014)
7. Landis, J.R., Koch, G.G.: The measurement of observer agreement for categorical data. Biometrics **33**, 159–174 (1977)
8. Medical Specialties. https://www.aamc.org/cim/specialty/exploreoptions/list/. Accessed 20 Apr 2018
9. Mikolov, T., Chen, K., Corrado, G., Dean, J.: Efficient estimation of word representations in vector space. ArXiv e-prints (2013)
10. Pennington, J., Socher, R., Manning, C.: GloVe: global vectors for word representation. In: Proceedings of the 2014 Conference on Empirical Methods in Natural Language Processing, EMNLP, pp. 1532–1543 (2014)
11. Pyysalo, S., Ginter, F., Moen, H., Salakoski, T., Ananiadou, S.: Distributional semantics resources for biomedical text processing. In: Proceedings of the 5th International Symposium on Languages in Biology and Medicine (2013)
12. Sensoy, M., Kandemir, M., Kaplan, L.: Evidential deep learning to quantify classification uncertainty. arXiv preprint arXiv:1806.01768 (2018)

Contender: Leveraging User Opinions for Purchase Decision-Making

Tiago de Melo[1,2]([envelope]), Altigran S. da Silva[2], Edleno S. de Moura[2],
and Pável Calado[3]

[1] Universidade do Estado do Amazonas, Manaus, Brazil
`tmelo@uea.edu.br`
[2] Universidade Federal do Amazonas, Manaus, Brazil
`{tiago.melo,alti,edleno}@icomp.ufam.edu.br`
[3] Universidade de Lisboa, Lisboa, Portugal
`pavel.calado@tecnico.ulisboa.pt`

Abstract. User opinions posted on e-commerce websites are a valuable source to support purchase making-decision. Unfortunately, it is not generally feasible for an ordinary buyer to examine a large set of reviews on a given product for useful information on certain attributes. We present a system named Contender that can summarize product reviews aligned to the attributes of these products. Contender is implemented as an Android app for smartphones.

Keywords: Opinion mining · User reviews · Experimentation

1 Introduction

When making purchasing decisions, customers usually rely on information from two types of sources: product specifications provided by manufacturers and reviews posted by other customers. Both kinds of information are often available on e-commerce websites. However, in some competitive markets, such as cell phones, many manufacturers make products with very similar characteristics. For example, the products *Moto Z Play* and *Samsung Galaxy S7 Edge* have the same screen size, operating system, camera resolution and so on. In addition, these products have almost the same price. Therefore, there is almost no difference between them. Especially in these cases, user reviews play an important role in purchase decision-making.

Unfortunately, it is not generally feasible for an ordinary buyer to examine a large set of reviews to manually compare similar products. For example, there are more than 2,000 reviews on *Samsung Galaxy S7 Edge* posted on the Amazon.com website. This problem has been addressed by techniques to identify comparative sentences from reviews [9–11]. These techniques have some important drawbacks. First, these are unable to identify the targets in comparative sentences [9]. Second, comparative sentences comprise only 10% of the total opinionated text [12].

© Springer Nature Switzerland AG 2019
L. Azzopardi et al. (Eds.): ECIR 2019, LNCS 11438, pp. 230–235, 2019.
https://doi.org/10.1007/978-3-030-15719-7_30

Finally, comparative sentences are most common in forum discussions than in user reviews [10], which are prevalent in e-commerce websites. Another approach to compare opinions on products is to leverage the ratings of the products. However, the usefulness of the star-based ratings in the reviews is limited for potential buyers, since a rating represents an average for the product as whole and can combine both positive and negative evaluations of many distinct attributes. In addition, this kind of evaluation does not convey any information about why users like a product or which characteristics they like the most. A user looking to buy a cell phone may want to know what user reviews say on the battery or screen, not just the general rating of the product.

A traditional way of organizing a large number of product reviews is to create *opinions summaries* which provide a condensed list of product aspects and their corresponding opinions, where the most common approach is called *aspect-based opinion summarization* [3]. However, current techniques are inadequate to properly support customer queries on specific product characteristics [4–6]. This is because, in current methods, opinions are arbitrarily clustered by aspects, causing these clusters to not necessarily align with the commonly expected product characteristics. For example, there may be several opinion clusters that refer to the screen: a cluster of opinions about resolution and color, another that mixes glossiness with size, and so on. Thus, the customer still has to carry out the non-trivial task of identifying which groups of aspects refer to the same aspect of interest. In fact, Zha et al. [7] reported that, for the *iPhone 3GS*, more than three hundred aspects were found in the reviews. Summarizing this information would generate hundreds of clusters, without identifying which specific ones refer to the actual cell phone screen.

To address the problem described above, particularly in the cell phone domain, we design and implement a system named *Contender*. This system can summarize product reviews aligned to the specification of attributes of these products. The system handles with real user reviews from e-commerce websites as a data source. Our system extracts the opinions of the reviews and then maps these opinions to the attributes defined for the product specifications. We adapt the unsupervised method described by Poria et al. [1] to opinion extraction task and use deep learning techniques to map opinions to attributes. We implement Contender as an Android app for smartphones and it is available at http://tiagodemelo.info/contender.

2   Contender Overview

Figure 1 presents an overview of Contender. The system has two modules: *pre-processing* (offline) and *user session* (online). In the *pre-processing* module, the specifications and opinions of products are extracted from websites and then the opinions are aligned (mapped) to the product attributes of specification. The *user session* provides an interface so that user can compare two products. We highlight and explain in detail the following major steps: crawler, product extraction, opinion extraction, indexing, searching, and ranking.

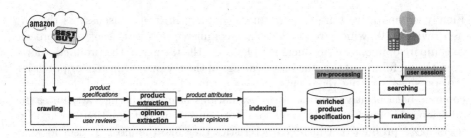

Fig. 1. Contender overview.

Crawling. The system collects web pages of product specifications and user reviews from e-commerce websites such as Amazon and BestBuy.

Product Extraction. Extraction of product attributes and their corresponding values from web pages collected. Notice that reviews may also include opinions that do not refer to a specific attribute of a product, but to the product as a whole. For example, the sentence *"Samsung is a trust device"* expresses a positive opinion for the product as a whole, and not for one of its attributes. Therefore, we should map this opinion to the target product. To enable this, we create a new attribute, called `General`, whose value is simply the product title. Furthermore, opinions may also target attributes that are not represented in the product catalog. For example, the sentence *"easy to use"* expresses a opinion for a characteristic of the product that is not represented as an attribute in the original product catalog. To handle these cases we create an attribute called `Other`. This attribute has no value in the product catalog.

Opinion Extraction. Let $R = \{r_1, r_2, \ldots, r_m\}$ be a set of reviews on a product p_i, where each review $r \in R$ contains a set of sentences $ST = \{st_1, st_2, \ldots, st_n\}$ and a numeric rating score which takes a value between 1 to 5. Each opinion o extracted from a sentence $st \in ST$ is represented by a triple $\langle p_i, a, rs \rangle$, where p_i is the target product, a is the aspect of the target product on which the opinion has been given, and rs is the numeric rating score of r. For the aspect identification task, we apply the unsupervised method proposed by Poria et al. [1].

Indexing. The system maps each opinion extracted from the reviews in R to specific attributes of the target product. This is the core contribution of our work in Contender. Our key insight is that there is a strong relationship between product attributes and the text that describes an opinion. Specifically, we represent product attributes by *attribute descriptors* and the opinionated text are represented by the aspect expressions, that is, by the text portion from the reviews that defines the aspect [8]. An attribute descriptor is a set whose elements are all the terms found in the name of attribute plus the terms found on the value of this attribute, for a given product. For example, the descriptor for the attribute `Screen` of a cell phone may be { "screen", "5.5", "TFT LCD"}.

To match descriptors and aspect expressions, we proposed a strategy that uses a similarity function. If they are deemed similar, the corresponding opinion is mapped to their corresponding attribute. An aspect expression α and a descriptor Δ are considered as similar, if there is some pair of terms $\langle w_1, w_2 \rangle \in \alpha \times \Delta$, such that $\cos(E_{w_1}, E_{w_2}) > \tau$, where E_w is the embedding of term w and τ is a threshold value. Currently, we use $\tau = 0.39$, a value we calculated empirically.

In this function, we use continuous vector representations, also known as *word embeddings* [2], for each word representation. More specifically, we train word embeddings on a dataset[1] with more than 1.2 million of reviews collected from Amazon.com. Our model uses skip-gram with negative sampling, a context window of five words, and 500 dimensions. We evaluated our strategy in a dataset manually annotated by us and it achieved 0.83 of F_1 when applied to the task of mapping opinions to specific attributes of the target product.

As an example of opinion mapping, the sentence *"the resolution is amazing"* has the aspect expression *resolution* and it is mapped to the attribute `Screen` because the similarity score obtained between *resolution* and `Screen` are higher when compared to the score of *resolution* with the other product attributes.

Searching. The app allows the user to search for two cell phones to be compared. For this, the user simply type the names of the cell phones. Searches by names are case insensitive. We have also provided an autocomplete search feature.

Ranking. The app presents a ranking of the product attributes. For this, it firsts aggregates the opinions retrieved in the searching. Then, it calculates an average of the opinion scores for each attribute. We adopt the same score normalization as Amazon and BestBuy, which consider 1–5 stars.

3   Demonstration

Figure 2 displays screenshots of the main modules of Contender. When the user opens the app, the system shows two search boxes where the user can type the names of devices. As shown in Fig. 2(a), a user inputs the names of two mobiles in the search box: *Motorola Moto X4* and *Samsung Galaxy S7 Edge*. Our system shows the results of the input query in three distinct modules: *Specs*, *Reviews*, and *Charts*.

Specs displays the specifications of each product provided by the manufacturers and made available by the e-commerce sites (Fig. 2(b)). *Reviews* displays user reviews for each product. The reviews can be filtered by the product attributes. For example, if the user selects `Screen`, the system will only display the reviews on that attribute (Fig. 2(c)). *Charts* displays four different types of charts so that the user can make a comparison between the two products: (a) bar chart; (b) radar chart; (c) pie chart; (d) rating score. For example, the Fig. 2(d) displays the attribute scores for each of the products. This is only possible because the system can align opinions to product attributes.

---

[1] Dataset will be made available on request.

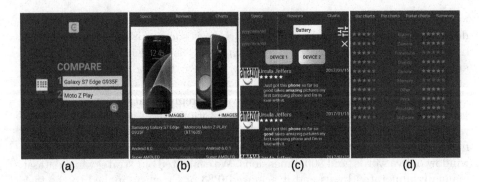

Fig. 2. Main modules of Contender.

4   System Setting

The current version uses a review corpus collected from Amazon and BestBuy, with 15,512 products indexed. The number of user reviews currently is over 2 millions, but this number is increasing, because our crawler daily collect new user reviews. Contender's backend is running on a host with the following setting 2.4 GHz Intel Xeon 2 vCPU processor, 2 GB of memory, and 50 GB of SSD.

5   Conclusion and Future Work

We presented a novel system for comparing two products at the attribute level granularity based on user opinions. To the best of our knowledge, our work is the first that address this problem. In future work, we plan to expand the system to other product categories, including, for example, cameras and laptops.

References

1. Poria, S., et al.: A rule-based approach to aspect extraction from product reviews. In: Proceedings of the Second Workshop on Natural Language Processing for Social Media (2014)
2. Mikolov, T., Sutskever, I., Chen, K., Corrado, G.S., Dean, J.: Distributed representations of words and phrases and their compositionality. In: Advances in Neural Information Processing Systems (2013)
3. Moussa, M.E., et al.: A survey on opinion summarization techniques for social media. Future Comput. Inform. J. **3**, 82–109 (2018)
4. Amplayo, R.K., Song, M.: An adaptable fine-grained sentiment analysis for summarization of multiple short online reviews. Data Knowl. Eng. **110**, 54–67 (2017)
5. Rakesh, V., et al.: A sparse topic model for extracting aspect-specific summaries from online reviews. In: Proceedings of the WWW Conference (2018)
6. Zhou, X., Wan, X., Xiao, J.: CMiner: opinion extraction and summarization for Chinese microblogs. IEEE Trans. Knowl. Data Eng. **28**, 1650–1663 (2016)

7. Zha, Z.-J., Yu, J., Tang, J., Wang, M., Chua, T.-S.: Product aspect ranking and its applications. IEEE Trans. Knowl. Data Eng. **26**, 1211–1224 (2014)
8. Liu, B.: Sentiment Analysis: Mining Opinions, Sentiments, and Emotions. Cambridge University Press, Cambridge (2015)
9. Varathan, K.D., Giachanou, A., Crestani, F.: Comparative opinion mining: a review. J. Assoc. Inf. Sci. Technol. **68**, 811–829 (2017)
10. Ganapathibhotla, M., Liu, B.: Mining opinions in comparative sentences. In: Proceedings of the 22nd International Conference on Computational Linguistics, vol. 1, pp. 241–248 (2008)
11. Jindal, N., Liu, B.: Identifying comparative sentences in text documents. In: SIGIR (2006)
12. Kessler, W., Kuhn, J.: Detection of product comparisons-how far does an out-of-the-box semantic role labeling system take you? In: Proceedings of the Conference on Empirical Methods in Natural Language Processing, pp. 1892–1897 (2013)

Rethinking 'Advanced Search': A New Approach to Complex Query Formulation

Tony Russell-Rose[1(✉)], Jon Chamberlain[2], and Udo Kruschwitz[2]

[1] UXlabs Ltd., Brunel House, 340 Firecrest Ct, Centre Park,
Warrington WA1 1RG, UK
`tgr@uxlabs.co.uk`
[2] School of Computer Science and Electronic Engineering, University of Essex,
Wivenhoe Park, Colchester, Essex CO4 3SQ, UK
`{jchamb,udo}@essex.ac.uk`

Abstract. Knowledge workers such as patent agents, recruiters and media monitoring professionals undertake work tasks where search forms a core part of their duties. In these instances, the search task often involves the formulation of complex queries expressed as Boolean strings. However, creating effective Boolean queries remains an ongoing challenge, often compromised by errors and inefficiencies. In this demo paper, we present a new approach to query formulation in which concepts are expressed on a two-dimensional canvas and relationships are articulated using direct manipulation. This has the potential to eliminate many sources of error, makes the query semantics more transparent, and offers new opportunities for query refinement and optimisation.

Keywords: Query formulation · Advanced search · Boolean ·
Search visualisation · Professional search

1 Introduction

Many knowledge workers rely on the effective use of search applications in the course of their professional duties [6]. Patent agents, for example, depend on accurate prior art search as the foundation of their due diligence process [10]. Similarly, recruitment professionals rely on Boolean search as the basis of the candidate sourcing process [8], and media monitoring professionals routinely manage thousands of Boolean expressions on behalf their client briefs [12].

The traditional solution is to formulate complex Boolean expressions consisting of keywords, operators and search commands, such as that shown in Fig. 1. However, the practice of using Boolean strings to articulate complex information needs suffers from a number of fundamental shortcomings [9]. First, it is poor at communicating structure: without some sort of physical cue such as indentation, parentheses and other delimiters can become lost among other alphanumeric characters. Second, it scales poorly: as queries grow in size, readability becomes progressively degraded. Third, they are error-prone: even if syntax checking is

© Springer Nature Switzerland AG 2019
L. Azzopardi et al. (Eds.): ECIR 2019, LNCS 11438, pp. 236–240, 2019.
https://doi.org/10.1007/978-3-030-15719-7_31

provided, it is still possible to place parentheses incorrectly, changing the semantics of the whole expression.

(cv OR "cirriculum vitae" OR resume OR "resumé") (filetype:doc OR filetype:pdf OR filetype:txt) (inurl:profile OR inurl:cv OR inurl:resume OR initile:profile OR intitle:cv OR initile:resume) ("project manager" OR "it project manager" OR "program* manager" OR "data migration manager" OR "data migration project manager") (leinster OR munster OR ulster OR connaught OR dublin) -template -sample -example -tutorial -builder -"writing tips" -apply -advert -consultancy

---

Fig. 1. An example from the Boolean Search Strings Repository

To mitigate these issues, many professionals rely on previous examples of best practice. Recruitment professionals, for example, draw on repositories such as the Boolean Search Strings Repository[1] and the Boolean String Bank[2]. However, these repositories store content as unstructured text strings, and as such their true value as source of experimentation and learning may never be fully realized.[3]

2dSearch[4] offers an alternative approach. Instead of formulating Boolean strings, queries are expressed by combining objects on a two-dimensional canvas and relationships are articulated using direct manipulation. This eliminates many sources of syntactic error, makes the query semantics more transparent, and offers further opportunities for query refinement and optimisation.

2 Related Work

The application of data visualisation to search query formulation can offer significant benefits, such as fewer zero-hit queries, improved query comprehension and better support for exploration of an unfamiliar database [3]. An early example is that of Anick et al. [1], who developed a two-dimensional graphical representation of a user's natural language query that supported reformulation via direct manipulation. Fishkin and Stone [2] investigated the application of direct manipulation techniques to database query formulation, using a system of 'lenses' to refine and filter the data. Jones [4] developed a query interface to the New Zealand Digital Library which uses Venn diagrams and integrated query result previews.

---

[1] https://booleanstrings.ning.com/forum/topics/boolean-search-strings-repository, accessed 10 Oct 2018.
[2] https://scoperac.com/booleanstringbank, accessed 10 Oct 2018.
[3] http://booleanblackbelt.com/2016/01/the-most-powerful-boolean-search-operator, accessed 10 Oct 2018.
[4] https://2dsearch.com, accessed 24 Oct 2018.

A further example is Yi et al. [13], who applied a 'dust and magnet' metaphor to multivariate data visualization. Nitsche and Nürnberger [5] developed a system based on a radial user interface that supports phrasing and interactive visual refinement of vague queries. A further example is Boolify[5], which provides a drag and drop interface to Google. More recently, de Vries et al. [11] developed a system which utilizes a visual canvas and elementary building blocks to allow users to graphically configure a search engine. 2dSearch differs from the prior art in offering a database-agnostic approach with automated query suggestions and support for optimising, sharing and re-using query templates and best practices.

3   Design Concept

At the heart of 2dSearch is a graphical editor which allows the user to formulate queries as objects on a two-dimensional canvas. Concepts can be simple keywords or attribute: value pairs representing controlled vocabulary terms or database-specific search operators. Concepts can be combined using Boolean (and other) operators to form higher-level groups and then iteratively nested to create expressions of arbitrary complexity. Groups can be expanded or collapsed on demand to facilitate transparency and readability.

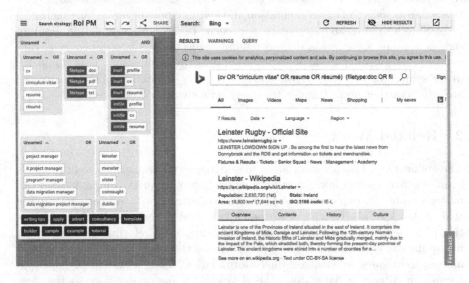

Fig. 2. The 2dSearch app showing query canvas (left) and search results pane (right). (Color figure online)

The application consists of two panes (see Fig. 2): a query canvas and a search results pane (which can be resized or detached in a separate window). The canvas can be resized or zoomed, and features an 'overview' widget to allow users to navigate to elements that may be outside the current viewport. Adopting design

---

[5] https://www.kidzsearch.com/boolify/, accessed 23 Oct 2018.

cues from Google's Material Design language[6], a sliding menu is offered on the left, providing file I/O and other options. This is complemented by a navigation bar which provides support for document-level functions such as naming and sharing queries.

Although 2dSearch supports creation of complex queries from a blank canvas, its value is most readily understood by reference to an example such as that of Fig. 1, which is intended to find social profiles for data migration project managers located in Dublin. Although relatively simple, this query is still difficult to interpret, optimise or debug. However, when opened with 2dSearch, it becomes apparent that the overall expression consists of a conjunction of OR clauses (nested blocks) with a number of specialist search operators (dark blue) and negated terms (white on black). To edit the expression, the user can move terms using direct manipulation or create new groups by combining terms. They can also cut, copy, delete, and lasso multiple objects. If they want to understand the effect of one group in isolation, they can execute it individually. Conversely, if they want to remove one element from consideration, they can disable it. In each case, the effects of each operation are displayed in real time in the adjacent search results pane.

2dSearch functions as a meta-search engine, so is in principle agnostic of any particular search technology or platform. In practice however, to execute a given query, the semantics of the canvas content must be mapped to the API of the underlying database. This is achieved via an abstraction layer or set of 'adapters' for common search platforms such as Bing, Google, PubMed, Google Scholar, etc. These are user selectable via a drop-down control.

Support for query optimisation is provided via a 'Messages' tab on the results pane. For example, if the user tries to execute via Bing a query string containing operators specific to Google, an alert is shown listing the unknown operators. 2dSearch also identifies redundant structure (e.g. spurious brackets or duplicate elements) and supports comparison of canonical representations. Query suggestions are provided via an NLP services API which utilises various Python libraries (for word embedding, keyword extraction, etc.) and SPARQL endpoints (for linked open data ontology lookup) [7].

4   Summary and Further Work

2dSearch is a framework for search query formulation in which information needs are expressed by manipulating objects on a two-dimensional canvas. Transforming logical structure into physical structure mitigates many of the shortcomings of Boolean strings. This eliminates syntax errors, makes the query semantics more transparent and offers new ways to optimise, save and share best practices. In due course, we hope to engage in a formal, user-centric evaluation, particularly in relation to traditional query builders. We are currently engaging in an outreach programme and invite subject matter experts to work with us in building repositories of curated (or user generated) examples and templates.

---

[6] https://material.io.

Adopting a database-agnostic approach presents challenges, but it also offers the prospect of a *universal* framework in which information needs can be articulated in a *generic* manner and the task of mapping to an underlying database can be delegated to platform-specific adapters. This could have profound implications for the way in which professional search skills are taught, learnt and applied.

References

1. Anick, P.G., Brennan, J.D., Flynn, R.A., Hanssen, D.R., Alvey, B., Robbins, J.M.: A direct manipulation interface for boolean information retrieval via natural language query. In: Proceedings of the 13th Annual International ACM SIGIR Conference on Research and Development in Information Retrieval. SIGIR 1990, pp. 135–150. ACM, New York, NY, USA (1990). https://doi.org/10.1145/96749.98015
2. Fishkin, K., Stone, M.C.: Enhanced Dynamic Queries Via Movable Filters, pp. 415–420. ACM Press, New York (1995)
3. Goldberg, J.H., Gajendar, U.N.: Graphical condition builder for facilitating database queries. U.S. Patent No. 7,383,513. 3 (2008)
4. Jones, S.: Graphical query specification and dynamic result previews for a digital library. In: Proceedings of the 11th Annual ACM Symposium on User Interface Software and Technology, UIST 1998, pp. 143–151. ACM, New York, NY, USA (1998). https://doi.org/10.1145/288392.288595
5. Nitsche, M., Nürnberger, A.: QUEST: querying complex information by direct manipulation. In: Yamamoto, S. (ed.) HIMI 2013. LNCS, vol. 8016, pp. 240–249. Springer, Heidelberg (2013). https://doi.org/10.1007/978-3-642-39209-2_28
6. Russell-Rose, T., Chamberlain, J., Azzopardi, L.: Information retrieval in the workplace: a comparison of professional search practices. Inf. Process. Manag. **54**(6), 1042–1057 (2018). https://doi.org/10.1016/j.ipm.2018.07.003
7. Russell-Rose, T., Gooch, P.: 2dsearch: a visual approach to search strategy formulation. In: Proceedings of DESIRES: Design of Experimental Search & Information REtrieval Systems. DESIRES 2018 (2018)
8. Russell-Rose, T., Chamberlain, J.: Real-world expertise retrieval: the information seeking behaviour of recruitment professionals. In: Ferro, N., et al. (eds.) ECIR 2016. LNCS, vol. 9626, pp. 669–674. Springer, Cham (2016). https://doi.org/10.1007/978-3-319-30671-1_51
9. Russell-Rose, T., Chamberlain, J.: Searching for talent: the information retrieval challenges of recruitment professionals. Bus. Inf. Rev. **33**(1), 40–48 (2016)
10. Tait, J.I.: An introduction to professional search. In: Paltoglou, G., Loizides, F., Hansen, P. (eds.) Professional Search in the Modern World. LNCS, vol. 8830, pp. 1–5. Springer, Cham (2014). https://doi.org/10.1007/978-3-319-12511-4_1
11. de Vries, A.P., Alink, W., Cornacchia, R.: Search by strategy. In: Proceedings of the Third Workshop on Exploiting Semantic Annotations in Information Retrieval, pp. 27–28. ACM (2010)
12. Pazer, J.W.: The importance of the boolean search query in social media monitoring tools. DragonSearch white paper (2013). https://www.dragon360.com/wp-content/uploads/2013/08/social-media-monitoring-tools-boolean-search-query.pdf. (Accessed 22 Mar 2018)
13. Yi, J.S., Melton, R., Stasko, J., Jacko, J.A.: Dust & magnet: multivariate information visualization using a magnet metaphor. Inf. Vis. **4**(4), 239–256 (2005). https://doi.org/10.1057/palgrave.ivs.9500099

node-indri: Moving the Indri Toolkit to the Modern Web Stack

Felipe Moraes(✉) and Claudia Hauff(✉)

{f.moraes,c.hauff}@tudelft.nl

Delft University of Technology, Delft, The Netherlands

Abstract. We introduce `node-indri`, a `Node.js` module that acts as a wrapper around the `Indri` toolkit, and thus makes an established IR toolkit accessible to the modern web stack. `node-indri` exposes many of `Indri`'s functionalities and provides direct access to document content and retrieval scores for web development (in contrast to, for instance, the `Pyndri` wrapper). This setup reduces the amount of *glue code* that has to be developed and maintained when researching search interfaces, which today tend to be developed with specific JavaScript libraries such as `React.js`, `Angular.js` or `Vue.js`. The `node-indri` repository is open-sourced at https://github.com/felipemoraes/node-indri.

1 Introduction

The information retrieval (IR) field is aided by numerous efficient search engine implementations, aimed at research and industry, such as `Indri` [1], `Lucene`[1], `Terrier` [2] and `Anserini` [3]. In the data science field, some of these efforts have evolved into frameworks such as `Elasticsearch`[2] and `Terrier-Spark` [4]. Recently, in order to enable data scientists to make use of `Indri` as part of their workflow, Van Gysel et al. [5] have made `Indri` accessible to the Python ecosystem (via `Pyndri`).

In this paper, we make `Indri` accessible to the modern web stack. Many modern web applications and frameworks make use of `Node.js`.[3] A significant advantage of this framework is the single programming language on the client and server-side (JavaScript), which simplifies development; in addition, `Node.js` is highly scalable [6]. In order to design and evaluate web search interfaces, a backend, implemented in `Node.js`, requires access to a search system. One option is to call `Indri` via system calls. However, the disadvantage of system calls via shell commands is the extra layer of communication with the operating system.

Here, we present an alternative, `node-indri`, a `Node.js` module implemented with an easy-to-use API. It provides access to basic `Indri` functionalities such as search with relevance feedback and document scoring. Importantly, `node-indri` is implemented in a non-blocking manner. We here discuss `node-indri`'s module

---

[1] http://lucene.apache.org/.

[2] https://www.elastic.co/products/elasticsearch.

[3] https://nodejs.org/.

© Springer Nature Switzerland AG 2019
L. Azzopardi et al. (Eds.): ECIR 2019, LNCS 11438, pp. 241–245, 2019.
https://doi.org/10.1007/978-3-030-15719-7_32

features and we compare its efficiency with `Indri` and `Pyndri` on the Aquaint and ClueWeb12 corpora with a load of 10K queries. We find that `node-indri` can be efficiently used in modern web backend development with comparable efficiency to `Indri` and `Pyndri`.

2   The `node-indri` Module

`node-indri`'s development started with the need to make `Indri`'s state-of-the-art relevance feedback models accessible to students, that (i) tend to have little experience with C++, but are familiar with modern web programming paradigms and (ii) are not IR experts and thus struggle to make sense of `Indri`'s internals.

2.1   Functionalities

Figure 1 shows the three layers of abstraction of `node-indri`. Our module exposes `Indri` features through the `Searcher`, `Reader`, and `Scorer` classes. These classes are implemented in C++ with the help of Native Abstractions for `Node.js`[4], a series of macros that abstract away the differences between the V8 and `libuv` API versions (which together form the core of the `Node.js` framework and are written in C++).

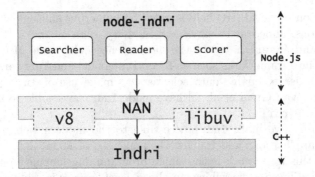

Fig. 1. Overview of `node-indri` layers of abstractions and their implementation languages and platforms.

In Table 1, we list the arguments of `node-indri`'s three classes. Each class has at most two methods with arguments that depend on the functionalities exposed from `Indri`. The last argument is a callback function implementing the error-first pattern. In this manner, `node-indri` is an asynchronous module, with most of these functions assessing lower-level system functionalities through

---

[4] https://github.com/nodejs/nan.

Table 1. Overview of the arguments necessary for node-indri's method calls. Our API is simple and includes only one method per class. The last argument is always a callback that is executed when the data has been retrieved. Underlined are the required parameters.

Searcher.search	Reader.getDocument	Scorer.scoreDocuments	Scorer.retrieveTopKScores
query, page, results-per-page, feedback-docs, callback	docid, callback	query, docs, callback	query, number-of-results, callback

libuvl. This in turn means that the methods are executed in Node.js' thread pool, making node-indri naturally parallel.

The models' hyperparameter settings (e.g. μ in the case of language modeling with Dirichlet smoothing) are manually set via a configuration file. We now discuss the goal of each of the three classes node-indri makes available to its users in turn:

Searcher. This class exposes the functionalities of Indri's QueryEnvironment and RMExpander classes through the method search which returns a list of search results in a paginated manner. When a Searcher object is instantiated, it takes a configuration object as argument (these settings include the retrieval models' hyperparameters and flags of the type of data to return). When a call to search() is made and no feedback documents are provided as argument, the standard query likelihood model is employed, otherwise RM3 is [7]. Depending on the configuration settings, the returned result list may contain document snippets (as provided by Indri's SnippetBuilder), document scores, document text and other metadata.

Reader. This class exposes the functionalities of an Indri index through the method getDocument in order to return a document's meta- and text data.

Scorer. This class provides access to the retrieval scores of a list of documents via the method scoreDocuments. In addition, it provides retrieveTopKScores to retrieve the scores and document ids of the top ranked documents for a query.

2.2   Use Cases

We have used node-indri as search results' provider in the backend of a large-scale collaborative search system, SearchX [8]. SearchX's backend supports the inclusion of many IR backends such as Elasticsearch and Bing API calls. In order to include node-indri as one of the supported backends, we implemented Searcher to provide search results (with or without snippets) in a pagination manner leveraging feedback documents, Reader to enable the rendering of a document's content when a user has clicked on it, and Scorer to enable our backend to have direct access to documents' scores for reranking purposes.

3   Efficiency Analysis

We now present an efficiency analysis of node-indri, comparing it to Indri and Pyndri. We indexed two standard test corpora—Aquaint and ClueWeb12B—with Indri and measured the execution time for 10k queries of the TREC 2007 Million Query track[5] across the three toolkits. As retrieval model we fixed language modeling with Dirichlet smoothing; up to 1000 documents were returned per query. We use Indri's IndriRunQuery application for this purpose; for Pyndri and node-indri we implemented scripts to achieve the same behaviour. Specifically, Pyndri's results were obtained with a Python script implemented with multiprocessing. In node-indri, we make use of Promises.all. We limit the execution to 20 threads for all three toolkits[6]. Table 2 presents the overall query execution time of the three toolkits.

Our results show that node-indri has execution times comparable to Indri and Pyndri. We can thus use node-indri efficiently in modern web backend development. We find for a small collection such as Aquaint (1 million documents), all three toolkits to obtain very similar execution times (between 25 and 29 s). In contrast, for a larger collection such as ClueWeb12B (50 million documents), the execution times differ to some extent: Indri takes on average 27 min to process 10K queries, while node-indri and Pyndri take 34 and 37 min respectively. This is expected, as both node-indri and Pyndri have additional overhead due to the frameworks they are built upon.

Table 2. Overview of retrieval efficiency (in seconds) across two corpora. 10K queries from the TREC 2017 Million Queries track were executed 20 times. Reported are the average (standard deviation) execution time of the batches.

	Aquaint	ClueWeb12B
Indri	29 s (0.30 s)	1645 s (20 s)
Pyndri	25 s (1.22 s)	2262 s (340 s)
node-indri	25 s (0.58 s)	2058 s (338 s)

4   Conclusions

We have introduced node-indri, a Node.js module to enable users with a good web development background (but minimal C++ knowledge) to efficiently implement search applications. We have described how node-indri exposes Indri's functionalities and how it is currently being used in the backend of a large-scale collaborative search system, that has been successfully tested with several hundred users. Furthermore, we compared node-indri's efficiency in the batch setting with Indri's and Pyndri's.

---

[5] https://trec.nist.gov/data/million.query07.html.
[6] More details and benchmark code are included in the GitHub repository.

Acknowledgements. This work was funded by NWO projects LACrOSSE (612.001.605) and SearchX (639.022.722). The authors would like to thank Harry Scells for his input.

References

1. Strohman, T., Metzler, D., Turtle, H., Croft, W.B.: Indri: a language model-based search engine for complex queries. In: ICIA (2005)
2. Ounis, I., Amati, G., Plachouras, V., He, B., Macdonald, C., Lioma, C.: Terrier: a high performance and scalable information retrieval platform. In: OSIR (2006)
3. Yang, P., Fang, H., Lin, J.: Anserini: enabling the use of Lucene for information retrieval research. In: SIGIR (2017)
4. Macdonald, C.: Combining terrier with Apache Spark to create agile experimental information retrieval pipelines. In: SIGIR (2018)
5. Van Gysel, C., Kanoulas, E., de Rijke, M.: Pyndri: a Python interface to the Indri search engine. In: Jose, J.M., et al. (eds.) ECIR 2017. LNCS, vol. 10193, pp. 744–748. Springer, Cham (2017). https://doi.org/10.1007/978-3-319-56608-5_74
6. Tilkov, S., Vinoski, S.: Node.js: using JavaScript to build high-performance network programs. IEEE Internet Comput. **14**, 80–83 (2010)
7. Lavrenko, V., Croft, W.B.: Relevance based language models. In: SIGIR (2001)
8. Putra, S.R., Moraes, F., Hauff, C.: Searchx: Empowering collaborative search research. In: SIGIR (2018)

PaperHunter: A System for Exploring Papers and Citation Contexts

Michael Färber[1]([✉])([iD]), Ashwath Sampath[1], and Adam Jatowt[2]

[1] Department of Computer Science, University of Freiburg,
Freiburg im Breisgau, Germany
michael.faerber@cs.uni-freiburg.de, ashwath92@gmail.com
[2] Department of Social Informatics, Kyoto University, Kyoto, Japan
adam@kuis.db.kyoto-u.ac.jp

Abstract. In this paper, we present a system that allows researchers to search for papers and in-text citations in a novel way. Specifically, our system allows users to search for the textual contexts in which publications are cited (so-called *citation contexts*), given either the cited paper's title or the cited paper's author name. To better assess the citations qualitatively, our system displays indications about the so-called *citation polarity*, i.e., whether the authors wrote about the cited publication in a positive, neutral, or negative way. Our system is based on all computer science papers from arXiv.org and can be used by computer science researchers to reflect on their appearance within the scientific community as well as by researchers studying citations.

Keywords: Scientific papers · Bibliometrics · Citation context · Citation polarity

1 Motivation

Researchers in all scientific fields are nowadays confronted with vast amounts of scientific papers published within a short period [1–3]. As a consequence, scientists are increasingly dependent on publication search engines [3], such as Google Scholar, to search – typically via keywords – for the metadata of relevant papers and the papers themselves. However, to our knowledge, no system lets users search the citation relationships between papers directly.

We present such a system here. Besides the usual ability to search for papers' full metadata and papers themselves, it provides the following exclusive features:

1. Given a paper's title, the system lets users search for the text passages in which the paper is cited (i.e., citation contexts). Thus, this search functionality can be used to analyze how the paper is cited in other papers.
2. Given an author's name, the system lets users search for the text passages in which the author's publications are cited. Also, the system provides the metadata for these citing papers. This search functionality allows users to analyze how the community perceives an author.

© Springer Nature Switzerland AG 2019
L. Azzopardi et al. (Eds.): ECIR 2019, LNCS 11438, pp. 246–250, 2019.
https://doi.org/10.1007/978-3-030-15719-7_33

3. To allow the user to quickly recognize how the citations are used within the text passages, our system presents indications about the so-called *citation polarity*, i.e., whether the authors of the text passage wrote about the cited publication in a positive, neutral, or negative way.

Several user groups can benefit from using our system:

1. *Ordinary researchers* might be interested to know *in which papers, in which contexts, and in which ways they or people in their environment (e.g., colleagues, competitors) are cited.* Having this information will allow researchers to react accordingly. For instance, aspects concerning the cited papers that other researchers have mentioned incorrectly or imprecisely, or that are missing, can be clarified in future papers and in communication between authors. Thus, our search functionalities will allow for more nuanced scientific exchanges between researchers.
2. *Bibliometrics, scientometrics, and social analysis researchers* who focus on analyzing citations and measure the impact of citations can use our search system to gain new insights concerning the usage of citations.
3. *Practitioners*, such as software developers, can use our system to determine how methods and data sets (published via papers) have been used.

Our demonstration system is available online at http://paperhunter.net. Also, the source code is available online as open source code.[1]

2   System Design

We use the *arXiv CS data set* [4] as our database. This data set contains metadata about all arXiv.org papers in the field of computer science published before the end of 2017 as well as the contents of these publications in plaintext. Overall, this data set contains about 16 million sentences from about 90,000 papers and is said to be one of the few data sets containing the full texts of papers of such a size and cleanliness [4]. Note that in the data set the formulas have been replaced by placeholders and in-text citation markers have been replaced by identifiers. Separate files contain the mappings of these identifiers to the cited papers' metadata (including the authors' names, title, venue, and year).

We index the arXiv CS data set using Apache Solr.[2] When a user searches for papers or citation contexts, our system retrieves all the result items according to the default TF-IDF scoring and then returns the top n results based on the papers' publication date. This is done due to users typically being interested in more recent papers. n is provided by the user (by default, $n = 100$). We use Python to process the data and Django to create the user interface.

In the following, we present the different search capabilities provided by our system. Note that the first three search functionalities can be considered basic functions and are prevalent in existing related works. The last two search possibilities are novel, exclusive search functionalities proposed in this paper.

---

[1] https://github.com/michaelfaerber/paperhunter.
[2] https://lucene.apache.org/solr/.

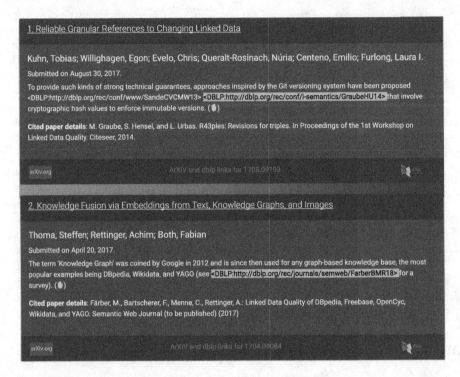

Fig. 1. Example of searching for citation contexts given a cited paper's title (here, the partial title "linked data quality").

1. Search for papers given a phrase. Given a phrase (here within the field of computer science), this search functionality allows users to search for all papers on arXiv that contain this phrase in the body text. Thus, this functionality is particularly suitable when papers with very specific keywords (e.g., "semantic cognition", "knowledge base completion", and "stochastic pooling") and not only abstract research topics need to be retrieved.

2. Search for papers given a paper's (incomplete) title. Given the title of a paper, this search functionality allows users to search for its full metadata. Also, only parts of the paper's title can be provided as inputs. For instance, searching for "linked data quality" allows users to search for all papers on that topic.

3. Search for papers given an author's name. Given an author's name, this functionality allows users to retrieve the full metadata of all papers written by this author.

4. Search for citation contexts given the cited paper's title. Given any paper's title, this search functionality allows users to retrieve all sentences from the bodies of arXiv papers in which the specified paper is cited (see Fig. 1). If a publication is cited several times within a paper, then all citation contexts of this paper are grouped together. To allow a quick assessment of the retrieved citation contexts by the user, an icon is presented next to each citation context. This icon indicates the citation context's polarity [5,6] and can be positive (i.e., the cited paper is praised), neutral, or negative. The citation polarity values were determined offline by using Athar et al.'s approach [7] and data set for training.

5. *Search for citation contexts given the cited paper's author.* Here, users can search for the citation contexts (plus the papers' metadata and the links to arXiv.org) in which papers written by the given author were cited. For instance, by searching for "Tim Berners-Lee", our search engine retrieves all contexts in which papers written by Tim Berners-Lee have been cited. We also provide the citation polarity indication for this search functionality.

3 Related Work

Search Engines for Papers and Citation Contexts. Paper search systems, such as *Google Scholar*, can be used to retrieve papers and obtain papers' metadata. The keywords provided by the user are thereby matched with the papers' metadata and with the papers' contents. Consequently, such systems essentially cover the first three functionalities. However, to the best of our knowledge, no system has been presented that allows users to search specifically for citation contexts, as enabled in our system. Note that citation contexts have already been analyzed with respect to various aspects [8–12]. However, we are not aware of available systems that provide the citation polarity information [5,7] of citation contexts.

Data Sets with Papers' Contents and Citation Contexts. Besides the arXiv CS data set, the CiteSeerX [13] and the Microsoft Academic Graph (MAG) [14] are also considerably large data sets that contain already extracted citation contexts. However, the version of CiteSeerX available online is not up-to-date. MAG's citation contexts are partially noisy, and the exact positions of citations within the citation contexts are not provided.

4 Conclusion

In this paper, we presented a system that allows searching for papers and citation contexts in a novel way. As our system is based on the large collection of computer science papers on arXiv.org, it can be used by any computer science researcher. Also, scientometrics researchers can use the system to explore how citations are embedded in papers. In terms of future work, we plan on improving the ranking of papers and citation contexts and to include a component for recommending papers that are related to the retrieved citation contexts.

Acknowledgements. This research has been supported by the Research Innovation Fund of the University of Freiburg (#2100189801), by JSPS KAKENHI grants (#17H01828, #18K19841), and by the MIC/SCOPE (#171507010) grant.

References

1. Bornmann, L., Mutz, R.: Growth rates of modern science: a bibliometric analysis based on the number of publications and cited references. J. Assoc. Inf. Sci. Technol. 66(11), 2215–2222 (2015)
2. Fortunato, S., et al.: Science of science. Science 359(6379) (2018). http://science.sciencemag.org/content/359/6379/eaao0185
3. Ware, M., Mabe, M.: The STM report: an overview of scientific and scholarly journal publishing. (2015)
4. Färber, M., Thiemann, A., Jatowt, A.: A high-quality gold standard for citation-based tasks. In: Proceedings of the International Conference on Language Resources and Evaluation, LREC 2018 (2018)
5. Abu-Jbara, A., Ezra, J., Radev, D.R.: Purpose and polarity of citation: towards NLP-based bibliometrics. In: Proceedings of the 2013 Conference of the North American Chapter of the Association of Computational Linguistics, NAACL 2013, pp. 596–606 (2013)
6. Ghosh, S., Das, D., Chakraborty, T.: Determining sentiment in citation text and analyzing its impact on the proposed ranking index. In: Proceedings of the 17th International Conference on Computational Linguistics and Intelligent Text Processing, CICLing 2016, pp. 292–306 (2016)
7. Athar, A.: Sentiment analysis of citations using sentence structure-based features. In: Proceedings of the 49th Annual Meeting of the Association for Computational Linguistics, NAACL 2011, pp. 81–87 (2011)
8. Alvarez, M.H., Gómez, J.M.: Survey about citation context analysis: tasks, techniques, and resources. Nat. Lang. Eng. 22(3), 327–349 (2016)
9. Tahamtan, I., Bornmann, L.: Core elements in the process of citing publications: conceptual overview of the literature. J. Inf. 12(1), 203–216 (2018)
10. Ding, Y., Zhang, G., Chambers, T., Song, M., Wang, X., Zhai, C.: Content-based citation analysis: the next generation of citation analysis. JASIST 65(9), 1820–1833 (2014)
11. Bornmann, L., Daniel, H.: What do citation counts measure? A review of studies on citing behavior. J. Doc. 64(1), 45–80 (2008)
12. Todeschini, R., Baccini, A.: Handbook of Bibliometric Indicators: Quantitative Tools for Studying and Evaluating Research. Wiley, Hoboken (2016)
13. Caragea, C., et al.: CiteSeer^x: a scholarly big dataset. In: de Rijke, M., et al. (eds.) ECIR 2014. LNCS, vol. 8416, pp. 311–322. Springer, Cham (2014). https://doi.org/10.1007/978-3-319-06028-6_26
14. Sinha, A., et al.: An overview of microsoft academic service (MAS) and applications. In: Proceedings of the 24th International Conference on World Wide Web, WWW 2015, pp. 243–246 (2015)

Interactive System for Automatically Generating Temporal Narratives

Arian Pasquali[1,2]([✉]), Vítor Mangaravite[1], Ricardo Campos[1,3],
Alípio Mário Jorge[1,2], and Adam Jatowt[4]

[1] LIAAD – INESCTEC, Porto, Portugal
{arrp,vima}@inesctec.pt
[2] FCUP, University of Porto, Porto, Portugal
amjorge@fc.up.pt
[3] Polytechnic Institute of Tomar - Smart Cities Research Center, Tomar, Portugal
ricardo.campos@ipt.pt
[4] Kyoto University, Kyoto, Japan
adam@dl.kuis.kyoto-u.ac.jp

Abstract. In this demo, we present a tool that allows to automatically generate temporal summarization of news collections. Conta-me Histórias (Tell me stories) is a friendly user interface that enables users to explore and revisit events in the past. To select relevant stories and temporal periods, we rely on a key-phrase extraction algorithm developed by our research team, and event detection methods made available by the research community. Additionally, we offer the engine as an open source package that can be extended to support different datasets or languages. The work described here stems from our participation at the Arquivo.pt 2018 competition, where we have been awarded the first prize.

Keywords: Information retrieval · Temporal summarization

1 Introduction

During the last decade, we have been witnessing an ever-growing number of online content posing new challenges for those who aim to understand a given event. This exponential growth of the volume of data, together with the phenomenon of media bias, fake news and filter bubbles, has contributed to the creation of new challenges in information access and transparency. For instance, following the media coverage of long-lasting events like wars, migration or economic crises can be oftentimes confusing and demanding. One possible solution is the adoption of timelines to support story-telling as a way to organize the different phases of complex events. Media outlets use this type of solution very often. However, manually building such timelines can be very laborious and time-consuming for journalists. Besides, it simply does not scale. One possible approach to overcome this problem is to automatically summarize large amount of news into consistent narratives, an active topic of research in the academic community [1,6] with the proposals of innovative and creative solutions [7,8].

© Springer Nature Switzerland AG 2019
L. Azzopardi et al. (Eds.): ECIR 2019, LNCS 11438, pp. 251–255, 2019.
https://doi.org/10.1007/978-3-030-15719-7_34

2   Proposed Solution

In this demo paper, we propose a user-friendly interface that allows running queries on news sources and exploring the results in a summarized and temporally organized manner with the help of an interactive timeline. Our project named Conta-me Histórias (Tell me stories) results from the participation on the Arquivo.pt 2018 contest, where we have achieved the first rank among 27 competing teams[1].

Given a user query, our system automatically identifies relevant dates and the most important headlines to illustrate the story. Figure 1 gives an overview of the framework, which can be described in 5 simple steps: (1) News Retrieval; (2) Term Weighting; (3) Identifying Relevant Time Intervals; (4) Computing Headline Scores; (5) Deduplication.

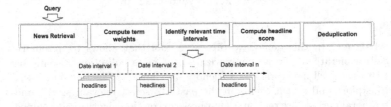

Fig. 1. System architecture

News Retrieval. The first step in the pipeline is to run the query against any data source of interest. The only requirements are that the result set must contain a list of items with a headline, timestamp, URL and optionally a source name. The source code for our temporal summarization framework, as well as examples on how to adapt for different data sources are available online[2].

Term Weighting. We then calculate each term weight through an adapted version of YAKE! [2,3] keyword extractor method (Best Short Paper at ECIR'18), which relies on statistical features to select the most important key-phrases of a document. In our approach, every headline is treated as an independent document. We calculate a number of term statistics, such as frequency within the entire result set, average frequency, standard deviation and positional features. All these features enable the identification of common and rare terms with much higher accuracy than TF-IDF. This step produces a term dictionary that will be used later in the pipeline to identify n-grams candidates. A more thorough discussion of the details of YAKE!, may be found in the above referred papers.

Identifying Relevant Time Intervals. To select relevant time periods, we applied a strategy that forces the system to select intervals with at least one

---

[1] http://sobre.arquivo.pt/en/arquivo-pt-2018-award-winners/.
[2] https://github.com/LIAAD/TemporalSummarizationFramework.

peak of occurrence. This strategy tries to identify main events related to the query assuming such events result in many headlines in a short period of time. We begin by dividing the timespan into 60 equi-width intervals (partitions). These partitions are used to aggregate the frequencies in order to find peaks of occurrences. The interval boundaries are then given by the fewer partition (*smallest peak*) among each pair of peaks; We apply the function *argrelextrema* from the Scipy's signal processing module[3] in order to find each relative peak of occurrences. Figure 2 illustrates time interval selection for the query *"Guerra na Siria"* (War at Syria). In this case, the system identified seven important time intervals (between 2010 and 2016): the red lines represent interval boundaries, while the blue ones highlight number of news aggregated by date.

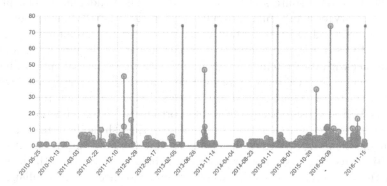

Fig. 2. Relevant time intervals detection for the query *"War at Syria"*

Computing Headline Scores. After determining the relevant time periods, we now aim to determine the most important headlines for each temporal interval. It is important to mention that it is not a simple concatenation of headlines, as we will only present the most important ones. In order to proceed with this summarization process, we look at each individual term of the headline and compute an aggregated value based on its individual term weight. A fully detailed description of the underlying scientific approach and the evaluation methodology on this particular aspect can be found in our recent work [3].

De-duplication. Finally, we eliminate similar key-phrases based on Levenshtein similarity measure. We specify a threshold where we ignore key-phrases that are more than 80% similar. When comparing a pair of strings we keep the longer one, assuming it carries more information than the shorter version. This threshold is a parameter and can be fine tuned for different cases. For each time interval, we then select the top 20 key-phrases ordered by their relevance. It also is possible to experiment with different methods like Jaccard, Monge-Elkan and Jaro-Winkler. A formal evaluation on this and other parameters are left for future work.

---

[3] https://docs.scipy.org/doc/scipy/reference/signal.html.

3   Demonstration

Users can interact with our demo either through Conta-me Histórias[4] or Tell
me Stories[5]. The former allows users to explore the Portuguese Web Archive [5]
through their free-text search API which enable users to explore their archived
results from a period that is mostly concentrated on 2010 to the present. To guarantee the plurality and the diversity of the information, we consider news from 24
popular Portuguese media outlets. The latter is built on top of the Signal Media
Dataset [4], a one-million news articles collection (mainly English, but also non-
English and multi-lingual articles) which were originally collected rom a variety
of news sources (such as Reuters) for a period of 1 month (1–30 September 2015).

Our proposed solution can be easily adapted to other scenarios including different kinds of data sources (e.g. social media posts, academic papers, propri-
etary repository, etc) and languages since it is mostly language independent. This
may be understood as an important contribution for anyone interested in having access to a summarized temporal view of their data. In its current form, users
may interact with the system through an interface where several options, including the specification of a query (a free text field) and a time interval (last five, ten,
twenty or thirty years) are offered to the users. Below we present the results for
the query War at Syria obtained from the Portuguese Web Archive project. The
results (which were translated from Portuguese to English) helps us to understand
this long-lasting long-lasting conflict, highlighting the most important headlines
since 2010, when the first popular protests reached the streets and the news. We
can see the brutal repression and the escalation of violence against civilians. Subsequently, that year, a famous journalist died in an attack, other countries got
involved in the conflict and news about humanitarian crises appeared. We also
confirm events where civilians suffered attacks like the infamous air strike and
bombing on a maternity unit. This piece of example illustrates how this kind of
tool can help to understand the escalation of violence in Syria, helping users to
explore relevant dates, headlines and actors in the story.

From	To	Top headlines
5/2010	7/2011	Syrian officials launch tear gas against protesters Security forces shoot at protesters New York Times journalist with the Pulitzer died of an Asthma attack in Syria
8/2011	3/2012	Assad promises elections in February in Syria US withdraws ambassador from Syria for security reasons NATO says goodbye to Libya and the world turns to Syria
7/2012	12/2012	Meeting of senior officials in Geneva failed agreement to end violence in Syria Russia delivers three war helicopters to Syria Red Cross says Syria is in civil war
7/2016	11/2016	Maternity unit among hospitals bombed in Idlib air strikes Russian helicopter shot down in Syria. Turkish army enters Syria

---

[4] http://contamehistorias.pt/arquivopt.
[5] http://signal.tellmestories.pt.

An additional available experimental interface[6] was made available such that interested researchers may experimentally test our demo with several different options, such as deduplication and event detection methods.

Acknowledgements. This work is partially funded by the ERDF through the COMPETE 2020 Programme within project POCI-01-0145-FEDER-006961, and by National Funds through the FCT as part of project UID/EEA/50014/2013.

References

1. Aslam, J.A., Ekstrand-Abueg, M., Pavlu, V., Diaz, F., Sakai, T.: TREC 2013 temporal summarization. In: TREC (2013)
2. Campos, R., Mangaravite, V., Pasquali, A., Jorge, A.M., Nunes, C., Jatowt, A.: YAKE! Collection-independent automatic keyword extractor. In: Pasi, G., Piwowarski, B., Azzopardi, L., Hanbury, A. (eds.) ECIR 2018. LNCS, vol. 10772, pp. 806–810. Springer, Cham (2018). https://doi.org/10.1007/978-3-319-76941-7_80
3. Campos, R., et al.: A text feature based automatic keyword extraction method for single documents. In: Pasi, G., Piwowarski, B., Azzopardi, L., Hanbury, A. (eds.) ECIR 2018. LNCS, vol. 10772, pp. 684–691. Springer, Cham (2018). https://doi.org/10.1007/978-3-319-76941-7_63
4. Corney, D., Albakour, D., Martinez, M., Moussa, S.: What do a million news articles look like? In: Proceedings of the First International Workshop on Recent Trends in News Information Retrieval co-located with 38th European Conference on Information Retrieval (ECIR 2016), Padua, Italy, 20 March 2016, pp. 42–47 (2016)
5. Gomes, D., Cruz, D., Miranda, J., Costa, M., Fontes, S.: Search the past with the Portuguese web archive. In: 22nd International World Wide Web Conference, Rio de Janeiro, Brasil (2013)
6. Jorge, A.M., et al.: Report on the first international workshop on narrative extraction from texts (Text2Story 2018). In: SIGIR Forum, vol. 52, no. 1, pp. 150–152. ACM Press (2018)
7. Schubotz, T., Krestel, R.: Online temporal summarization of news events. In: 2015 IEEE/WIC/ACM International Conference on Web Intelligence and Intelligent Agent Technology (WI-IAT) vol. 1, pp. 409–412 (2015)
8. Tran, G., Alrifai, M., Herder, E.: Timeline summarization from relevant headlines. In: Hanbury, A., Kazai, G., Rauber, A., Fuhr, N. (eds.) ECIR 2015. LNCS, vol. 9022, pp. 245–256. Springer, Cham (2015). https://doi.org/10.1007/978-3-319-16354-3_26

---

[6] http://labs.tellmestories.pt.

CLEF Organizers Lab Track

Early Detection of Risks on the Internet: An Exploratory Campaign

David E. Losada[1]([✉]), Fabio Crestani[2], and Javier Parapar[3]

[1] Centro Singular de Investigación en Tecnoloxías da Información (CiTIUS),
Universidade de Santiago de Compostela, Santiago de Compostela, Spain
david.losada@usc.es
[2] Faculty of Informatics, Universitá della Svizzera italiana (USI),
Lugano, Switzerland
fabio.crestani@usi.ch
[3] Information Retrieval Lab, University of A Coruña, A Coruña, Spain
javierparapar@udc.es

Abstract. This paper summarizes the activities related to the CLEF
lab on early risk prediction on the Internet (eRisk). eRisk was initiated in 2017 as an attempt to set the experimental foundations of early
risk detection. The first edition essentially focused on a pilot task on
early detection of signs of depression. In 2018, the lab was enlarged
and included an additional task oriented to early detection of signs of
anorexia. We review here the main lessons learned and we discuss our
plans for 2019.

1 Introduction

The main goal of eRisk, a CLEF lab on early risk detection [3,4], is to explore
issues of evaluation methodology, performance metrics and other issues related
to building testbeds for early risk detection. Early risk detection can be useful
in different areas, particularly those related to health and safety. For instance,
warning alerts can be given when a predator starts contacting a child for sexual
purposes, or when an offender publishes antisocial threats on Social Media. eRisk
intends to pioneer a new interdisciplinary area of research whose results would
be potentially applicable to detect potential paedophiles, stalkers, individuals
with a latent tendency to fall into the hands of criminal organizations, people
with suicidal inclinations, or people susceptible to depression.

The lab views early risk prediction as a process of accumulation of evidence
where alerts should be made when there is enough evidence about a certain
type of risk. For example, the pieces of evidence could be Social Media posts
submitted at various times. A common characteristic of the campaigns run so
far is that the pilot tasks worked with stream data and the participating teams
had to find a balance between emitting early decisions· (based on just a few
pieces of evidence) and emitting *not-so-early* decisions (if they opt to wait and
analyze more pieces of evidence). Although the collection building strategies and

L. Azzopardi et al. (Eds.): ECIR 2019, LNCS 11438, pp. 259–266, 2019.
https://doi.org/10.1007/978-3-030-15719-7_35

performance metrics are generic and potentially applicable to the usage scenarios described above, all previous editions of eRisk have focused on data related to psychological disorders.

The rest of the paper discusses previous results and sketches our plans for eRisk 2019.

2   Previous Editions of eRisk

eRisk 2017 included a pilot task on early detection of depression. This exploratory task was based on the test collection and metrics presented in [2]. The interactions between depression and natural language use is a challenging problem and by sharing this collection with other teams we expected to instigate fruitful discussions on these issues. The 2017 participants employed a wide range of techniques in information access and closely related fields, such as Natural Language Processing, Machine Learning, and Information Retrieval. This pilot task was moderately successful. More than 30 teams registered for this task and got access to the data. The challenge was demanding because it involved ten different releases of data, and, after each release, the teams had one week to submit their results. Furthermore, eRisk was new to all CLEF participants and, thus, the teams were not familiar with the task. As a result, only 8 teams were able to follow this tight and novel schedule. We got 30 different contributions (system variants) from the 8 contributing teams.

In 2018, we proposed two campaign-like tasks: task 1 was a continuation of the pilot task that ran in 2017 (early detection of signs of depression) and task 2 was new (early detection of signs of anorexia). Both tasks had the same structure and evaluation design. Compared with eRisk 2017, eRisk 2018 received increased attention. In 2018, we had 41 registered participants. We got 45 submissions (system variants) for Task 1 and 35 submissions (system variants) for Task 2. There were 11 active teams that engaged into the eRisk tasks. These numbers suggest that the lab is slowly becoming an experimental reference for early detection technologies. In 2019, we expect to increase participation because many groups are now familiar with eRisk and its tight schedule, and some of them have already worked with the data (although they could not make it to send the required results in the previous years).

2.1   Tasks

The tasks consisted of sequentially processing pieces of evidence –in the form of writings (post or comments) posted by Social Media users– and learn to detect early signs of risk as soon as possible. Texts had to be processed by the participating systems in the order they were created. In this way, systems that effectively perform this task could be employed to sequentially track user publications in blogs, social networks, or other types of online media. Table 1 reports the main statistics of the collections utilized in eRisk 2017 and eRisk 2018.

Table 1. Main statistics of the train and test collections used in the eRisk 2017 and 2018 tasks (depression and anorexia).

	Train		Test	
eRisk 2017 - depression task				
	Depressed	*Control*	*Depressed*	*Control*
Num. subjects	83	403	52	349
Num. submissions (posts & comments)	30,851	264,172	18,706	217,665
Avg num. of submissions per subject	371.7	655.5	359.7	623.7
Avg num. of days from first to last submission	572.7	626.6	608.31	623.2
Avg num. words per submission	27.6	21.3	26.9	22.5
eRisk 2018 - depression task				
	Depressed	*Control*	*Depressed*	*Control*
Num. subjects	135	752	79	741
Num. submissions (posts & comments)	49,557	481,837	40,665	504,523
Avg num. of submissions per subject	367.1	640.7	514.7	680.9
Avg num. of days from first to last submission	586.43	625.0	786.9	702.5
Avg num. words per submission	27.4	21.8	27.6	23.7
eRisk 2018 - anorexia task				
	Anorexia	*Control*	*Anorexia*	*Control*
Num. subjects	20	132	41	279
Num. submissions (posts & comments)	7,452	77,514	17,422	151,364
Avg num. of submissions per subject	372.6	587.2	424.9	542.5
Avg num. of days from first to last submission	803.3	641.5	798.9	670.6
Avg num. words per submission	41.2	20.9	35.7	20.9

Reddit was the main source of data for our experimental tasks. It is an open-source platform where community members submit content, vote submissions, and publications are organized by areas of interests (*subreddits*). Reddit has a large community of members (*redditors*) and many of the members have a large history of previous submissions (covering several years). It also contains substantive contents about different medical conditions, such as depression or eating disorders. Reddit's terms and conditions allow to use its contents for research purposes[1].

The test collections used in eRisk 2017 and eRisk 2018 have the same format as the collection described in [2]. It is a collection of publications (posts or comments) done by redditors. For each task, there were two classes of users: the positive class (depression or anorexia, respectively) and a negative class (con-

---

[1] Reddit privacy policy states explicitly that the posts and comments redditors make are not private and will still be accessible after the redditor's account is deleted. Reddit does not permit unauthorized commercial use of its contents or redistribution, except as permitted by the doctrine of fair use. This research is an example of fair use.

trol group). The positive class was extracted following the approach proposed by Coppersmith et al. [1]. These authors proposed an automatic method to identify people diagnosed with depression in Twitter. We have adapted this estimation method to Reddit as follows. Self-expressions related to diagnoses can be obtained by running specific searches against Reddit (e.g. "I was diagnosed with anorexia"). Next, we manually reviewed the matched posts to verify that they were really genuine. Our confidence on the quality of these assessments is high. In Reddit, there are many support communities for people suffering from different disorders and it is often the case that redditors go there and are very explicit about their problems and medical condition. Although this method requires manual intervention, it is a simple and effective way to extract a large group of people that explicitly declare having being diagnosed with a given disorder. The manual reviews were strict. Expressions like "I have anorexia", "I think I have anorexia", or "I am anorexic" did not qualify as explicit expressions of a diagnosis. We only included a redditor into the positive group when there was a clear and explicit mention of a diagnosis (e.g., "In 2013, I was diagnosed with anorexia nervosa", "After struggling with anorexia for many years, yesterday I was diagnosed").

For each user, the collection contains his sequence of submissions (in chronological order) and this sequence was split into 10 chunks. The first chunk has the oldest 10% of the submissions, the second chunk has the second oldest 10%, and so forth. Each task was organized into two different stages:

- **Training stage.** Initially, the teams that participated in the task had access to some training data. In this stage, we released the entire history of submissions done by a set of training users. All chunks of all training users were sent to the participants and the actual class of each training user was provided.
- **Test stage.** The test stage had 10 releases of data (one release per week). The first week we gave the 1st chunk of data to the teams (oldest submissions of all test users), the second week we gave the 2nd chunk of data (second oldest submissions of all test users), and so forth. After each release, the teams had to process the data and, before the next week, each team had to choose between: (a) emitting a decision on the user (i.e. positive or negative), or (b) making no decision (i.e. waiting to see more chunks). This choice had to be made for each user in the test split. If the team emitted a decision then the decision was considered as final. The systems were evaluated based on the accuracy of the decisions and the number of chunks required to take the decisions (see below).

2.2 Evaluation Metrics for Early Risk Detection

The evaluation of the tasks considered standard classification metrics, such as F1, Precision and Recall (computed with respect to the positive class) and the early risk detection measure proposed in [2]. The standard classification measures evaluate the teams' estimations with respect to golden truth judgments. We included them in our experimental evaluation because these metrics are well-known and easily interpretable. However, they are time-unaware and do not penalize late decisions. In order to reward early alerts, we employed ERDE, an error measure for early risk detection [2] for which the fewer writings required to make the alert, the better.

ERDE (*early risk detection error*) takes into account the correctness of the (binary) decision and the delay, which is measured by counting the number (k) of distinct submissions (posts or comments) seen before taking the decision. For instance, imagine a user u who posted a total number of 150 posts or comments (15 submissions per chunk). If a team's system emitted a decision for user u after the third chunk of data then the delay k would be 45.

Another important factor is that data are unbalanced (many more negative cases than positive cases) and, thus, the evaluation measure needs to weight different errors in a different way. Consider a binary decision d taken by a team's system with delay k. Given golden truth judgments, the prediction d can be a true positive (TP), true negative (TN), false positive (FP) or false negative (FN). Given these four cases, the ERDE measure is defined as:

$$ERDE_o(d, k) = \begin{cases} c_{fp} & \text{if } d = \text{positive AND ground truth} = \text{negative (FP)} \\ c_{fn} & \text{if } d = \text{negative AND ground truth} = \text{positive (FN)} \\ lc_o(k) \cdot c_{tp} & \text{if } d = \text{positive AND ground truth} = \text{positive (TP)} \\ 0 & \text{if } d = \text{negative AND ground truth} = \text{negative (TN)} \end{cases}$$

How to set c_{fp} and c_{fn} depends on the application domain and the implications of FP and FN decisions. In evaluating the systems, we fixed c_{fn} to 1 and c_{fp} was set according to the proportion of positive cases in 2017's test data (e.g. we set c_{fp} to 0.1296). The factor $lc_o(k)(\in [0, 1])$ represents a cost associated to the delay in detecting true positives. We set c_{tp} to c_{fn} (i.e. c_{tp} was set to 1) because late detection can have severe consequences (as a late detection is considered as equivalent to not detecting the case at all). The function $lc_o(k)$ is a monotonically increasing function of k (sigmoid). The latency cost factor was only used for the true positives because we understand that late detection is not an issue for true negatives. True negatives are non-risk cases that, of course, would not demand early intervention (i.e. these cases just need to be effectively filtered out from the positive cases). The systems must therefore focus on early detecting risk cases and detecting non-risk cases (regardless of when these non-risk cases are detected).

2.3 Performance Results

In general, the effectiveness of the submitted systems was weak (particularly, for the depression task, see Table 2, which reports the results obtained in the last edition of eRisk). This suggests that the depression task is really challenging and we still need further research on the intriguing aspects of early risk detection. Most of the teams focused on classification aspects (i.e. how to learn effective classifiers from the training data) and no much attention was paid to the tradeoff

Table 2. Performance results achieved by the eRisk 2018 participants (depression and anorexia tasks).

	Depression					Anorexia				
	$ERDE_5$	$ERDE_{50}$	F1	P	R	$ERDE_5$	$ERDE_{50}$	F1	P	R
FHDO-BCSGA	9.21%	6.68%	0.61	0.56	0.67	12.17%	7.98%	0.71	0.67	0.76
FHDO-BCSGB	9.50%	**6.44%**	**0.64**	0.64	0.65	11.75%	6.84%	0.81	0.84	0.78
FHDO-BCSGC	9.58%	6.96%	0.51	0.42	0.66	13.63%	9.64%	0.55	0.47	0.66
FHDO-BCSGD	9.46%	7.08%	0.54	0.64	0.47	12.15%	**5.96%**	0.81	0.75	**0.88**
FHDO-BCSGE	9.52%	6.49%	0.53	0.42	0.72	11.98%	6.61%	**0.85**	0.87	0.83
LIIRA	9.46%	7.56%	0.50	0.61	0.42	12.78%	10.47%	0.71	0.81	0.63
LIIRB	10.03%	7.09%	0.48	0.38	0.67	13.05%	10.33%	0.76	0.79	0.73
LIIRC	10.51%	7.71%	0.42	0.31	0.66					
LIIRD	10.52%	7.84%	0.42	0.31	0.66					
LIIRE	9.78%	7.91%	0.55	0.66	0.47					
LIRMMA	10.66%	9.16%	0.49	0.38	0.68	13.65%	13.04%	0.54	0.52	0.56
LIRMMB	11.81%	9.20%	0.36	0.24	0.73	14.45%	12.62%	0.52	0.41	0.71
LIRMMC	11.78%	9.02%	0.35	0.23	0.71	16.06%	15.02%	0.42	0.28	0.78
LIRMMD	11.32%	8.08%	0.32	0.22	0.57	17.14%	14.31%	0.34	0.22	0.76
LIRMME	10.71%	8.38%	0.37	0.29	0.52	14.89%	12.69%	0.41	0.32	0.59
PEIMEXA	10.30%	7.22%	0.38	0.28	0.62	12.70%	9.25%	0.46	0.39	0.56
PEIMEXB	10.30%	7.61%	0.45	0.37	0.57	12.41%	7.79%	0.64	0.57	0.73
PEIMEXC	10.07%	7.35%	0.37	0.29	0.51	13.42%	10.50%	0.43	0.37	0.51
PEIMEXD	10.11%	7.70%	0.39	0.35	0.44	12.94%	9.86%	0.67	0.61	0.73
PEIMEXE	10.77%	7.32%	0.35	0.25	0.57	12.84%	10.82%	0.31	0.28	0.34
RKMVERIA	10.14%	8.68%	0.52	0.49	0.54	12.17%	8.63%	0.67	0.82	0.56
RKMVERIB	10.66%	9.07%	0.47	0.37	0.65	12.93%	12.31%	0.46	0.81	0.32
RKMVERIC	9.81%	9.08%	0.48	**0.67**	0.38	12.85%	12.85%	0.25	0.86	0.15
RKMVERID	9.97%	8.63%	0.58	0.60	0.56	12.89%	12.89%	0.31	0.80	0.20
RKMVERIE	9.89%	9.28%	0.21	0.35	0.15	12.93%	12.31%	0.46	0.81	0.32
UDCA	10.93%	8.27%	0.26	0.17	0.53					
UDCB	15.79%	11.95%	0.18	0.10	**0.95**					
UDCC	9.47%	8.65%	0.18	0.13	0.29					
UDCD	12.38%	8.54%	0.18	0.11	0.61					
UDCE	9.51%	8.70%	0.18	0.13	0.29					
UNSLA	**8.78%**	7.39%	0.38	0.48	0.32	12.48%	12.00%	0.17	0.57	0.10
UNSLB	8.94%	7.24%	0.40	0.35	0.46	**11.40%**	7.82%	0.61	0.75	0.51
UNSLC	8.82%	6.95%	0.43	0.38	0.49	11.61%	7.82%	0.61	0.75	0.51
UNSLD	10.68%	7.84%	0.45	0.31	0.85	12.93%	9.85%	0.79	**0.91**	0.71
UNSLE	9.86%	7.60%	0.60	0.53	0.70	12.93%	10.13%	0.74	0.90	0.63
UPFA	10.01%	8.28%	0.55	0.56	0.54	13.18%	11.34%	0.72	0.74	0.71
UPFB	10.71%	8.60%	0.48	0.37	0.70	13.01%	11.76%	0.65	0.81	0.54
UPFC	10.26%	9.16%	0.53	0.48	0.61	13.17%	11.60%	0.73	0.76	0.71
UPFD	10.16%	9.79%	0.42	0.42	0.42	12.93%	12.30%	0.60	0.86	0.46
UQAMA	10.04%	7.85%	0.42	0.32	0.62					
TBSA	10.81%	9.22%	0.37	0.29	0.52	13.65%	11.14%	0.67	0.60	0.76
TUA1A	10.19%	9.70%	0.29	0.31	0.27	–	–	0.00	0.00	0.00
TUA1B	10.40%	9.54%	0.27	0.25	0.28	19.90%	19.27%	0.25	0.15	0.76
TUA1C	10.86%	9.51%	0.47	0.35	0.71	13.53%	12.57%	0.36	0.42	0.32
TUA1D	–	–	0.00	0.00	0.00					

between accuracy and delay. Only a couple of teams tried to define temporal models that incorporate some sort of sophisticated estimation of the evolution of the disorders. A full description and analysis of the results can be found in the lab overviews [3,4] and working note proceedings. Another important outcome of eRisk 2017 and eRisk 2018 is related to the evaluation measures. How to define appropriate metrics for early risk prediction is a challenge by itself and eRisk labs have already instigated the development of new early prediction metrics [5,6].

3 Conclusions and Future Work

eRisk will continue at CLEF 2019. Our plan is to organize up to three different tasks. The first task will be a continuation of 2018's eRisk task on early detection of signs of anorexia. We will use the eRisk 2018 data as training data, and new anorexia and non-anorexia test cases will be collected and included into the 2019 test split. The second task will follow a slightly different format. First, we will provide no training data. In this way, the participants will be encouraged to design predictive methods (e.g. based on search) that require no labelled examples. Most of the systems implemented for the previous eRisk editions were heavily dependent on supervised learning techniques. In 2019, we want to explore some tasks where training data are not available. Second, this task will focus on self-harm problems and, for each individual, the algorithms would be given only the history of the postings *before* the individual entered into the self-harm community. An individual who is active on a self-harm community perhaps has already done some sort of self-harm to his body. We want algorithms that detect the cases earlier on (and not when the cases are explicit and the individual is already engaging in a support forum). As a consequence, the participants would only be given the texts posted by the affected individuals before they first engaged in the self-harm community (before their first post in this community).

eRisk 2019 will also include a third task on searching for signs of depression. We have collected new data on depression that will consist not only on postings submitted by the depressed users but also on standard questionnaires that estimate their level of depression. Participants will be asked to automatically fill the depression questionnaire based on the user's postings. In this way, we can evaluate how good the algorithms are at detecting multiple elements or symptoms associated with depression.

Acknowledgements. We thank the support obtained from the Swiss National Science Foundation (SNSF) under the project "Early risk prediction on the Internet: an evaluation corpus", 2015.

This research has received financial support from the Galician Ministry of Education (grants ED431C 2018/29, ED431G/08 and ED431G/01 –"Centro singular de investigación de Galicia"–). All grants were co-funded by the European Regional Development Fund (ERDF/FEDER program).

References

1. Coppersmith, G., Dredze, M., Harman, C.: Quantifying mental health signals in Twitter. In: ACL Workshop on Computational Linguistics and Clinical Psychology (2014)
2. Losada, D.E., Crestani, F.: A test collection for research on depression and language use. In: Fuhr, N., et al. (eds.) CLEF 2016. LNCS, vol. 9822, pp. 28–39. Springer, Cham (2016). https://doi.org/10.1007/978-3-319-44564-9_3
3. Losada, D.E., Crestani, F., Parapar, J.: eRISK 2017: CLEF lab on early risk prediction on the internet: experimental foundations. In: Jones, G.J.F., et al. (eds.) CLEF 2017. LNCS, vol. 10456, pp. 346–360. Springer, Cham (2017). https://doi.org/10.1007/978-3-319-65813-1_30
4. Losada, D.E., Crestani, F., Parapar, J.: Overview of eRisk: early risk prediction on the internet. In: Bellot, P., et al. (eds.) CLEF 2018. LNCS, vol. 11018, pp. 343–361. Springer, Cham (2018). https://doi.org/10.1007/978-3-319-98932-7_30
5. Sadeque, F., Xu, D., Bethard, S.: Measuring the latency of depression detection in social media. In: Proceedings of the Eleventh ACM International Conference on Web Search and Data Mining, WSDM 2018, pp. 495–503. ACM, New York (2018)
6. Trotzek, M., Koitka, S., Friedrich, C.M.: Utilizing neural networks and linguistic metadata for early detection of depression indications in text sequences. CoRR, abs/1804.07000 (2018)

CLEF eHealth 2019 Evaluation Lab

Liadh Kelly[1]([✉]), Lorraine Goeuriot[2], Hanna Suominen[3,4,5,6], Mariana Neves[7], Evangelos Kanoulas[8], Rene Spijker[9,10], Leif Azzopardi[11], Dan Li[8], Jimmy[12], João Palotti[13,14], and Guido Zuccon[12]

[1] Maynooth University, Maynooth, Ireland
liadh.kelly@mu.ie
[2] Univ. Grenoble Alpes, Grenoble, France
Lorraine.Goeuriot@imag.fr
[3] Research School of Computer Science, The Australian National University (ANU),
Canberra, ACT, Australia
hanna.suominen@anu.edu.au
[4] Data61, Commonwealth Scientific and Industrial Research Organisation (CSIRO),
Canberra, ACT, Australia
[5] Faculty of Science and Technology, University of Canberra,
Canberra, ACT, Australia
[6] Department of Future Technologies, University of Turku, Turku, Finland
[7] German Federal Institute for Risk Assessment (BfR), Berlin, Germany
mariana.lara-neves@bfr.bund.de
[8] Informatics Institute, University of Amsterdam, Amsterdam, Netherlands
E.Kanoulas@uva.nl
[9] Cochrane Netherlands, Julius Center for Health Sciences and Primary Care,
UMC Utrecht, Utrecht University, Utrecht, Netherlands
R.Spijker-2@umcutrecht.nl
[10] Medical Library, Amsterdam Public Health, Amsterdam UMC,
University of Amsterdam, Amsterdam, Netherlands
[11] Computer and Information Sciences, University of Strathclyde, Glasgow, UK
leif.azzopardi@strath.ac.uk
[12] The University of Queensland, St Lucia, QLD, Australia
jimmy@uq.net.au, g.zuccon@uq.edu.au
[13] Vienna University of Technology, Vienna, Austria
palotti@ifs.tuwien.ac.at
[14] Qatar Computing Research Institute, Doha, Qatar
jpalotti@hbku.edu.qa

Abstract. Since 2012 CLEF eHealth has focused on evaluation resource building efforts around the easing and support of patients, their next-of-kins, clinical staff, and health scientists in understanding, accessing, and authoring eHealth information in a multilingual setting. This year's lab offers three tasks: Task 1 on multilingual information extraction; Task 2 on technology assisted reviews in empirical medicine; and Task 3 on consumer health search in mono- and multilingual settings. Herein, we describe the CLEF eHealth evaluation series to-date and then present the 2019 tasks, evaluation methodology, and resources.

LK, LG & HS co-chair the CLEF eHealth lab and contributed equally to this paper.
MN, EK & RS & LA & DL, and J & JP & GZ lead 2019 lab Tasks 1–3, respectively.

L. Azzopardi et al. (Eds.): ECIR 2019, LNCS 11438, pp. 267–274, 2019.
https://doi.org/10.1007/978-3-030-15719-7_36

Keywords: eHealth · Medical informatics · Information extraction · Information storage and retrieval · Information management · Systematic reviews

1   Introduction

In today's information overloaded society it is increasingly difficult to retrieve and digest valid and relevant information to make health-centered decisions. *Electronic Health* (eHealth) content is becoming available in a variety of forms ranging from patient records and medical dossiers, scientific publications, and health-related websites to medical-related topics shared across social networks. Laypeople, clinicians, and policy-makers need to easily retrieve, and make sense of this content to support their decision making.

Information retrieval (IR) systems have been commonly used as a means to access health information available online. However, the reliability, quality, and suitability of the information for the target audience varies greatly while high recall or coverage, that is finding all relevant information about a topic, is often as important as high precision, if not more. Furthermore, the information seekers in the health domain also experience difficulties in expressing their information needs as search queries.

CLEF eHealth[1], established as a lab workshop in 2012 as part of the *Conference and Labs of the Evaluation Forum* (CLEF), has offered since 2013 evaluation labs in the fields of layperson and professional health information extraction, management, and retrieval with the aims of bringing together researchers working on related information access topics and providing them with datasets to work with and validate the outcomes. More specifically, these labs and their subsequent workshops target

1. developing processing methods and resources (e.g., dictionaries, abbreviation mappings, and data with model solutions for method development and evaluation) in a multilingual setting to enrich difficult-to-understand eHealth texts and provide personalized reliable access to medical information, and provide valuable documentation;
2. developing an evaluation setting and releasing evaluation results for these methods and resources;
3. contributing to the participants and organizers' professional networks and interaction with all interdisciplinary actors of the ecosystem for producing, processing, and consuming eHealth information.

In this paper we overview the CLEF eHealth evaluation lab series to-date [5, 6,10,11,19,20] and present this year's evaluation lab challenges.

---

[1] https://sites.google.com/site/clefehealth/ (last accessed on 19 October 2018).

2 CLEF eHealth—Past and Future

In 2012, the CLEF eHealth workshop was organized to prepare for evaluation labs. Its outcome was the identification of the need for an evaluation lab focusing on patient-centric health language processing. The subsequent CLEF eHealth tasks offered from 2013–2018 can be broadly categorized as information extraction, management, and retrieval focused. In 2019, we offer information extraction and retrieval challenges (described in Sect. 3). Here we describe the growth path of these challenges.

2.1 Information Extraction from Clinical Text

The CLEF eHealth tasks on *information extraction* (IE) began in 2013 by considering English only but evolved by 2018 to considering more and more languages. In 2013, the focus of the information extraction task was on named entity recognition, normalization of disorders, and normalization of acronyms/abbreviations. In 2014, we extended the challenge with a focus on disorder attribute identification and normalization from clinical text. In 2015 and 2016, we supplemented the tasks by aiming to release nurses' time from documentation to patient communication by considering first clinical speech recognition to capture the verbal shift-change handover and then information extraction to pre-fill a handover form from the speech recognized text by automatically identifying relevant text-snippets for each slot of the form.

To continue this evolution from a widely studied corpus type (written in English) towards a larger variety of corpora by considering spoken English in the handover tasks, we introduced a multilingual challenge in 2015, which considered information extraction from French clinical texts. This challenge was grown in the subsequent years [12,13]. In last year's lab [14] we began the evolution of the multilingual element task towards the inclusion of other European languages, such as Hungarian and Italian. In this year's task we continue this evolution.

Our goal in the coming years is to offer an information extraction task using comparable corpora in several languages in order to challenge participants with the issue of language adaptation and to encourage the development of systems that are able to address a multilingual setting or can easily be tuned to specialize to specific languages.

2.2 Information Retrieval and Personalization

In 2013 and 2014 the focus of the information retrieval task was on evaluating the effectiveness of search engines to support people when searching for information about known conditions, for example, to answer queries like "thrombocytopenia treatment corticosteroids length", with multilingual queries added in the 2014 challenge [2–4]. This task aimed to model the scenario of a patient being discharged from hospital and wanting to seek more information about diagnosed conditions or prescribed treatments.

In 2015 the information retrieval task changed to focus on studying the effectiveness of search engines to support individuals' queries issued for self-diagnosis purposes, and again offered a multilingual queries challenge [15]. In addition, we began adding personalization elements to the challenge on an incremental basis by assessing the readability of information and taking this into account in the evaluation framework.

This individualized information retrieval approach was continued in the 2016 and 2017 labs [16,21] and we also introduced gradual shifts from an ad-hoc search paradigm (that of a single query and a single document ranking) to a session based search paradigm. Along these lines we also revised how relevance is measured for evaluation purposes, taking into account instead whole-of-session usefulness. In 2018 [7] we continued this evolution, and introduced query intent elements.

Our next goals are as follows: (1) to further progress the evaluation methodology for session based and query intent search paradigms that we laid the foundations for in the previous years, and (2) to introduce spoken query elements and supporting evaluation methodology.

2.3   Technology Assisted Reviews

The *Technology Assisted Reviews* (TARs) task, organized for the first time in 2017 and continued in 2018 [8,9], was a high-recall IR task in English that aimed at evaluating search algorithms that seek to identify all studies relevant for conducting a systematic review in empirical medicine. The task had a focus on *Diagnostic Test Accuracy* (DTA) reviews. The typical process of searching for scientific publications to conduct a systematic review consists of three stages:

1. specifying a number of inclusion criteria that characterize the articles relevant to the review and constructing a complex Boolean Query to express them,
2. screening the abstracts and titles that result from the Boolean query, and
3. screening the full documents that passed the Abstract and Title Screening.

The 2017 task focused on the second stage of the process, that is, Abstract and Title Screening. Building on this the 2018 task focused on the first stage (*subtask 1*) and second stage (*subtask 2*) of the process, that is, Boolean Search and Abstract and Title Screening. The task built two benchmark collections and implemented a number of evaluation metrics to automatically assess the quality of methods on these collection, all of which have been made available at https://github.com/CLEF-TAR.

Directions to take to further build the task in the coming years include the following: (1) developing metrics to evaluate systems on the ranking and thresholding tasks, (2) increasing the labelled data offered with the challenge, and (3) providing an infrastructure to support running of participants' algorithms in house, thus allowing for use of full text articles and live, iterative active learning technique development.

3   CLEF eHealth 2019 Tasks

Continuing the CLEF eHealth growth path from 2013–2018, in 2019 CLEF eHealth offers three tasks. Specifically, Task 1 on Multilingual Information Extraction, Task 2 on TARs in Empirical Medicine, and Task 3 on Consumer Health Search.

3.1   Task 1. Multilingual Information Extraction

This task builds upon the previous CLEF eHealth IE tasks. This year's task continues to explore the automatic assignment of the *International Classification of Diseases* (ICD-10) codes to health-related documents with the focus on the German language and on *Non-Technical Summaries* (NTSs) of animal experiments. Specifically, in 2019, participants are challenged with the semantic indexing of NTSs using codes from the German version of the ICD-10. The NTSs are short summaries which are currently publicly available in the Animal-TestInfo database[2], as part of the approval procedure for animal experiments in Germany [1]. The database currently contains more than 8,000 NTSs, which have been manually indexed by domain experts, and that was used to generate a training dataset. The task can be treated as a named entity recognition and normalization task, but also as a text classification task. Only fully automated means are allowed, that is, human-in-the-loop approaches are not permitted.

3.2   Task 2. Technology Assisted Reviews in Empirical Medicine

This task builds on the TAR task first introduced in 2017. The task is a ranking and classification task (similar to the 2017 and 2018 version), and includes two subtasks: (1) No Boolean Query and (2) Title and Abstract Screening. For the former users are provided with a set of topics and parts of the systematic review protocol. The goal of the participants is to rank PubMed abstracts and titles and provide a threshold on the ranking. For the latter users are provided with a set of topics, the original Boolean query used by the researchers that conducted the systematic review, and the results of that query. The goal of the participants is to rank PubMed abstracts and titles and provide a threshold on the ranking.

3.3   Task 3. Consumer Health Search

This task builds on the CLEF eHealth information retrieval tasks that have ran since the onset of CLEF eHealth. The main components of the *Consumer Health Search* (CHS) task are the document collection, the set of topics, and the system evaluation. This year's challenge uses the new document collection introduced in last year's challenge, consisting of over 5 million Web pages. It is a compilation of Web pages of selected domains acquired from the CommonCrawl[3]. User

---

[2] https://www.animaltestinfo.de/ (last accessed on 18 October 2018).
[3] http://commoncrawl.org/ (last accessed on 19 October 2018).

stories for query (and query variant) generation are created using the discharge summaries and forum posts we used in previous years of the task. For the first time, queries are also offered as spoken queries, with automatic speech-to-text transcripts provided. The challenge is structured into 5 subtasks, specifically: ad-hoc search, personalization search, query variations, multilingual search, and search intent.

4   CLEF eHealth Contributions

In its seven years of existence, the CLEF eHealth series has offered a recurring contribution to the creation and dissemination of text analytics resources, methods, test collections, and evaluation benchmarks in order to ease and support patients, their next-of-kins, clinical staff, and health scientists in understanding, accessing, and authoring eHealth information in a multilingual setting. In 2012–2017 alone it has attracted over 700 teams to register their interest in its 15 tasks, leading to 130 task submissions, 180 papers, and their 1, 300 citations for the 741 included authors from 33 countries across the world [18].

The annual workshops and evaluation labs offered by CLEF eHealth have matured and established their presence over the years. In total, 70 unique teams registered their interest and 28 teams took part in the 2018 tasks (14 in Task 1, 7 in Task 2 and 7 in Task 3). In comparison, in 2017, 2016, 2015, 2014, and 2013, the number of team registrations was 67, 116, 100, 220, and 175, respectively and the number of participating teams was 32, 20, 20, 24, and 53 [5,6,10,11,19,20].

Given the significance of the tasks, all problem specifications, test collections, and text analytics resources associated with the lab have been made available to the wider research community through our CLEF eHealth website[4].

5   Conclusion

In this paper, we have provided an overview of the CLEF eHealth evaluation lab series and presented the 2019 lab tasks. The CLEF eHealth workshop series was established in 2012 as a scientific workshop with an aim of establishing an evaluation lab [17]. This ambition was realized in the CLEF eHealth evaluation lab, which has ran since 2013. This annual lab offers shared tasks in the eHealth space each year in the domain of medical information retrieval, management and extraction [5,6,10,11,19,20].

The CLEF eHealth 2019 lab offers three shared tasks: Task 1 on multilingual information extraction to extend the 2018 task on French, Hungarian, and Italian corpora to German; Task 2 on technologically assisted reviews in empirical medicine building on the 2018 task in English; and Task 3 on patient-centered IR in mono- and multilingual settings that builds on the 2013–18 IR tasks. Test collections generated by each of the three CLEF eHealth 2019 tasks offer a specific

---

[4] https://sites.google.com/site/clefehealth/datasets (last accessed on 18 October 2018).

task definition, implemented in a dataset distributed together with an implementation of relevant evaluation metrics to allow for direct comparability of the results reported by systems evaluated on the collections.

Acknowledgements. We gratefully acknowledge the people involved in the CLEF eHealth labs as participants or organizers. We also acknowledge the many organizations that have supported CLEF eHealth labs since 2012. The CLEF eHealth 2019 evaluation lab is supported in part by the CLEF Initiative, Data61/CSIRO, a Google Faculty Research Award, and ARC DECRA grant DE180101579.

References

1. Bert, B., et al.: Rethinking 3R strategies: digging deeper into AnimalTestInfo promotes transparency in in vivo biomedical research. PLoS Biol. **15**(12), 1–20 (2017). https://doi.org/10.1371/journal.pbio.2003217
2. Goeuriot, L., et al.: ShARe/CLEF eHealth evaluation lab 2013, task 3: information retrieval to address patients' questions when reading clinical reports. CLEF 2013 Online Working Notes 8138 (2013)
3. Goeuriot, L., et al.: An analysis of evaluation campaigns in ad-hoc medical information retrieval: CLEF eHealth 2013 and 2014. Springer Inf. Retr. J. **21**(6), 507–540 (2018)
4. Goeuriot, L., et al.: ShARe/CLEF eHealth evaluation lab 2014, task 3: user-centred health information retrieval. In: CLEF 2014 Evaluation Labs and Workshop: Online Working Notes, Sheffield, UK (2014)
5. Goeuriot, L., et al.: Overview of the CLEF eHealth evaluation lab 2015. In: Mothe, J., et al. (eds.) CLEF 2015. LNCS, vol. 9283, pp. 429–443. Springer, Cham (2015). https://doi.org/10.1007/978-3-319-24027-5_44
6. Goeuriot, L., et al.: CLEF 2017 eHealth evaluation lab overview. In: Jones, G.J.F., et al. (eds.) CLEF 2017. LNCS, vol. 10456, pp. 291–303. Springer, Cham (2017). https://doi.org/10.1007/978-3-319-65813-1_26
7. Jimmy, Zuccon, G., Palotti, J., Goeuriot, L., Kelly, L.: Overview of the CLEF 2018 consumer health search task. In: Working Notes of Conference and Labs of the Evaluation (CLEF) Forum. CEUR Workshop Proceedings (2018)
8. Kanoulas, E., Li, D., Azzopardi, L., Spijker, R.: CLEF 2017 technologically assisted reviews in empirical medicine overview. In: Working Notes of Conference and Labs of the Evaluation (CLEF) Forum. CEUR Workshop Proceedings (2017)
9. Kanoulas, E., Li, D., Azzopardi, L., Spijker, R.: CLEF 2018 technologically assisted reviews in empirical medicine overview. In: Working Notes of Conference and Labs of the Evaluation (CLEF) Forum. CEUR Workshop Proceedings (2018)
10. Kelly, L., Goeuriot, L., Suominen, H., Névéol, A., Palotti, J., Zuccon, G.: Overview of the CLEF eHealth evaluation lab 2016. In: Fuhr, N., et al. (eds.) CLEF 2016. LNCS, vol. 9822, pp. 255–266. Springer, Cham (2016). https://doi.org/10.1007/978-3-319-44564-9_24
11. Kelly, L., et al.: Overview of the ShARe/CLEF eHealth evaluation lab 2014. In: Kanoulas, E., et al. (eds.) CLEF 2014. LNCS, vol. 8685, pp. 172–191. Springer, Cham (2014). https://doi.org/10.1007/978-3-319-11382-1_17
12. Névéol, A., et al.: Clinical information extraction at the CLEF eHealth evaluation lab 2016. In: Balog, K., Cappellato, L., Ferro, N., Macdonald, C. (eds.) CLEF 2016 Working Notes. CEUR Workshop Proceedings (CEUR-WS.org) (2016). http://ceur-ws.org/Vol-1609/. ISSN 1613-0073

13. Névéol, A., et al.: CLEF eHealth 2017 multilingual information extraction task overview: ICD10 coding of death certificates in English and French. In: CLEF 2017 Online Working Notes. CEUR-WS (2017)
14. Névéol, A., et al.: CLEF eHealth 2018 multilingual information extraction task overview: ICD10 coding of death certificates in French, Hungarian and Italian. In: CLEF 2018 Online Working Notes. CEUR-WS (2018)
15. Palotti, J., et al.: CLEF eHealth evaluation lab 2015, task 2: retrieving information about medical symptoms. In: CLEF 2015 Online Working Notes. CEUR-WS (2015)
16. Palotti, J., et al.: CLEF 2017 task overview: the IR task at the eHealth evaluation lab. In: Working Notes of Conference and Labs of the Evaluation (CLEF) Forum. CEUR Workshop Proceedings (2017)
17. Suominen, H.: CLEF 2012 Working Notes. CEUR Workshop Proceedings (CEUR-WS.org). Ed. by Forner, P., Karlgren, J., Womser-Hacker, C., Ferro, N. (2012). http://ceur-ws.org/Vol-1178/. ISSN 1613-0073
18. Suominen, H., Kelly, L., Goeuriot, L.: Scholarly influence of the conference and labs of the evaluation forum ehealth initiative: review and bibliometric study of the 2012 to 2017 outcomes. JMIR Res. Protoc. 7(7), e10961 (2018). https://doi.org/10.2196/10961
19. Suominen, H., et al.: Overview of the CLEF ehealth evaluation lab 2018. In: Bellot, P., et al. (eds.) CLEF 2018. LNCS, vol. 11018, pp. 286–301. Springer, Heidelberg (2018). https://doi.org/10.1007/978-3-319-98932-7_26
20. Suominen, H., et al.: Overview of the ShARe/CLEF eHealth Evaluation Lab 2013. In: Forner, P., Müller, H., Paredes, R., Rosso, P., Stein, B. (eds.) CLEF 2013. LNCS, vol. 8138, pp. 212–231. Springer, Heidelberg (2013). https://doi.org/10.1007/978-3-642-40802-1_24
21. Zuccon, G., et al.: The IR task at the CLEF eHealth evaluation lab 2016: user-centred health information retrieval. In: CLEF 2016 Evaluation Labs and Workshop: Online Working Notes, CEUR-WS, September 2016

LifeCLEF 2019: Biodiversity Identification and Prediction Challenges

Alexis Joly[1], Hervé Goëau[2(✉)], Christophe Botella[1,3], Stefan Kahl[7],
Marion Poupard[4], Maximillien Servajean[8], Hervé Glotin[4], Pierre Bonnet[2],
Willem-Pier Vellinga[5], Robert Planqué[5], Jan Schlüter[4], Fabian-Robert Stöter[1],
and Henning Müller[6]

[1] Inria, LIRMM, Montpellier, France
[2] CIRAD, UMR AMAP, Montpellier, France
herve.goeau@cirad.fr
[3] INRA, UMR AMAP, Montpellier, France
[4] AMU, Univ. Toulon, CNRS, ENSAM, LSIS UMR 7296, IUF, Toulon, France
[5] Xeno-canto Foundation, Groningen, The Netherlands
[6] HES-SO, Sierre, Switzerland
[7] Chemnitz University of Technology, Chemnitz, Germany
[8] LIRMM, Université Paul Valéry, University of Montpellier, CNRS,
Montpellier, France

Abstract. Building accurate knowledge of the identity, the geographic
distribution and the evolution of living species is essential for a sustainable development of humanity, as well as for biodiversity conservation.
However, the burden of the routine identification of plants and animals
in the field is strongly penalizing the aggregation of new data and knowledge. Identifying and naming living plants or animals is actually almost
impossible for the general public and often a difficult task for professionals and naturalists. Bridging this gap is a key challenge towards enabling
effective biodiversity information retrieval systems. The LifeCLEF evaluation campaign, presented in this paper, aims at boosting and evaluating
the advances in this domain since 2011. In particular, the 2019 edition
proposes three data-oriented challenges related to the identification and
prediction of biodiversity: (i) an image-based plant identification challenge, (ii) a bird sounds identification challenge and (iii) a location-based
species prediction challenge based on spatial occurrence data and environmental tensors.

Keywords: Biodiversity · Informatics · Machine learning ·
Species identification · Species prediction · Plant identification ·
Bird identification · Species distribution model

1 Introduction

Identifying organisms is a key for accessing information related to the uses and
ecology of species. This is an essential step in recording any specimen on earth

© Springer Nature Switzerland AG 2019
L. Azzopardi et al. (Eds.): ECIR 2019, LNCS 11438, pp. 275–282, 2019.
https://doi.org/10.1007/978-3-030-15719-7_37

to be used in ecological studies. Unfortunately, this is difficult to achieve due to the level of expertise necessary to correctly record and identify living organisms (for instance plants are one of the most difficult groups to identify with an estimated number of 400,000 species). This *taxonomic gap* has been recognized since the Rio Conference of 1992, as one of the major obstacles to the global implementation of the Convention on Biological Diversity. Among the diversity of methods used for species identification, Gaston and O'Neill [2] discussed in 2004 the potential of automated approaches typically based on machine learning and multimedia data analysis. They suggested that, if the scientific community is able to (i) overcome the production of large training datasets, (ii) more precisely identify and evaluate the error rates, (iii) scale up automated approaches, and (iv) detect novel species, it will then be possible to initiate the development of a generic automated species identification system that could open up vistas of new opportunities for theoretical and applied work in biological and related fields.

Since the question raised by Gaston and O'Neill [2], *automated species identification: why not?*, a lot of work has been done on the topic (*e.g.* [1,4,5,11,13,16,17]) and it is still attracting much research today, in particular in deep learning [3,6,14]. In order to measure the progress made in a sustainable and repeatable way, the LifeCLEF[1] research platform was created in 2014 as a continuation of the plant identification task [10] that was run within the ImageCLEF lab[2] the three years before [7–9]. LifeCLEF enlarged the evaluated challenge by considering animals in addition to plants, and audio and video contents in addition to images. In 2018, a new challenge dedicated to the location-based prediction of species was finally introduced (GeoLifeCLEF).

2   PlantCLEF Challenge

2.1   Methodology

The plant identification challenge of CLEF has been run since 2011, offering today a seven-year follow-up of the progress made in image-based plant identification. From the beginning, it mainly relied on real-world collaborative data and the evaluation protocol was defined in collaboration with biologists so as to reflect realistic usage scenarios. In particular, it considers the problem of classifying plant observations based on several images of the same individual plant rather than considering a classical image classification task. Indeed, it is usually required to observe several organs of a plant to identify it accurately (*e.g.* the flower, the leaf, the fruit, the stem, etc.). As a consequence, the same individual plant is often photographed several times by the same observer resulting in contextually similar pictures and/or near-duplicates. To avoid bias, it is crucial to consider such image sets as a single plant observation that should not be split across the training and the test set. In addition to the raw pictures, plant observations are usually associated with contextual and social data. This

---

[1] http://www.lifeclef.org/.
[2] http://www.imageclef.org/.

includes geo-tags or location names, time information, author names, collaborative ratings, vernacular names (common names), picture type tags, etc. Within all PlantCLEF challenges, the use of this additional information was considered as part of the problem because it was judged as potentially useful for a real-world usage scenario.

The data that was shared within the PlantCLEF challenge was considerably enriched along the years. The number of species was increased from 71 species in 2011 to 10,000 species in 2017 and 2018 (illustrated by more than 1 million images). This durable scaling-up was made possible thanks to the close collaboration of LifeCLEF with several important actors in the digital botany domain, in particular the TelaBotanica network of expert and amateur botanists (about 40K members) and the Pl@ntNet citizen science platform (million of users).

2.2 Main Outcomes of the Previous Edition

The main novelty of the 2018 edition of PlantCLEF was to involve 9 of the best expert botanists of the French flora who accepted to compete with AI algorithms on a difficult subset of the whole test set. The results confirmed that identifying plants from images is a difficult task, even for some of the highly skilled specialists who accepted to participate in the experiment. Images only contain a partial information of the plant and that it is often not sufficient to determine the right species with certainty. Regarding the performance of the automated approaches, the results showed that there is still a margin of progression but that it is becoming tighter and tighter. The best system was able to correctly classify 84% of the test samples, better than 5 of the 9 experts.

2.3 PlantCLEF 2019

The main novelty of the 2019 edition of PlantCLEF will be to extend the challenge to the flora of *data deficient* regions, *i.e.* regions having the richest biodiversity (tropical ones) but for which data availability is much lower than northern countries. Indeed, it is estimated that there is over 391K species of vascular plants on earth, much beyond the 10K species of PlantCLEF 2018 that are among the most common ones. The additional data will be aggregated in two ways. For the training set, we will mainly rely on raw web data collected by querying popular image search engines with the binomial latin name of the targeted species. We actually did show in previous editions of LifeCLEF that training deep learning models on such noisy big data is as effective as training models on cleaner but smaller expert data. For the test set, on the other hand, we will use expert data without any uncertainty. More precisely, we will rely on 3 collections of expert botanists who accepted to share their unpublished observations for the challenge. One is a collection of trees, shrubs, herbs and ferns from French Guyana (wet evergreen Amazonian forest). The second one is a specialized collection of pictures related to epiphytic orchids, mainly from Laos. And the third one is a collection of endemic species of South Africa. The main evaluation measure for the challenge will be the Mean Reciprocal Rank.

3   BirdCLEF Challenge

3.1   Methodology

The bird identification challenge of LifeCLEF, initiated in 2014 in collaboration with Xeno-Canto, considerably increased the scale of the seminal challenges. The first bird challenge ICML4B [4] initiated in 2012 by DYNI/SABIOD had only 35 species, but received 400 runs. The next at MLSP had only 15 species, the third (NIPS4B [5] in 2013 by SABIOD) had 80 species. Meanwhile, Xeno-canto, launched in 2005, hosts bird sounds from all continents and daily receives new recordings from some of the remotest places on Earth. It currently archives with 379472 recordings, 9779 species of birds, making it one of the most comprehensive collections of bird sound recordings worldwide, and certainly the most comprehensive collection shared under Creative Commons licenses. For the first Bird-CLEF challenge, it was decided to not consider the whole Xeno-Canto dataset but to rather focus on a specific region, *i.e.* the Amazonian rain forest because it is one of the richest in the world in terms of biodiversity but also one of the most endangered. The geographical extent and the number of species were progressively increased over the years so as to reach 1000 species in 2015/2016, and 1500 in 2017/2018. By nature, the Xeno-Canto data as well as the BirdCLEF subset has a massive class imbalance. For instance, the 2017 dataset contains 48,843 recordings in total, with a minimum of four recordings for *Laniocera rufescens* and a maximum of 160 recordings for *Henicorhina leucophrys*.

In 2016, the BirdCLEF challenge was extended to *soundscape* recordings in addition to the classical *mono-directional* Xeno-Canto recordings. This enables more passive monitoring scenarios such as setting up a network of static recorders that would continuously capture the surrounding sound environment. One of the limitations of this new content, however, was that the vocalizing birds were not localized in the recordings. Thus, to allow a more accurate evaluation, new *time-coded soundscapes* were introduced within the BirdCLEF 2017 and 2018 challenges. In total, 6.5 h of recordings were collected in the Amazonian forests and were manually annotated by two experts including a native of the Amazon forest, in the form of time-coded segments with associated species name.

3.2   Main Outcomes of the Previous Edition

The best system of the 2018 edition of the BirdCLEF challenge achieved an impressive Mean Average Precision score of 0.83 on the mono-directional recordings. This performance could probably even be improved by a few points by combining it with a metadata-based prediction model, as shown by the second best participant to the challenge. This means that the technology is now mature enough for this scenario. Concerning the soundscapes recordings however, we did not observe any significant improvement over the performance of the 2017 edition. Recognizing many overlapping birds remains a hard problem and none of the efforts made by the participants to tackle it provided observable improvement.

3.3   BirdCLEF 2019

The 2019 edition of the BirdCLEF challenge will mainly focus on the soundscape scenario that remains very challenging whereas the mono-directional identification task is now better solved. Two tasks will be evaluated, (i) the recognition of all specimens singing in a long sequence (up to one hour) of raw soundscapes that can contain tens of birds singing simultaneously, and (ii) source separation or source count estimation in complex soundscapes that were recorded using multiple microphones. Therefore, two new corpus of soundscapes will be added to the existing soundscape dataset: (i) 100+ hours of manually annotated soundscapes recorded using 30 field recorders between January and June of 2017 in Ithaca, NY, USA. (ii) 50 h of four-channel or stereophonic binaural recordings acquired in Papa New Guinea in november 2017 at high sampling rate (96 kHz SR) and high dynamics (24 bits) [15]. For this purpose we designed binaural or quadriphonic recording stations, specifically for localisation in azimuth and elevation of singing birds, in order to help in a second stage the recognition of the species. These recordings contain some endemic bird species that had never been recorded before. The evaluation measure used for the species detection task will be the classification mean Average Precision (c-mAP [12]). The evaluation measure used for the count estimation task is the mean absolute count error.

4   GeoLifeCLEF Challenge

4.1   Methodology

Predicting the shortlist of species that are likely to be observed at a given geographical location should significantly help to reduce the candidate set of species to be identified. However, none of the attempt to do so within previous Life-CLEF editions successfully used this information. The GeoLifeCLEF challenge was specifically created in 2018 to tackle this problem through a standalone task. More generally, automatically predicting the list of species that are likely to be observed at a given location might be useful for many other scenarios in biodiversity informatics. It could facilitate biodiversity inventories through the development of location-based recommendation services (typically on mobile phones) as well as the involvement of non-expert nature observers. It might also serve educational purposes thanks to biodiversity discovery applications providing functionalities such as contextualized educational pathways.

The challenge relies on a large data set of 291,392 occurrences of around 3K plant species, each occurrence being associated to a location, a species name and a multi-channel image characterizing the local environment. Indeed, it is usually not possible to learn a species distribution model directly from spatial positions because of the limited number of occurrences and the sampling bias. What is usually done in ecology is to predict the distribution on the basis of a representation in the environmental space, typically a feature vector composed of climatic variables (average temperature at that location, precipitation, etc.) and other variables such as soil type, land cover, distance to water, etc. The originality of GeoLifeCLEF is to generalize such niche modeling approach to the use of an image-based environmental representation space. Instead of learning a model from environmental feature vectors, the goal of the task will be to learn a model from k-dimensional image patches, each patch representing the value of an environmental variable in the neighborhood of the occurrence.

4.2 Main Outcomes of the Previous Edition

The main outcome of the first edition of GeoLifeCLEF was that Convolutional Neural Networks models learned on environmental tensors revealed to be the most performing method. They performed better than boosted classification trees that were known as providing state-of-the-art performance for environmental modelling. However, the achieved performance is still low with regard to the targeted scenario and there is a large room of improvement and research opportunities regarding such models, like appropriately integrating neighbours species correlations in the model, using external expert information about related species like taxonomic or phylogenetic classification, or correcting for observer reporting bias.

4.3 GeoLifeCLEF 2019

The 2019 edition of the challenge will tackle some of the methodological weaknesses that were revealed by the pilot 2018 edition. In particular, we will rely on the top-30 accuracy instead of the Mean Average Precision as the main evaluation metric. This will allow to better take into account the fact that many species co-exist at small spatial scales (under the meter), much lower than the accuracy of the geo-coordinates in the data set. We will also produce a new dataset fixing some issues of the previous one related to the incompleteness of some environmental variables and the spatial degradation of some occurrences. More precisely, the training set will be composed of nearly one million geo-locations of plant species living on the French territory (coming from two main platforms: (i) the Global Biodiversity Information Facility and (ii), the Pl@ntNet participatory application). For the test set, on the other hand, we will rely solely on expert data without any uncertainty coming from the French national conservatories. Regarding the environmental variables, we will provide about 30 rasters of data covering the whole French territory (related to climatology, altitude, soil type, land cover, distance to water, etc.). We will also provide tools to extract environmental tensors from that rasters (at the positions of the plant occurrences in the training and test sets).

5   Timeline and Registration Instructions

All information about the timeline and the participation to the challenges is provided on the LifeCLEF 2019 web pages[3]. The system used to run the challenges (registration, submission, leaderboard, etc.) is the crowdAI platform[4].

6   Discussion and Conclusion

Boosting research on biodiversity informatics in the long term is crucial in terms of societal impact. Researchers are actually often opportunistic regarding the choice of a dataset and an interesting related challenge. And so are end-users regarding the use of applications emerging from that research. To fully reach its objective, an evaluation campaign such as LifeCLEF requires a long-term research effort so as to (i) encourage non-incremental contributions, (ii) measure consistent performance gaps, (iii) progressively scale-up the problem and (iv), enable the emergence of a strong community. The 2019-th edition of the lab will support this vision but will still include a set of consistent novelties:

- The historical BirdCLEF subtask related to monospecies recordings will be stopped in order to concentrate all efforts on the most challenging subtask of recognizing birds in soundscapes and on a new subtask relying on polyphonic recordings.
- We will go deeper in the comparison of automated approaches with human expertise by extending the PlantCLEF task to more complex taxonomic groups, in particular the floras of several tropical countries that are known only by a few specialists who will participate in the evaluation.
- The evaluation methodology of the GeoLifeCLEF challenge will be improved according to the feedback of the first edition and the dataset will be enriched with more diverse and more precise plant occurrences.

References

1. Cai, J., Ee, D., Pham, B., Roe, P., Zhang, J.: Sensor network for the monitoring of ecosystem: bird species recognition. In: 3rd International Conference on Intelligent Sensors, Sensor Networks and Information, ISSNIP 2007 (2007). https://doi.org/10.1109/ISSNIP.2007.4496859
2. Gaston, K.J., O'Neill, M.A.: Automated species identification: why not? Philos. Trans. R. Soc. Lond. B: Biol. Sci. **359**(1444), 655–667 (2004)
3. Ghazi, M.M., Yanikoglu, B., Aptoula, E.: Plant identification using deep neural networks via optimization of transfer learning parameters. Neurocomputing **235**, 228–235 (2017)
4. Glotin, H., Clark, C., LeCun, Y., Dugan, P., Halkias, X., Sueur, J.: Proceedings of the 1st Workshop on Machine Learning for Bioacoustics - ICML4B, ICML, Atlanta USA (2013)

---

[3] https://www.imageclef.org/lifeclef2019.
[4] https://www.crowdai.org.

5. Glotin, H., LeCun, Y., Artiéres, T., Mallat, S., Tchernichovski, O., H.X.: Proceedings of the Neural Information Processing Scaled for Bioacoustics, from Neurons to Big Data. NIPS International Conference, Tahoe USA (2013). http://sabiod.org/nips4b

6. Goeau, H., Bonnet, P., Joly, A.: Plant identification based on noisy web data: the amazing performance of deep learning (LifeCLEF 2017). In: CLEF 2017 Conference and Labs of the Evaluation Forum, pp. 1–13 (2017)

7. Goëau, H., et al.: The ImageCLEF 2013 plant identification task. In: CLEF, Valencia, Spain (2013)

8. Goëau, H., et al.: The ImageCLEF 2011 plant images classification task. In: CLEF 2011 (2011)

9. Goëau, H., et al.: ImageCLEF 2012 plant images identification task. In: CLEF 2012, Rome (2012)

10. Goëau, H., et al.: The ImageCLEF plant identification task 2013. In: Proceedings of the 2nd ACM International Workshop on Multimedia Analysis for Ecological Data, pp. 23–28. ACM (2013)

11. Joly, A., et al.: Interactive plant identification based on social image data. Ecol. Inform. **23**, 22–34 (2014)

12. Joly, A., et al.: Overview of LifeCLEF 2018: a large-scale evaluation of species identification and recommendation algorithms in the era of AI. In: Bellot, P., et al. (eds.) CLEF 2018. LNCS, vol. 11018, pp. 247–266. Springer, Cham (2018). https://doi.org/10.1007/978-3-319-98932-7_24

13. Lee, D.J., Schoenberger, R.B., Shiozawa, D., Xu, X., Zhan, P.: Contour matching for a fish recognition and migration-monitoring system. In: Optics East, pp. 37–48. International Society for Optics and Photonics (2004)

14. Lee, S.H., Chan, C.S., Remagnino, P.: Multi-organ plant classification based on convolutional and recurrent neural networks. IEEE Trans. Image Process. **27**(9), 4287–4301 (2018)

15. Poupard, M., Glotin, H., Lengagne, T., Bougrain-Dubourg, A., Bedu, A.L.: Multichannel soundscape recordings of wild versus anthropised new guinea - a first acoustic repertory of some endemic species. LIS DYNI CNRS Toulon Research Report (2018)

16. Towsey, M., Planitz, B., Nantes, A., Wimmer, J., Roe, P.: A toolbox for animal call recognition. Bioacoustics **21**(2), 107–125 (2012)

17. Trifa, V.M., Kirschel, A.N., Taylor, C.E., Vallejo, E.E.: Automated species recognition of antbirds in a mexican rainforest using hidden markov models. J. Acoust. Soc. Am. **123**, 2424 (2008)

CENTRE@CLEF 2019

Nicola Ferro[1], Norbert Fuhr[2], Maria Maistro[1,3(✉)], Tetsuya Sakai[4], and Ian Soboroff[5]

[1] University of Padua, Padua, Italy
{ferro,maistro}@dei.unipd.it
[2] University Duisburg-Essen, Duisburg, Germany
norbert.fuhr@uni-due.de
[3] University of Copenhagen, Copenhagen, Denmark
mm@di.ku.dk
[4] Waseda University, Tokyo, Japan
tetsuyasakai@acm.org
[5] National Institute of Standards and Technology (NIST), Gaithersburg, USA
ian.soboroff@nist.gov

Abstract. Reproducibility of experimental results has recently become a primary issue in the scientific community at large, and in the information retrieval community as well, where initiatives and incentives to promote and ease reproducibility are arising. In this context, CENTRE is a joint CLEF/TREC/NTCIR lab which aims at raising the attention on this topic and involving the community in a shared reproducibility exercise. In particular, CENTRE focuses on three objectives, e.g. replicability, reproducibility and generalizability, and for each of them a dedicated task is designed. We expect that CENTRE may impact on the validation of some key achievement in IR, help in designing shared protocols for reproducibility, and improve the understanding on generalization across collections and on the additivity issue.

1 Introduction

Reproducibility is becoming a primary concern in many areas of science [13,18] as well as in computer science, as also witnessed by the recent ACM policy on result and artefact review and badging. Also in *Information Retrieval (IR)* replicability and reproducibility of the experimental results are becoming a more and more central discussion item in the research community [2,6,8,9,14,17,19]. We now commonly find questions about the extent of reproducibility of the reported experiments in the review forms of all the major IR conferences, such as SIGIR, CHIIR, ICTIR and ECIR, as well as journals, such as ACM TOIS. We also witness the raise of new activities aimed at verifying the reproducibility of the results: for example, the "Reproducibility Track" at ECIR since 2015 hosts papers which replicate, reproduce and/or generalize previous research results.

Nevertheless, it has been repeatedly shown that the best TREC systems still outperforms off-the-shelf open source systems [2–4,16,17]. This is due to many

---

http://www.centre-eval.org/clef2019/.

L. Azzopardi et al. (Eds.): ECIR 2019, LNCS 11438, pp. 283–290, 2019.
https://doi.org/10.1007/978-3-030-15719-7_38

different factors, among which are the lack of tuning on a specific collection when using default configuration, and the lack of specifications about advanced components and resources adopted by the best systems.

It has been also shown that additivity is an issue, since adding a component on top of a weak or strong base does not produce the same level of gain [4,16]. This poses a serious challenge when off-the-shelf open source systems are used as stepping stone to test a new component on top of them, because the gain might appear bigger starting from a weak baseline.

Moreover, as also emerged from a recent survey within the SIGIR community [10] while there is a very positive attitude towards reproducibility and it is considered very important from a scientific point of view, there are many obstacles to it, such as the effort required to put it into practice, the lack of rewards for achieving it, the possible barriers for new and inexperienced groups, and, last but not least, the (somehow optimistic) researcher's perception that their own research is already reproducible.

Finally, the other side of reproducibility is the generalizability of the experimental results which plays an important role for future research. Indeed, both a Dagstuhl Perspectives Workshop [7] and the recent SWIRL III strategic workshop [1] have put on the IR research agenda the need to develop both better explanatory models of IR system performance and new predictive models, able to anticipate the performance of IR systems in new operational conditions.

This paper is organized as follows: Sect. 2 presents the objectives and scope of CENTRE, Sect. 3 describes the tasks proposed at CENTRE@CLEF 2019 and provides details about the measures used to evaluate the submitted runs, finally Sect. 4 reports some observations and lessons learnt from CENTRE@CLEF 2018, which were useful to design the 2019 edition.

2   Aims and Scope

Overall, the above considerations stress the need and urgency for a systematic approach to reproducibility and generalizability in IR. Therefore, the goal of *CLEF NTCIR TREC REproducibility (CENTRE)* at CLEF 2019 is to run a joint CLEF/NTCIR/TREC task on challenging participants:

- to replicate and reproduce best results of best/most interesting systems in previous editions of CLEF/NTCIR/TREC by using standard open source IR systems;
- to contribute back to the community the additional components and resources developed to reproduce the results in order to improve existing open source systems;
- to start exploring the generalizability of our findings and the possibility of predicting IR system performances.

We targeted evaluation campaigns to run CENTRE since we need third-party-ness with respect to the original developers of a technique, thus the author of the method should not attempt in reproducing it. Moreover, the critical mass involved in an evaluation campaign is needed for sharing the effort, achieving enough coverage and getting multiple independent checks for the same techniques. Indeed, if a system is reproduced by more than one single group, they can possibly discover more issues concerning a given technique and they can get as close as possible to actually reproducing it. Finally, we need to develop a common and shared protocol for reproducibility, to this end the experimental results and the developed components should be publicly accessible and an evaluation campaign represents one of the best venues to achieve this purpose.

We designed CENTRE as a joint CLEF/NTCIR/TREC task to further promote the possibility for third-party-ness, asking members of a community to reproduce what has been developed in another community. Moreover, we can simultaneously cover almost all the geographical areas, synchronously progressing the IR community at large towards reproducibility and the participants have the possibility to report their results in a globally shared task, at the closest and more convenient venue among CLEF/NTCIR/TREC. Finally, this is also an experiment to understand how a closer cooperation among CLEF/NTCIR/TREC might work.

3   CENTRE@CLEF2019 Tasks

In this edition of the lab, we target three specific objectives, according to the ACM badging terminology, which may need to be slightly adapted to the IR context:

Replicability (different team, same experimental setup): we use the collections, topics and ground-truth on which the methods and solutions have been developed and evaluated.
Reproducibility (different team, different experimental setup): we use a different experimental collection, but in the same domain, from those used to originally develop and evaluate a solution;
Generalizability (different team, different experimental setup): use sub-collections or different collections, but in the same domain.

For each of the aforementioned objectives, we designed a different task. Therefore, CENTRE@CLEF 2019 offers the following three tasks:

- Task 1 - Replicability: the task focuses on the replicability of selected methods on the same experimental collections;
- Task 2 - Reproducibility: the task focuses on the reproducibility of selected methods on the different experimental collections;
- Task 3 - Generalizability: the task focuses on collection performance prediction and the goal is to rank (sub-)collections on the basis of the expected performance over them.

3.1 Replicability and Reproducibility

Tasks 1 and 2 are the same tasks as in the CENTRE@CLEF2018 edition[1], targeting selected runs from CLEF/NTCIR/TREC on the same collections for replicability and on different collections for reproducibility. According to the discussion and feedback from attendees at CLEF 2018, we modified and changed the set of the targeted runs with respect to those used during the 2018 edition.

In particular, two valid suggestions were proposed by the participants in CENTRE@CLEF2018. First, we promote a partnership with the ECIR 2020 reproducibility track. To this end we encourage a collaboration among CENTRE participants, who reproduced the same algorithm. The outcome of this collaboration will be a joint paper, summarizing their reproducibility efforts and findings, which can be submitted at the ECIR 2020 reproducibility track. If enough teams will reproduce the same algorithm, the outcome paper will be even strengthened by the different perspectives and strategies adopted in the reproducibility process. We hope that this might represent a reward and a further incentive to participate in CENTRE. Furthermore, from the scheduling point of view, this partnership with ECIR is particularly well timed, since CENTRE deadlines are around May/June, while ECIR is early October. Thus participants will have the possibility to gather during CLEF, in early September, and to jointly finalize their paper.

Second, we select the replication and replicability targets among the best systems submitted at the labs of CLEF 2018. We decided to choose among those labs that will continue with the same task at CLEF 2019. This should motivate prospective participants in developing a baseline, since they would anyway need to do it in order to participate in their preferred lab. Moreover, this should also be useful for lab organizers, since they will be provided with state-of-the-art baselines available for their lab. We have already polled some lab organizers, who gave us their support in this respect.

Therefore, for the replicability and reproducibility activities, we select, among the methods/systems submitted to the CLEF tasks last year, the top performing and most impacting ones. In addition, we select methods/systems from TREC and NTCIR, following the same approach.

Each participating group will be challenged to replicate and/or reproduce one or more of the selected systems by only using standard open source IR systems, like Lucene, Terrier, and others, and they will submit one or more runs, in TREC format, representing the output of their reproduced systems. Participating groups will have to develop and integrate into the open source IR systems all the missing components and resources needed to replicate/reproduce the selected systems and they need to contribute back to open source all the developed components, resources, and configuration via a common repository, e.g. on Bitbucket.

We evaluate the quality of the replicated runs from two points of views: effectiveness and ranking. Effectiveness evaluates how close are the performance

---

[1] http://www.centre-eval.org/clef2018/.

scores of the reproduced systems to those of the original ones. This is measured using the *Root Mean Square Error (RMSE)* between the new and original *Average Precision (AP)* scores as follows:

$$\text{RMSE} = \sqrt{\frac{1}{m} \sum_{i=1}^{m} \left(AP_{\text{orig},i} - AP_{\text{replica},i} \right)^2} \tag{1}$$

where m is the total number of topics, $AP_{\text{orig},i}$ is the AP score of the original target run on topic t_i and $AP_{\text{replica},i}$ is the AP score of the replicated run on topic t_i.

Since different result lists may produce the same effectiveness score, we also measure how close are the ranked results lists of the replicated systems to those of the original ones. This is measured using the correlation coefficient Kendall's τ between the original and replicated run:

$$\tau_i(\text{orig}, \text{replica}) = \frac{P - Q}{\sqrt{(P + Q + T)(P + Q + U)}}$$

$$\bar{\tau}(\text{orig}, \text{replica}) = \frac{1}{m} \sum_{i=1}^{m} \tau_i(\text{orig}, \text{replica}) \tag{2}$$

where P is the total number of concordant pairs (document pairs that are ranked in the same order in both vectors) Q the total number of discordant pairs (document pairs that are ranked in opposite order in the two vectors), T and U are the number of ties, respectively, in the first and in the second ranking.

Evaluating the quality of the reproduced runs is less straightforward since there is no original run that can be used as a comparison point. Therefore, the idea is to compare the difference with respect to the improvement, in terms of AP, of a baseline run in both collections.

3.2   Generalizability

For the generalizability task, participants needs to rank document collections by the expected performance over them. The task is divided in three phases: training, test, and validation.

During the training phase participants are given topics, ground-truth, and a set of sub-collections (e.g. some newspaper collections from ad-hoc CLEF). They need to work on a selected method (e.g. a specific system as Lucene with BM25, ...) to allow for comparability across participants. Moreover, if they wish, they can also work on their own preferred method. The aim of this phase is to identify features of collections and methods that allow participants to rank and predict collections.

Then, during the test phase, the participants are given different sets of sub-collections (e.g. newspaper from ad-hoc CLEF in a different language) and they have to rank these collections with respect to the mandatory method and their own method.

Finally, the validation phase is conducted after the submission. We provide the topics and the ground-truth on the test sub-collections which are needed to verify how the different methods perform. Note that CLEF topics in different languages are translations one of each other and this should minimize the impact of the topic effect on the prediction. Indeed, generalizing a method through different topics should not be too hard, since topics are related and what differs is just the language used to describe them.

We evaluate the quality of the rankings and predictions of the generalizability task with *Mean Absolute Error (MAE)*, defined as follows:

$$\text{MAE} = \frac{1}{n} \sum_{j=1}^{n} |AP_{\text{orig},j} - AP_{\text{predict},j}| \tag{3}$$

where n is the number of sub-collections and $AP_{\text{predict},i}$ is the score of the predicted ranking. Furthermore, we use RMSE, as in Eq. (1), between the predicted and actual performance on the given collections.

4   Lessons Learnt from CENTRE@CLEF2018

CENTRE has been run for the first time at CLEF 2018 and variants of it are running at TREC 2018 and NTCIR-14 (due June 2019). The CENTRE@CLEF2018 edition [11,12] had 17 registered participants, but only 1 actually submitted results, Technical University of Wien (TUW) [15]. TUW failed to replicate the targeted bilingual run, indeed, AP_{orig} was 0.0667, while AP_{replica} was 0.0030, RMSE computed with Eq. (1) was 0.1132 and Kendall's τ in Eq. (2) was $-5.69 \cdot 10^{-04}$.

This leads to two observations. First, it indicates that engaging participants is a critical issue and that the community needs to be involved more in reproducibility. Second, replicability, reproducibility, and generalizability are still very hard to achieve, showing once more that reproducibility represents a serious limit for the advancement of research.

These issues were presented during the CENTRE session at CLEF 2018. We discussed with attendees measures for attracting more participation at the task and for lowering their barriers of entry. Thus, for CENTRE@CLEF 2019 we select the target systems among the best systems submitted at CLEF 2018 and we start the partnership with ECIR 2020 reproducibility track.

In addition to these incentives, we are contacting the colleagues who have master courses in IR to consider CENTRE tasks as part of the students assignments they already do. We already had positive feedback and availability from some colleagues.

Finally, we also hope that the new task on generalizability can raise the participation in the lab.

Acknowledgments. AMAOS (Advanced Machine Learning for Automatic Omni-Channel Support). Funded by: Innovationsfonden, Denmark.

References

1. Allan, J., et al.: Research frontiers in information retrieval - report from the third strategic workshop on information retrieval in lorne (SWIRL 2018). SIGIR Forum **52**(1), 34–90 (2018)

2. Arguello, J., Crane, M., Diaz, F., Lin, J., Trotman, A.: Report on the SIGIR 2015 workshop on reproducibility, inexplicability, and generalizability of results (RIGOR). SIGIR Forum **49**(2), 107–116 (2015)

3. Armstrong, T.G., Moffat, A., Webber, W., Zobel, J.: Has adhoc retrieval improved since 1994? In: Allan, J., Aslam, J.A., Sanderson, M., Zhai, C., Zobel, J. (eds.) Proceedings of the 32nd Annual International ACM SIGIR Conference on Research and Development in Information Retrieval (SIGIR 2009), pp. 692–693. ACM Press, New York (2009)

4. Armstrong, T.G., Moffat, A., Webber, W., Zobel, J.: Improvements that don't add up: ad-hoc retrieval results since 1998. In: Cheung, D.W.L., Song, I.Y., Chu, W.W., Hu, X., Lin, J.J. (eds.) Proceedings of the 18th International Conference on Information and Knowledge Management (CIKM 2009), pp. 601–610. ACM Press, New York (2009)

5. Cappellato, L., Ferro, N., Nie, J.Y., Soulier, L. (eds.): CLEF 2018 Working Notes. CEUR Workshop Proceedings (CEUR-WS.org) (2018). http://ceur-ws.org/Vol-2125/. ISSN 1613–0073

6. Ferro, N., et al.: Report on ECIR 2016: 38th european conference on information retrieval. SIGIR Forum **50**(1), 12–27 (2016)

7. Ferro, N., et al.: Manifesto from Dagstuhl perspectives workshop 17442 - building a predictive science for performance of information retrieval, recommender systems, and natural language processing applications. Dagstuhl Manifestos, Schloss Dagstuhl-Leibniz-Zentrum für Informatik, Germany **7**(1) (2018)

8. Ferro, N., Fuhr, N., Rauber, A.: Introduction to the special issue on reproducibility in information retrieval: evaluation campaigns, collections, and analyses. ACM J. Data Inf. Qual. (JDIQ) **10**(3), 9:1–9:4 (2018)

9. Ferro, N., Fuhr, N., Rauber, A.: Introduction to the special issue on reproducibility in information retrieval: tools and infrastructures. ACM J. Data Inf. Qual. (JDIQ) **10**(4), 1–4 (2018)

10. Ferro, N., Kelly, D.: SIGIR initiative to implement ACM artifact review and badging. SIGIR Forum **52**(1), 4–10 (2018)

11. Ferro, N., Maistro, M., Sakai, T., Soboroff, I.: CENTRE@CLEF2018: overview of the replicability task. In: Cappellato et al. [5]

12. Ferro, N., Maistro, M., Sakai, T., Soboroff, I.: Overview of CENTRE@CLEF 2018: a first tale in the systematic reproducibility realm. In: Bellot, P., et al. (eds.) CLEF 2018. LNCS, vol. 11018, pp. 239–246. Springer, Cham (2018). https://doi.org/10. 1007/978-3-319-98932-7_23

13. Freire, J., Fuhr, N., Rauber, A. (eds.): Report from Dagstuhl Seminar 16041: Reproducibility of Data-Oriented Experiments in e-Science. Dagstuhl Reports, vol. 6, no. 1, Schloss Dagstuhl-Leibniz-Zentrum für Informatik, Germany (2016)

14. Fuhr, N.: Some common mistakes in IR evaluation, and how they can be avoided. SIGIR Forum **51**(3), 32–41 (2017)

15. Jungwirth, M., Hanbury, A.: Replicating an experiment in cross-lingual information retrieval with explicit semantic analysis. In: Cappellato et al. [5]

16. Kharazmi, S., Scholer, F., Vallet, D., Sanderson, M.: Examining additivity and weak baselines. ACM Trans. Inf. Syst. (TOIS) **34**(4), 23:1–23:18 (2016)

17. Lin, J., et al.: Toward reproducible baselines: the open-source IR reproducibility challenge. In: Ferro, N., et al. (eds.) ECIR 2016. LNCS, vol. 9626, pp. 408–420. Springer, Cham (2016). https://doi.org/10.1007/978-3-319-30671-1_30
18. Munafò, M.R., et al.: A manifesto for reproducible science. Nat. Hum. Behav. **1**, 0021:1–0021:9 (2017)
19. Zobel, J., Webber, W., Sanderson, M., Moffat, A.: Principles for robust evaluation infrastructure. In: Agosti, M., Ferro, N., Thanos, C. (eds.) Proceedings of the Workshop on Data Infrastructures for Supporting Information Retrieval Evaluation (DESIRE 2011), pp. 3–6. ACM Press, New York (2011)

A Decade of Shared Tasks in Digital Text Forensics at PAN

Martin Potthast[1], Paolo Rosso[2], Efstathios Stamatatos[3(✉)], and Benno Stein[4]

[1] Department of Computer Science, Leipzig University, Leipzig, Germany
[2] PRHLT Research Center, Universitat Politècnica de València, Valencia, Spain
[3] Department of Information and Communication Systems Engineering,
University of the Aegean, Samos, Greece
`stamatatos@aegean.gr`
[4] Web Technology and Information Systems, Bauhaus-Universität Weimar,
Weimar, Germany
`pan@webis.de`
`https://pan.webis.de`

Abstract. Digital text forensics aims at examining the originality and credibility of information in electronic documents and, in this regard, to extract and analyze information about the authors of these documents. The research field has been substantially developed during the last decade. PAN is a series of shared tasks that started in 2009 and significantly contributed to attract the attention of the research community in well-defined digital text forensics tasks. Several benchmark datasets have been developed to assess the state-of-the-art performance in a wide range of tasks. In this paper, we present the evolution of both the examined tasks and the developed datasets during the last decade. We also briefly introduce the upcoming PAN 2019 shared tasks.

1 Introduction

Digital Text Forensics is a text mining field examining authenticity and credibility issues of information included in electronic documents. It is closely related with text reuse and deception detection applications. But its main focus is on authorship analysis, aiming to reveal information about the author(s) of electronic documents. This is crucial in applications of cybersecurity, digital humanities, and social media analytics. Writing style, rather than topic information, is the primary factor in text forensics tasks [11].

PAN[1] is a series of shared tasks in digital text forensics, started in 2009, and held in conjunction with CLEF evaluation labs since 2010 [35,38]. During the last decade, PAN explored several text forensics tasks and attracted the attention of the international research community. A significant number of new evaluation datasets covering multiple languages and genres have been developed

---

[1] The acronym originates from the title of the first PAN workshop held at SIGIR-2007: Plagiarism analysis, Authorship identification, and Near-duplicate detection [36].

© Springer Nature Switzerland AG 2019
L. Azzopardi et al. (Eds.): ECIR 2019, LNCS 11438, pp. 291–300, 2019.
https://doi.org/10.1007/978-3-030-15719-7_39

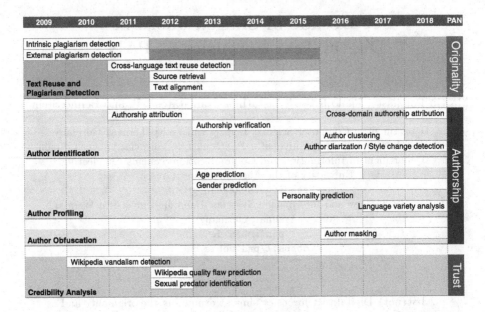

Fig. 1. Development of the most important digital text forensics tasks at PAN, starting at 2009. The tasks address three aspects: originality (top), authorship (middle), and trust (bottom). For each aspect various tasks have been suggested, varied, and further specialized.

and quickly established as reference benchmarks in this area. Since 2013, only software submissions are allowed in PAN tasks and all submitted software is evaluated on the specifically developed TIRA experimentation platform [26]. Apart from enabling reproducibility of results, the collected software can easily be tested on alternative datasets. In this paper, we present the evolution of main tasks organized by PAN during the last decade depicted in Fig. 1. In addition, we describe the datasets introduced by PAN to estimate the effectiveness and weaknesses of state-of-the-art methods.

2   Plagiarism Detection

Plagiarism, the unacknowledged use of another author's original work, is considered a problem in publishing, science, and education. Texts and other works of art have been plagiarized throughout history, but with the advent of the World Wide Web, text reuse and plagiarism have been observed at large scale. Looking for theory, concepts, and algorithms to detect text reuse, computer-based plagiarism detection breaks down this task into manageable parts: *"Given a text d and a reference collection D, does d contain a section s for which one can find a document d' ∈ D that contains a section s' such that under some retrieval model the similarity between s and s' is above a threshold?"*

The above definition presumes a closed world where a reference collection D is given, which is why this kind of analysis is called *external* plagiarism detection. Since D can be extremely large—possibly the entire indexed part of the World Wide Web—the respective research covers near-similarity search, near-duplicate detection, similarity hashing techniques, and indexes tailored to these problems In addition, situations where one would like to identify sections of plagiarized text if no reference collection is given can be imagined, a setting that is called *intrinsic* plagiarism detection. This problem is closely related to authorship verification: the goal of the former is to identify potential plagiarism by analyzing a document with respect to undeclared changes in writing style. In this regard, intrinsic plagiarism analysis can be understood as a more general form of the authorship verification problem: only a single document is given, and, one is faced with the problem of finding the suspicious sections. Both intrinsic plagiarism detection and authorship verification are one-class classification problems [37].

Against the above background the development of plagiarism detection tasks as shown in Fig. 1 (top ∼ "originality") becomes plausible: starting 2009, both intrinsic and external plagiarism detection were considered; over three years, the evaluation datasets have been improved and extended [21,23,30]. This experience and the improved problem understanding is also reflected in development of tailored detection measures such as "pladget", which combines precision, recall, and detection accuracy for plagiarized passages. While cross-language text reuse detection lost its importance with gaining popularity of machine translation and the Wikipedia-Based Multilingual Retrieval Model [29], it became clear that research for external plagiarism detection requires a two-fold strategy, adopted in the ensuing three years [24,25,27]: (1) finding promising candidates on the Web (the source retrieval task), and, (2) developing effective algorithms for fuzzy text matching (the text alignment task). Meanwhile, spin-off tasks at FIRE [6,8], also in the form of source code reuse detection [9,10], used the original tasks' setup to develop resources for other languages.

3   Author Identification

Author identification focuses on the personal style of the author(s) of electronic documents. The main assumption is that every author has her own stylistic "fingerprint" and that it is possible to identify the author(s) of a disputed document based on them [33]. There are several variations of this problem and PAN has explored many of them as shown in Fig. 1. In more detail, in *closed-set authorship attribution*, a well-defined list of suspects and samples of texts they authored are provided. The task is to identify the most likely author of a questioned document among them. In *open-set attribution*, the true author may not be included in the list of suspects. The first editions of PAN related to author identification focused on tasks already popular in the research community [33]. In the 2011 edition, a dataset using emails (extracted from the Enron corpus) and relatively large sets of candidate authors was developed [4]. In 2012, emphasis was put on smaller candidate sets and fiction in English [15]. Another important task is *author verification* where there is only one candidate author. This is an especially challenging task considered fundamental in authorship attribution [17]. PAN has spurred widespread interest in this task among the research community, obtaining rather high participation figures in verification tasks from 2013 to 2015 [11,26,34]. The developed datasets for these tasks cover four languages (Dutch, English, Greek, Spanish) and a variety of genres (e.g., newspaper articles, student essays, reviews, novels, textbooks).

PAN also explored tasks where no labeled (known authorship) documents are provided. One such task is *author clustering* where the goal is to group documents written by the same author given a document collection. Two editions of PAN in 2016 and 2017 introduced an evaluation framework that also considers a retrieval task (ranking document pairs by likelihood of common authorship) [28,31]. Three languages (English, Greek, and Spanish) and two genres (reviews and newspaper articles) are included in the developed datasets focusing on either full texts (2016 edition) or fragments (paragraphs) of texts (2017 edition). Another unsupervised task is *author diarization*, where the assumption that each document is written by a single author does not hold. The task aims to determine how many authors wrote the document and extract the authorial components. A few variations of this task have been included in recent PAN editions, moving from complicated ones (e.g., detection of the exact number of co-authors and their exact contribution) [28,31], which proved to be extremely difficult at present, to more basic ones (e.g. *style change detection*: distinguishing between single-author and multi-author documents) [35], which is more feasible with current technology. The datasets to support these tasks include synthetic multi-author documents in English (essays or Q&As) where topic is controlled [28,31,35].

More recently, PAN focused on a challenging, but quite realistic problem: *cross-domain authorship attribution*. In this task, the labeled and unlabeled documents differ with respect to topic, genre, or even language. Fanfiction, a large part of contemporary fiction written by non-professionals following a canon (e.g., a well-known novel or TV series), has been adopted to allow for controlling the domain of documents. Thus the target domain (fandom) is excluded from the training documents in a closed-set attribution framework. The datasets built for this task include five languages (English, French, Italian, Polish, and Spanish) [35].

4   Author Profiling

Author profiling aims at identifying personal traits of an author on the basis of her writing. Traits, such as gender, age, language variety, or personality, are of high interest for areas such as forensics, security, and also marketing. From a forensic linguistics perspective, one would like to be able to know the linguistic profile of the author of a harassing text message (language used by a certain type of people). From a security perspective, these technologies may allow to profile and identify criminals. From the marketing perspective, being able to identify personal traits from comments to blogs or reviews may provide advertisers with the possibility of better segmenting their audience, which is an important competitive advantage. Traditional investigations in computational linguistics [5] and social psychology [20] have been carried out mainly for English. Furthermore, pioneering research from Argamon et al. [5] and Holmes et al. [13] focused on formal and well-written texts. With the rise of social media, however, the focus has shifted to more informal usage found in blogs and forums [16,32].

Starting in 2013, PAN has been organizing author profiling-related tasks with several objectives as depicted in Fig. 1. We have covered different profiling aspects (age, gender, native language, language variety, personality), languages (Arabic, Dutch, English, Italian, Portuguese, Russian, Bengali, Hindi, Kannada, Malayo, Tamil, and Telugu), and genres (blogs, reviews, social media, and Twitter). The first edition was organized with the aim of investigating *age and gender identification* in a social media realistic scenario [11]. We collected thousands of social media posts in English and Spanish with a high variety of topics. With respect to age, we considered three classes following previous work by Schler et al. [32]: 10s (13–17), 20s (23–27) and 30s (33–47). Furthermore, we wanted to test the robustness of the systems when dealing with fake age profiles such as those induced by sexual predators. Therefore, we included texts from the previous year's shared task on sexual predator identification [14]. In the second edition [26], we extended the task to other genres besides social media focusing on Twitter, blogs, and hotel reviews, in English and Spanish. We realized the difficulty of obtaining high-quality labeled data and proposed a methodology to annotate age and gender. In 2014, we opted for modeling age classes without gaps: 18–24; 25–34; 35–49; 50–64; 65+. Finally, the Twitter subcorpus was constructed in cooperation with RepLab [3] in order to address also the reputational perspective (e.g., profiling social media influencers, journalists, professionals, celebrities, among others).

In 2015 [34], besides age and gender identification, we introduced the task of *personality recognition* in Twitter. We maintained the age ranges defined in 2014 (except "50–64" and "65+" that were merged to "50–XX") and, besides English and Spanish, we included also Dutch and Italian (yet, only for gender and personality recognition). The objective of the shared task organized in 2016 [31] was to investigate the robustness of the systems in a cross-genre scenario. That is, training the systems in one genre and testing their performance in other genres. In particular, we provided Twitter data for training in English, Spanish, and Dutch. The approaches were then tested on blogs and social media genres in English and Spanish, and essays and reviews in Dutch. In 2017 [28], we introduced two novelties: *language variety identification* (together with gender), and Arabic and Portuguese languages (besides English and Spanish). This marked the first time a task has been organized covering gender and language variety identification combined. Language variety was addressed from a fine-grained and coarse-grained perspective, where varieties that are close, geographically, were grouped together (e.g., Canada and United States, Great Britain and Ireland, or New Zealand and Australia). Finally, in 2018 [35], gender identification on Twitter was approached from a multimodal perspective. Three languages have been considered: English, Spanish, and Arabic. Further spin-off profiling tasks were organized at FIRE [18,19].

5   Author Obfuscation

Author obfuscation (in particular, author masking as a special case) was launched in 2016 within the PAN task series: as the adversary task to authorship verification, it deals with preventing verification by altering a to-be-verified text. The underlying question is whether the authorial style of a text can be consistently manipulated. Though this task is of public interest and has various applications, only a handful of approaches have been proposed so far, and they achieved limited success only. We hope that this dedicated PAN task will push the research boundaries for both obfuscation and verification, and help to develop theoretical backgrounds and new evaluation frameworks: an obfuscation software is called *safe* if a forensic analysis does not reveal the original author of its obfuscated texts, it is called *sound* if its obfuscated texts are textually entailed with their originals, and it is called *sensible* if its obfuscated texts are inconspicuous.

6   Trust-Related Tasks

The PAN tasks related to trust (see Fig. 1 bottom) have foreshadowed today's challenges that the Web and, in particular, social media platforms provide to computer scientists, psycholinguists, and psychologists, among others. Driven by the ideal of social responsibility and the scientific curiosity of the limits of "detectability", different tasks have been devised and operationalized.

Wikipedia vandalism detection (2010–2011) addressed the intentional damage of Wikipedia articles: given a set of edits on Wikipedia articles, the task was to separate ill-intentioned edits from well-intentioned edits. Wikipedia quality flaw prediction (2012) can be considered as a generalization of the vandalism detection task, focusing on the prediction of quality flaws in Wikipedia articles. It was driven by the observation that the majority of quality flaws in Wikipedia is not caused due to malicious intentions but stem from edits by inexperienced authors; examples include poor writing style, unreferenced statements, or missing neutrality. Since, by nature, no representative "negative" training data can be provided (articles that are tagged to not suffer from vandalism, articles that are tagged to not contain a particular flaw), vandalism detection and quality flaw prediction in Wikipedia represent one-class classification problems.

The goal of the sexual predator identification task (2012) was to identify online predators: the participants were given chat logs involving two (or more) people for which they had to determine who is the one trying to convince the other(s) to provide some sexual favor.

7 Discussion

During the last decade, PAN contributed to focus the attention of the research community on specific digital text forensics tasks, built benchmark datasets, and estimated the effectiveness as well as the weaknesses of the state of the art. The developed datasets cover multiple genres and languages while the top-ranked PAN submissions have been used as baselines in subsequent research [12,22]. In addition, the evolution of tasks within PAN made the exploration of new tasks feasible. For example, author obfuscation is based on the results of the author verification tasks. PAN also achieved to highlight the close relationship among certain tasks. For example, an approach to authorship clustering can be based on a verification method [7].

The upcoming edition of PAN will focus on four tasks. Two new tasks are introduced—*bots and gender profiling*, whose aim is to discriminate between human and robot Twitter profiles and in case of humans to profile their gender, and *celebrity profiling*, whose aim is to profile celebrities with regard to how they present themselves in public, be it personally or via public relations staff. In addition, the *cross-domain authorship attribution* task based on fanfiction documents, introduced in 2018, will continue. However, this time the open-set attribution scenario is adopted. Finally, another variant of the *style change detection* task will be included, this time focusing on the exact number of co-authors in a multi-author document.

Acknowledgements. We are indebted to many colleagues and friends who contributed greatly to PAN's tasks: Maik Anderka, Shlomo Argamon, Alberto Barrón-Cedeño, Fabio Celli, Fabio Crestani, Walter Daelemans, Andreas Eiselt, Tim Gollub, Parth Gupta, Matthias Hagen, Teresa Holfeld, Patrick Juola, Giacomo Inches, Mike Kestemont, Moshe Koppel, Manuel Montes-y-Gómez, Aurelio Lopez-Lopez, Francisco Rangel, Miguel Angel Sánchez-Pérez, Günther Specht, Michael Tschuggnall, and Ben Verhoeven. Our special thanks go to PAN's sponsors throughout the years and not least to the hundreds of participants.

References

1. FIRE 2015 Working Notes Papers, 4–6 December, Gandhinagar, India (2015). http://www.uni-weimar.de/medien/webis/events/pan-at-fire-15
2. FIRE 2017 Working Notes Papers, 8–11 December, Bangalore, India (2017)
3. Amigó, E., et al.: Overview of RepLab 2014: author profiling and reputation dimensions for online reputation management. In: Kanoulas, E., et al. (eds.) CLEF 2014. LNCS, vol. 8685, pp. 307–322. Springer, Cham (2014). https://doi.org/10.1007/978-3-319-11382-1_24
4. Argamon, S., Juola, P.: Overview of the international authorship identification competition at PAN-2011. In: Petras, V., Forner, P., Clough, P. (eds.) Notebook Papers of CLEF 2011 Labs and Workshops, 19–22 September, Amsterdam, Netherlands (2011). http://www.clef-initiative.eu/publication/working-notes
5. Argamon, S., Koppel, M., Fine, J., Shimoni, A.R.: Gender, genre, and writing style in formal written texts. TEXT **23**, 321–346 (2003)

6. Asghari, H., Mohtaj, S., Fatemi, O., Faili, H., Rosso, P., Potthast, M.: Algorithms and corpora for Persian plagiarism detection. In: Majumder, P., Mitra, M., Mehta, P., Sankhavara, J. (eds.) FIRE 2016. LNCS, vol. 10478, pp. 61–79. Springer, Cham (2018). https://doi.org/10.1007/978-3-319-73606-8_5

7. Bagnall, D.: Authorship clustering using multi-headed recurrent neural networks-notebook for PAN at CLEF 2016. In: Balog, K., Cappellato, L., Ferro, N., Macdonald, C. (eds.) CLEF 2016 Evaluation Labs and Workshop - Working Notes Papers, 5–8 September, Évora, Portugal. CEUR Workshop Proceedings, CEUR-WS.org, September 2016. http://ceur-ws.org/Vol-1609/

8. Bensalem, I., Boukhalfa, I., Rosso, P., Abouenour, L., Darwish, K., Chikhi, S.: Overview of the AraPlagDet PAN@FIRE2015 shared task on Arabic plagiarism detection. In: FIRE 2015 Working Notes Papers, 4–6 December, Gandhinagar, India [1]

9. Flores, E., Rosso, P., Moreno, L., Villatoro-Tello, E.: On the detection of SOurce COde re-use. In: FIRE 2014 Working Notes Papers, 5–7 December, Bangalore, India, pp. 21–30, December 2014

10. Flores, E., Rosso, P., Villatoro-Tello, E., Moreno, L., Alcover, R., Chirivella, V.: PAN@FIRE: Overview of CL-SOCO track on the detection of cross-language SOurce COde re-use. In: FIRE 2015 Working Notes Papers, 4–6 December, Gandhinagar, India, pp. 1–5 [1]

11. Gollub, T., et al.: Recent trends in digital text forensics and its evaluation. In: Forner, P., Müller, H., Paredes, R., Rosso, P., Stein, B. (eds.) CLEF 2013. LNCS, vol. 8138, pp. 282–302. Springer, Heidelberg (2013). https://doi.org/10.1007/978-3-642-40802-1_28

12. Halvani, O., Graner, L., Vogel, I.: Authorship verification in the absence of explicit features and thresholds. In: Pasi, G., Piwowarski, B., Azzopardi, L., Hanbury, A. (eds.) ECIR 2018. LNCS, vol. 10772, pp. 454–465. Springer, Cham (2018). https://doi.org/10.1007/978-3-319-76941-7_34

13. Holmes, J., Meyerhoff, M.: The Handbook of Language and Gender. Blackwell Handbooks in Linguistics. Wiley, Hoboken (2003)

14. Inches, G., Crestani, F.: Overview of the international sexual predator identification competition at PAN-2012. In: Forner, P., Karlgren, J., Womser-Hacker, C. (eds.) CLEF 2012 Evaluation Labs and Workshop - Working Notes Papers, 17–20 September, Rome, Italy (2012). http://www.clef-initiative.eu/publication/working-notes

15. Juola, P.: An overview of the traditional authorship attribution subtask. In: Forner, P., Karlgren, J., Womser-Hacker, C. (eds.) CLEF 2012 Evaluation Labs and Workshop - Working Notes Papers, 17–20 September, Rome, Italy (2012). http://www.clef-initiative.eu/publication/working-notes

16. Koppel, M., Argamon, S., Shimoni, A.R.: Automatically categorizing written texts by author gender (2003)

17. Koppel, M., Schler, J., Argamon, S., Winter, Y.: The "fundamental problem" of authorship attribution. Engl. Stud. 93(3), 284–291 (2012)

18. Litvinova, T., Rangel, F., Rosso, P., Seredin, P., Litvinova, O.: Overview of the RusProfiling PAN at FIRE track on cross-genre gender identification in Russian. In: FIRE 2017 Working Notes Papers, 8–11 December, Bangalore, India [2]

19. Anand Kumar, M., Barathi Ganesh, H.B., Singh, S., Soman, K.P., Rosso, P.: Overview of the INLI PAN at FIRE-2017 track on Indian native language identification. In: FIRE 2017 Working Notes Papers, 8–11 December, Bangalore, India [2]

20. Pennebaker, J.W.: The Secret Life of Pronouns: What Our Words Say About Us. Bloomsbury, USA (2013)
21. Potthast, M., Barrón-Cedeño, A., Eiselt, A., Stein, B., Rosso, P.: Overview of the 2nd international competition on plagiarism detection. In: Braschler, M., Harman, D., Pianta, E. (eds.) Working Notes Papers of the CLEF 2010 Evaluation Labs, September 2010. http://www.clef-initiative.eu/publication/working-notes
22. Potthast, M., et al.: Who wrote the web? Revisiting influential author identification research applicable to information retrieval. In: Ferro, N., et al. (eds.) ECIR 2016. LNCS, vol. 9626, pp. 393–407. Springer, Cham (2016). https://doi.org/10.1007/978-3-319-30671-1_29
23. Potthast, M., Eiselt, A., Barrón-Cedeño, A., Stein, B., Rosso, P.: Overview of the 3rd international competition on plagiarism detection. In: Notebook Papers of the 5th Evaluation Lab on Uncovering Plagiarism, Authorship and Social Software Misuse (PAN), Amsterdam, The Netherlands, September 2011
24. Potthast, M., et al.: Overview of the 4th international competition on plagiarism detection. In: Forner, P., Karlgren, J., Womser-Hacker, C. (eds.) Working Notes Papers of the CLEF 2012 Evaluation Labs, September 2012. http://www.clef-initiative.eu/publication/working-notes
25. Potthast, M., et al.: Overview of the 5th international competition on plagiarism detection. In: Forner, P., Navigli, R., Tufis, D. (eds.) Working Notes Papers of the CLEF 2013 Evaluation Labs, September 2013. http://www.clef-initiative.eu/publication/working-notes
26. Potthast, M., Gollub, T., Rangel, F., Rosso, P., Stamatatos, E., Stein, B.: Improving the reproducibility of PAN's shared tasks: plagiarism detection, author identification, and author profiling. In: Kanoulas, E., et al. (eds.) CLEF 2014. LNCS, vol. 8685, pp. 268–299. Springer, Cham (2014). https://doi.org/10.1007/978-3-319-11382-1_22
27. Potthast, M., et al.: Overview of the 6th international competition on plagiarism detection. In: Cappellato, L., Ferro, N., Halvey, M., Kraaij, W. (eds.) Working Notes Papers of the CLEF 2014 Evaluation Labs. CEUR Workshop Proceedings, CLEF and CEUR-WS.org, September 2014. http://www.clef-initiative.eu/publication/working-notes
28. Potthast, M., Rangel, F., Tschuggnall, M., Stamatatos, E., Rosso, P., Stein, B.: Overview of PAN'17: author identification, author profiling, and author obfuscation. In: Jones, G.J.F., et al. (eds.) CLEF 2017. LNCS, vol. 10456, pp. 275–290. Springer, Cham (2017). https://doi.org/10.1007/978-3-319-65813-1_25
29. Potthast, M., Stein, B., Anderka, M.: A Wikipedia-based multilingual retrieval model. In: Macdonald, C., Ounis, I., Plachouras, V., Ruthven, I., White, R.W. (eds.) ECIR 2008. LNCS, vol. 4956, pp. 522–530. Springer, Heidelberg (2008). https://doi.org/10.1007/978-3-540-78646-7_51
30. Potthast, M., Stein, B., Eiselt, A., Barrón-Cedeño, A., Rosso, P.: Overview of the 1st international competition on plagiarism detection. In: Stein, B., Rosso, P., Stamatatos, E., Koppel, M., Agirre, E. (eds.) SEPLN 2009 Workshop on Uncovering Plagiarism, Authorship, and Social Software Misuse (PAN 2009), pp. 1–9. CEUR-WS.org, September 2009. http://ceur-ws.org/Vol-502
31. Rosso, P., Rangel, F., Potthast, M., Stamatatos, E., Tschuggnall, M., Stein, B.: Overview of PAN'16: new challenges for authorship analysis: cross-genre profiling, clustering, diarization, and obfuscation. In: Fuhr, N., et al. (eds.) CLEF 2016. LNCS, vol. 9822, pp. 332–350. Springer, Cham (2016). https://doi.org/10.1007/978-3-319-44564-9_28

32. Schler, J., Koppel, M., Argamon, S., Pennebaker, J.W.: Effects of age and gender on blogging. In: AAAI Spring Symposium: Computational Approaches to Analyzing Weblogs, pp. 199–205. AAAI (2006)

33. Stamatatos, E.: A survey of modern authorship attribution methods. J. Am. Soc. Inf. Sci. Technol. **60**, 538–556 (2009)

34. Stamatatos, E., Potthast, M., Rangel, F., Rosso, P., Stein, B.: Overview of the PAN/CLEF 2015 evaluation lab. In: Mothe, J., et al. (eds.) CLEF 2015. LNCS, vol. 9283, pp. 518–538. Springer, Cham (2015). https://doi.org/10.1007/978-3-319-24027-5_49

35. Stamatatos, E., et al.: Overview of PAN 2018: author identification, author profiling, and author obfuscation. In: Bellot, P., et al. (eds.) CLEF 2018. LNCS, vol. 11018, pp. 267–285. Springer, Cham (2018). https://doi.org/10.1007/978-3-319-98932-7_25

36. Stein, B., Koppel, M., Stamatatos, E. (eds.): SIGIR 2007 Workshop on Plagiarism Analysis, Authorship Identification, and Near-Duplicate Detection (PAN 2007). CEUR-WS.org (2007). http://www.uni-weimar.de/medien/webis/events/pan-07

37. Stein, B., Lipka, N., Prettenhofer, P.: Intrinsic plagiarism analysis. Lang. Resour. Eval. (LRE) **45**(1), 63–82 (2011)

38. Stein, B., Rosso, P., Stamatatos, E., Koppel, M., Agirre, E. (eds.): SEPLN 2009 Workshop on Uncovering Plagiarism, Authorship, and Social Software Misuse (PAN 2009). Universidad Politécnica de Valencia and CEUR-WS.org (2009). http://ceur-ws.org/Vol-502

ImageCLEF 2019: Multimedia Retrieval in Lifelogging, Medical, Nature, and Security Applications

Bogdan Ionescu[1(✉)], Henning Müller[2,17], Renaud Péteri[3],
Duc-Tien Dang-Nguyen[16], Luca Piras[5], Michael Riegler[6], Minh-Triet Tran[7],
Mathias Lux[8], Cathal Gurrin[4], Yashin Dicente Cid[2], Vitali Liauchuk[9],
Vassili Kovalev[9], Asma Ben Abacha[11], Sadid A. Hasan[10], Vivek Datla[10],
Joey Liu[10], Dina Demner-Fushman[11], Obioma Pelka[12],
Christoph M. Friedrich[12], Jon Chamberlain[13], Adrian Clark[13],
Alba García Seco de Herrera[13], Narciso Garcia[14], Ergina Kavallieratou[15],
Carlos Roberto del Blanco[14], Carlos Cuevas Rodríguez[14],
Nikos Vasillopoulos[15], and Konstantinos Karampidis[15]

[1] University Politehnica of Bucharest, Bucharest, Romania
bionescu@alpha.imag.pub.ro
[2] University of Applied Sciences Western Switzerland (HES-SO),
Sierre, Switzerland
[3] University of La Rochelle, La Rochelle, France
[4] Dublin City University, Dublin, Ireland
[5] University of Cagliari, Cagliari, Italy
[6] University of Oslo, Oslo, Norway
[7] University of Science, Ho Chi Minh City, Vietnam
[8] Klagenfurt University, Klagenfurt, Austria
[9] Institute for Informatics, Minsk, Belarus
[10] Philips Research Cambridge, Cambridge, USA
[11] National Library of Medicine, Bethesda, USA
[12] University of Applied Sciences and Arts, Dortmund, Germany
[13] University of Essex, Colchester, UK
[14] E.T.S. Ingenieros Telecomunicación, Madrid, Spain
[15] University of the Aegean, Mytilene, Greece
[16] University of Bergen, Bergen, Norway
[17] University of Geneva, Geneva, Switzerland

Abstract. This paper presents an overview of the foreseen ImageCLEF 2019 lab that will be organized as part of the Conference and Labs of the Evaluation Forum - CLEF Labs 2019. ImageCLEF is an ongoing evaluation initiative (started in 2003) that promotes the evaluation of technologies for annotation, indexing and retrieval of visual data with the aim of providing information access to large collections of images in various usage scenarios and domains. In 2019, the 17th edition of Image-CLEF will run four main tasks: (i) a *Lifelog* task (videos, images and other sources) about daily activities understanding, retrieval and summarization, (ii) a *Medical* task that groups three previous tasks (caption analysis, tuberculosis prediction, and medical visual question answering)

L. Azzopardi et al. (Eds.): ECIR 2019, LNCS 11438, pp. 301–308, 2019.
https://doi.org/10.1007/978-3-030-15719-7_40

with newer data, (iii) a new *Coral* task about segmenting and labeling collections of coral images for 3D modeling, and (iv) a new *Security* task addressing the problems of automatically identifying forged content and retrieve hidden information. The strong participation, with over 100 research groups registering and 31 submitting results for the tasks in 2018 shows an important interest in this benchmarking campaign and we expect the new tasks to attract at least as many researchers for 2019.

Keywords: Lifelogging retrieval and summarization ·
Medical retrieval · Coral image segmentation and classification ·
File forgery detection · ImageCLEF benchmarking ·
Annotated datasets

1   Introduction

The ImageCLEF evaluation campaign was started as part of the CLEF (Cross Language Evaluation Forum) in 2003 [4,5]. It has been held every year since then and delivered many results in the analysis and retrieval of images [15,17]. Medical tasks started in 2004 and have in some years been the majority of the tasks in ImageCLEF [14].

The objectives of ImageCLEF have always been the multilingual or language-independent analysis of visual content. A focus has often been on multimodal data sets, so combining images with structure information, free text or other information that helps in the decision making.

Since 2018 ImageCLEF uses the crowdAI[1] platform to distribute the data and received the submitted results. The system allows having an online leader board and gives the possibility to keep data sets accessible beyond competition, including a continuous submission to the leader board.

Over the years, ImageCLEF and also CLEF have shown a strong scholarly impact that was captured in [21,22]. This underlines the importance of evaluation campaigns for disseminating best scientific practices.

In the following, we introduce the four tasks that are going to run in the 2019 edition[2], namely: ImageCLEFlifelog, ImageCLEFmedical, ImageCLEFcoral, and ImageCLEFsecurity. A sample of some of the provided visual data is presented in Fig. 1.

2   ImageCLEFlifelog

An increasingly wide range of personal devices, such as smart-phones, video cameras as well as wearable devices that allow capturing pictures, videos, and audio clips for every moment of our lives have become available. Considering the huge volume of data created, there is a need for systems that can automatically

---

[1] http://www.crowdA.org/.
[2] https://www.imageclef.org/2019.

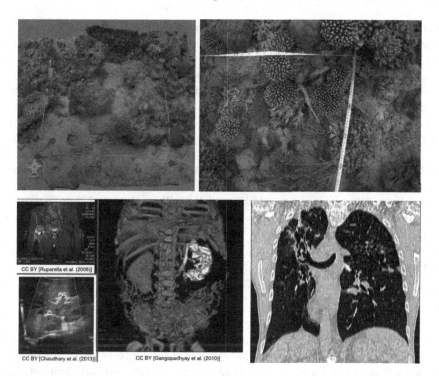

Fig. 1. Sample images from (left to right, top to bottom): ImageCLEFcoral, and Image-CLEFmedical, caption and tuberculosis tasks.

analyze the data in order to categorize, summarize and also query to retrieve the information the user may need.

The main goal of the Lifelog task since its first edition [6] has been to advance the state-of-the-art research in lifelogging as an application of information retrieval. As in the 2017 and 2018 editions, the 2019 task will be split into two related subtasks using a completely new rich multimodal data set. It consists of 42 days of data from two lifeloggers, namely: images (1,500–2,500 per day from wearable cameras), visual concepts (automatically extracted visual concepts with varying rates of accuracy), semantic content (semantic locations, semantic activities) based on sensor readings (via the Moves App) on mobile devices, biometrics information (heart rate, galvanic skin response, calories burn, steps, continual blood glucose, etc.), music listening history, computer usage (frequency of typed words via the keyboard and information consumed on the computer via Automatic Speech Recognition of on-screen activity on a per-minute basis). The copyright and ethical approval to release the data is held by one of the task organizers.

Subtask 1 (Puzzle): Solve my life puzzle. Given a set of lifelog images with associated metadata (e.g., biometrics, location, etc.), but no time stamps, the participants need to analyze these images and rearrange them in chronological order and predict the correct day (Monday or Sunday) and part of the day (morning, afternoon, or evening). The data set will be arranged into 75% training and 25% test data.

Subtask 2 (LMRT): Lifelog moment retrieval. This sub-task follows the success of the LMRT sub-task in ImageCLEFlifelog 2018 [7] with some minor adjustments. The participants have to retrieve a number of specific predefined activities in a lifelogger's life. For example, they need to return the relevant moments for the query "Find the moment(s) when I was shopping". Particular attention needs to be paid to the diversification of the selected moments with respect to the target scenario. The ground truth for this subtask was created using manual annotations.

For assessing performance, classic metrics will be deployed, e.g., precision, cluster recall (to account for the diversification), etc. In particular, the organizers would like to emphasize methods that allow interaction with real users (via Relevance Feedback - RF, for example) and the organizers will define appropriate evaluation measures.

3   ImageCLEFmedical

The medical tasks of ImageCLEF have started in 2004 and have been run almost every year since then [15]. In 2019 there will be three subtasks under the medical umbrella that will all be based on past tasks but concentrating on clean data and on bringing people of the three tasks together with fewer actual subtasks. The three tasks will be: *figure caption analysis* [8,13], *tuberculosis analysis* [8,13], and *visual question answering* [12].

The *caption analysis task* will use a new and manually curated data set of images from the biomedical literature, thus reducing variability in the data and making the extraction of concepts cleaner, as only clinical images are present and as there are also quality constraints on the captions. The Radiology Objects in Context (ROCO) [18] data set is used. It contains over 81,000 radiology images from the medical literature including caption information and a manual control of the image type. The task will concentrate on extracting Unified Medical Language System (UMLS) concepts and not the prediction of a precise caption. Trivial concepts will be removed and also concepts occurring in only a single image.

The *tuberculosis task* uses 3D image volumes (Computed Tomography with 3 mm slice thickness and around 150 slices per image volume) and clinical data to detect tuberculosis type and severity from these data. The multiple drug resistance task was dropped for 2019, as results were of limited quality. The other two tasks are clinically more interesting.

The *medical Visual Question Answering (VQA) task* is an exciting problem that combines natural language processing and computer vision techniques. Inspired by the recent success of visual question answering in the general domain, this year's task will focus on a new, larger and nicely manually curated dataset.

Given a medical image accompanied with a clinically relevant question, participating systems are tasked with answering the question based on the visual image content.

4   ImageCLEFcoral

Most coral reefs are in danger of being lost within the next 30 years, and with them the ecosystems they support [1]. This catastrophe will see the extinction of many marine species, such as shellfish, corals and many micro-organisms in the ocean. It also reduces reef fishery production, which is an important source of income and food source [2,19]. By monitoring the changes in the structural complexity and composition of coral reefs we can help prioritize conservation efforts. Autonomous Underwater Vehicles (AUV) can collect data for many hours at a time. However, the complexity of the images makes it impossible for human annotators to assess the contents of images on a large scale [3]. Advances in automatically annotating images for complexity and benthic composition have been promising [11,20], and we are interested in automatically identifying areas of interest and to label them appropriately for monitoring coral reefs.

Similar to previous ImageCLEF annotation tasks [9,10,23–25], the 2019 ImageCLEFcoral task will require participants to automatically annotate and localize a collection of images with types of benthic substrate, such as hard coral and sponge. The data for this task originates from a growing, large-scale collection of images taken from coral reefs around the world as part of a coral reef monitoring project with the Marine Technology Research Unit at the University of Essex (currently containing over 2TB of image data of benthic reef structure).

The annotated data set comprises several sets of overlapping images, each set taken in an area of underwater terrain. Each image will be labelled by experts for training and evaluation.

The performance of the algorithms will be evaluated using the PASCAL VOC[3] style metric of intersection over union (IoU) that calculates the area of intersection between the foreground in the proposed output localization and the ground-truth bounding box localization, divided by the area of their union.

5   ImageCLEFsecurity

File Forgery Detection (FFD) is a serious problem concerning digital forensics examiners. Fraud or counterfeits are common causes for altering files. Another example is a child predator who hides porn images by altering the image extension and in some cases by changing the image signature. Many proposals have been made to solve this problem and the most promising ones concentrate on the image content. It is also common that someone who wants to hide information in plain sight without being perceived might use steganography. Steganography is the practice of concealing a file, message, image or video within another file,

---

[3] http://host.robots.ox.ac.uk/pascal/VOC/.

message, image, or video. The most usual cover medium for hiding data are images. For more information, we refer the reader to [16].

The specific objective of this task is first to examine if an image has been forged and then if it could hide a text message. Last objective is to retrieve the potentially hidden message from the forged steganography images.

The participant takes the role of a professional digital forensic examiner collaborating with the police, who suspects that there is an ongoing fraud in the Central Bank. After obtaining a court order, police gain access to a suspect's computer in the bank with the purpose of looking for images proving the suspect guilty. However, police suspects that the suspect managed to change file extensions and signatures of some images, so that they look like PDF (Portable Document Format) files or other types. It is probable that the suspect has used steganography software to hide messages within the forged images that can reveal valuable information. The following subtasks are defined.

Subtask 1: perform detection of altered (forged) images (both extension and signature) and predict the actual type of the forged file. *Subtask 2:* identify the altered images that hide steganographic content. *Subtask 3:* retrieve the hidden messages (text) from the forged steganographic images.

The data set consists of about 9,000 forged images and pdfs, divided into 3 groups of 3,000 images each one. Every group of images is used for a specific task, where 2,000 images are for training and 1,000 for test. All participants have access to the training data set along with the ground truth. The participants will also have the opportunity to publish an extended version of their proposed methodology and experiments in a special issue of the Journal of Imaging.

6   Conclusions

In this paper we presented an overview of the upcoming ImageCLEF 2019 campaign. ImageCLEF has organized many tasks in a variety of domains over the past 17 years, from general stock photography, medical and biodiversity data to multimodal lifelogging. The focus has always been on language independent approaches and most often on multimodal data analysis. 2019 has a set of interesting tasks that are expected to again draw a large number of participants. A focus for 2019 has been on the diversity of applications and on creating clean data sets to provide a solid basis for the evaluations.

References

1. Birkeland, C.: Global status of coral reefs: in combination, disturbances and stressors become ratchets. In: World Seas: An Environmental Evaluation, pp. 35–56. Elsevier (2019)
2. Brander, L.M., Rehdanz, K., Tol, R.S., Van Beukering, P.J.: The economic impact of ocean acidification on coral reefs. Clim. Change Econ. **3**(01), 1250002 (2012)
3. Bullimore, R.D., Foster, N.L., Howell, K.L.: Coral-characterized benthic assemblages of the deep northeast atlantic: defining "coral gardens" to support future habitat mapping efforts. ICES J. Marine Sci. **70**(3), 511–522 (2013)

4. Clough, P., Müller, H., Sanderson, M.: The CLEF 2004 cross-language image retrieval track. In: Peters, C., Clough, P., Gonzalo, J., Jones, G.J.F., Kluck, M., Magnini, B. (eds.) CLEF 2004. LNCS, vol. 3491, pp. 597–613. Springer, Heidelberg (2005). https://doi.org/10.1007/11519645_59

5. Clough, P., Sanderson, M.: The CLEF 2003 cross language image retrieval track. In: Peters, C., Gonzalo, J., Braschler, M., Kluck, M. (eds.) CLEF 2003. LNCS, vol. 3237, pp. 581–593. Springer, Heidelberg (2004). https://doi.org/10.1007/978-3-540-30222-3_56

6. Dang-Nguyen, D.T., Piras, L., Riegler, M., Boato, G., Zhou, L., Gurrin, C.: Overview of ImageCLEFlifelog 2017: lifelog retrieval and summarization. In: Cappellato, L., Ferro, N., Goeuriot, L., Mandl, T. (eds.) CLEF 2017 Working Notes. CEUR Workshop Proceedings (CEUR-WS.org) (2017). http://ceur-ws.org/Vol-1866/. ISSN 1613–0073

7. Dang-Nguyen, D.T., Piras, L., Riegler, M., Zhou, L., Lux, M., Gurrin, C.: Overview of ImageCLEFlifelog 2018: daily living understanding and lifelog moment retrieval. In: Cappellato, L., Ferro, N., Goeuriot, L., Mandl, T. (eds.) CLEF 2018 Working Notes. CEUR Workshop Proceedings (CEUR-WS.org) (2018). http://ceur-ws.org/Vol-1866/. ISSN 1613–0073

8. Eickhoff, C., Schwall, I., de Herrera, A.G.S., Müller, H.: Overview of ImageCLEF-caption 2017 - the image caption prediction and concept extraction tasks to understand biomedical images. In: CLEF2017 Working Notes. CEUR Workshop Proceedings (CEUR-WS.org), 11–14 September 2017, Dublin, Ireland (2017). http://ceur-ws.org

9. Gilbert, A., et al.: Overview of the ImageCLEF 2015 scalable image annotation, localization and sentence generation task. In: CLEF Working Notes (2015)

10. Gilbert, A., et al.: Overview of the ImageCLEF 2016 scalable concept image annotation task. In: CLEF Working Notes, pp. 254–278 (2016)

11. Gonzalez-Rivero, M., et al.: The Catlin seaview survey–kilometre-scale seascape assessment, and monitoring of coral reef ecosystems. Aquat. Conserv.: Mar. Freshw. Ecosyst. **24**, 184–198 (2014)

12. Hasan, S.A., Ling, Y., Farri, O., Liu, J., Lungren, M., Müller, H.: Overview of the ImageCLEF 2018 medical domain visual question answering task. In: CLEF2018 Working Notes. CEUR Workshop Proceedings (CEUR-WS.org), 11–14 September 2018, Avignon, France (2018). http://ceur-ws.org

13. de Herrera, A.G.S., Eickhoff, C., Andrearczyk, V., Müller, H.: Overview of the ImageCLEF 2018 caption prediction tasks. In: CLEF2018 Working Notes. CEUR Workshop Proceedings (CEUR-WS.org), 10–14 September 2018, Avignon, France (2018). http://ceur-ws.org

14. Ionescu, B., et al.: Overview of ImageCLEF 2018: challenges, datasets and evaluation. In: Bellot, P., et al. (eds.) CLEF 2018. LNCS, vol. 11018, pp. 309–334. Springer, Cham (2018). https://doi.org/10.1007/978-3-319-98932-7_28

15. Kalpathy-Cramer, J., de Herrera, A.G.S., Demner-Fushman, D., Antani, S., Bedrick, S., Müller, H.: Evaluating performance of biomedical image retrieval systems: overview of the medical image retrieval task at ImageCLEF 2004–2014. Comput. Med. Imaging Graph. **39**, 55–61 (2015)

16. Karampidis, K., Kavallieratou, E., Papadourakis, G.: A review of image steganalysis techniques for digital forensics. J. Inf. Secur. Appl. 40 (2018). https://doi.org/10.1016/j.jisa.2018.04.005

17. Müller, H., Clough, P., Deselaers, T., Caputo, B. (eds.): ImageCLEF - Experimental Evaluation in Visual Information Retrieval. The Springer International Series on Information Retrieval, vol. 32. Springer, Berlin (2010). https://doi.org/10.1007/978-3-642-15181-1

18. Pelka, O., Koitka, S., Rückert, J., Nensa, F., Friedrich, C.M.: Radiology objects in COntext (ROCO): a multimodal image dataset. In: Stoyanov, D., et al. (eds.) LABELS/CVII/STENT -2018. LNCS, vol. 11043, pp. 180–189. Springer, Cham (2018). https://doi.org/10.1007/978-3-030-01364-6_20

19. Speers, A.E., Besedin, E.Y., Palardy, J.E., Moore, C.: Impacts of climate change and ocean acidification on coral reef fisheries: an integrated ecological-economic model. Ecol. Econ. **128**, 33–43 (2016)

20. Stokes, M., Deane, G.: Automated processing of coral reef benthic images. Limnol. Oceanogr. Methods **7**, 157–168 (2009)

21. Tsikrika, T., de Herrera, A.G.S., Müller, H.: Assessing the scholarly impact of ImageCLEF. In: Forner, P., Gonzalo, J., Kekäläinen, J., Lalmas, M., de Rijke, M. (eds.) CLEF 2011. LNCS, vol. 6941, pp. 95–106. Springer, Heidelberg (2011). https://doi.org/10.1007/978-3-642-23708-9_12

22. Tsikrika, T., Larsen, B., Müller, H., Endrullis, S., Rahm, E.: The scholarly impact of CLEF (2000–2009). In: Forner, P., Müller, H., Paredes, R., Rosso, P., Stein, B. (eds.) CLEF 2013. LNCS, vol. 8138, pp. 1–12. Springer, Heidelberg (2013). https://doi.org/10.1007/978-3-642-40802-1_1

23. Villegas, M., Paredes, R.: Overview of the ImageCLEF 2012 scalable web image annotation task. In: CLEF Working Notes (2012)

24. Villegas, M., Paredes, R.: Overview of the ImageCLEF 2014 scalable concept image annotation task. In: CLEF Working Notes, pp. 308–328. Citeseer (2014)

25. Villegas, M., Paredes, R., Thomee, B.: Overview of the ImageCLEF 2013 scalable concept image annotation subtask. In: CLEF Working Notes (2012)

CheckThat! at CLEF 2019: Automatic Identification and Verification of Claims

Tamer Elsayed[1], Preslav Nakov[2], Alberto Barrón-Cedeño[2],
Maram Hasanain[1(✉)], Reem Suwaileh[1], Giovanni Da San Martino[2],
and Pepa Atanasova[3]

[1] Qatar University, Doha, Qatar
{telsayed,maram.hasanain}@qu.edu.qa, rs081123@student.qu.edu.qa
[2] Qatar Computing Research Institute, HBKU, Doha, Qatar
{pnakov,albarron,gmartino}@hbku.edu.qa
[3] Sofia University, Sofia, Bulgaria
pepa.k.gencheva@gmail.com

Abstract. We introduce the second edition of the CheckThat! Lab, part of the 2019 Cross-Language Evaluation Forum (CLEF). CheckThat! proposes two complementary tasks. Task 1: predict which claims in a political debate should be prioritized for fact-checking. Task 2: rank Web-retrieved pages against a check-worthy claim based on their usefulness for fact-checking, extract useful passages from those pages, and then use them all to decide whether the claim is factually true or false. Checkthat! provides a full evaluation framework, consisting of data in English (derived from fact-checking sources) and Arabic (gathered and annotated from scratch) and evaluation based on mean average precision (MAP) for ranking and F_1 for classification tasks.

1 Overview

The current coverage of news in both the press and in social media has led to an unprecedented situation. Like never before, a statement in an interview, a press release, or a blog note can spread almost instantaneously. This proliferation speed leaves little time to double-check claims against the facts, which has proven critical in electoral campaigns, e.g., during the 2016 US presidential campaign in the USA and during Brexit. Indeed, some politicians were fast to notice that when it comes to shaping public opinion, facts were secondary and that appealing to emotions and beliefs worked better, especially in social media. It has been even proposed that this was marking the dawn of a *post-truth age*.

Investigative journalists and volunteers have been working hard trying to get to the root of a claim and to present solid evidence in favor or against it. However, manual fact-checking is very time-consuming, and thus automatic methods have been proposed as a way to speed-up the process. For instance, there has been work on checking the factuality/credibility of a claim, of a news article, or of an entire news outlet [2,4,6,8–11,13,15]. However, less attention has been paid to other steps of the fact-checking pipeline shown in Fig. 1, e.g.,

L. Azzopardi et al. (Eds.): ECIR 2019, LNCS 11438, pp. 309–315, 2019.
https://doi.org/10.1007/978-3-030-15719-7_41

check-worthiness estimation has been severely understudied as a problem [1,5, 7]. A typical fact-checking pipeline includes the following steps. First, check-worthy text fragments are identified. Then, documents that might be useful for fact-checking the claim [14] are retrieved from various sources, and supporting evidence is extracted. By comparing a claim against the retrieved evidence, a system can determine whether the claim is likely true or likely false (or unsure, if no supporting evidence either way could be found). This **CheckThat!** CLEF 2019 lab addresses these understudied aspects through two tasks:

Task 1: Check-Worthiness. The task aims at predicting which claim in a political debate should be prioritized for fact-checking.

Task 2: Evidence and Factuality. The task focuses on extracting evidence to support fact-checking a claim.

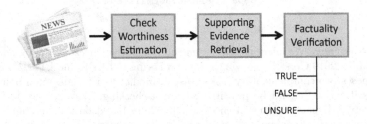

Fig. 1. Information verification pipeline: our two tasks cover all three steps.

2   Usage Scenarios

Automated systems for claim identification and verification could be very useful as supportive technology for investigative journalism. They provide assistance and guidance and save time. A system could automatically identify check-worthy claims and present them to the journalist as a ranking from more to less relevant. Additionally, for a claim, the system could identify documents that are *useful* for humans to manually fact-check and it could also estimate a *veracity score* supported by evidence extracted from such documents, which would help the journalist to focus on the most outstanding cases.

Another useful scenario, with the potential of impacting larger communities, would be helping the social media users who get a large flow of claims daily and want help in verifying them.

3   Target Audience

The main targets for nourishing the list of participants are the information retrieval, computational linguistics, and machine learning communities. We also hope that the lab attracts neighboring communities that would be interested in the problem, maybe from slightly different angles, e.g., social computing, social sciences, and investigative journalism.

4 Description of the Tasks

CheckThat! 2019[1] is a continuation of the evaluation lab at CLEF-2018 [12].[2] It is organized around two different tasks, which correspond to the three main blocks in the verification pipeline (Fig. 1): check-worthiness estimation (**Task 1**), and extracting supporting evidence and factuality verification (**Task 2**). We address these two tasks separately in order to ease the participation and to have independent evaluations and more meaningful comparisons of systems.

4.1 Task 1: Check-Worthiness

Task 1 is defined as follows: *Given a political debate or a transcribed speech, segmented into sentences, with speakers annotated, identify which sentence should be prioritized for fact-checking.* This is a ranking task and systems are required to produce a score per sentence, according to which the ranking will be performed. This task will be run in English.

Dataset. The training data for *Task 1* is ready. We selected four transcripts of the 2016 US election: one vice-presidential and three presidential debates. For each debate, we used the publicly-available manual analysis about it from nine reputable fact-checking sources (ABC News, Chicago Tribune, CNN, FactCheck.org, NPR, PolitiFact, The Guardian, The New York Times, and The Washington Post). This could include not just a statement about factuality, but any free text that journalists decided to add, e.g., links to biographies or behavioral analysis of the opponents and moderators. We converted this to binary annotation about whether a particular sentence was annotated for factuality by a given source.

The training dataset of four debates, contains a total of 5,415 annotated sentences in context, with 880 of them being identified as check-worthy by at least one of the sources. The agreement between the sources is not high. The reason for this is that different media aim at annotating sentences according to their own editorial line, rather than trying to be exhaustive in any way. This suggests that the task of predicting which sentence would contain check-worthy claims will be challenging. Thus, we focus on a ranking task rather than on absolute predictions.

The test set for *Task 1* will be created following the same approach as the training data: more debates will be collected with the same nine fact-checking sources to annotate the check-worthy claims. The volume of this data will be around 25% of the training set size.

Table 1 shows an excerpt from the first presidential debate in the American elections in 2016 together with the annotation flag (0 or 1) indicating whether each of the media has fact-checked the claim. The positive examples for *Task 1* will be those fact-checked by at least one source (those highlighted in blue in Table 1).

---

[1] http://alt.qcri.org/clef2019-checkthat/.
[2] http://alt.qcri.org/clef2018-factcheck/.

Table 1. Excerpt from the transcript of the first US Presidential Debate in 2016, annotated by nine sources: Chicago Tribune, ABC News, CNN, Washington Post, NPR, PolitiFact, The Guardian, The New York Times and FactCheck.org. Whether the media fact-checked the claim or not is indicated by a 1 or 0, respectively. The blue examples are the positive examples for *Task 1* (i.e., those with a positive number of sources that commented on the claim).

Speaker	Total	CT	ABC	CNN	WP	NPR	PF	TG	NYT	FC	Text
Clinton	0	0	0	0	0	0	0	0	0	0	So we're now on the precipice of having a potentially much better economy, but the last thing we need to do is to go back to the policies that failed us in the first place.
Clinton	6	1	1	0	0	1	1	0	1	1	Independent experts have looked at what I've proposed and looked at what Donald's proposed, and basically they've said this, that if his tax plan, which would blow up the debt by over $5 trillion and would in some instances disadvantage middle-class families compared to the wealthy, were to go into effect, we would lose 3.5 million jobs and maybe have another recession.
Clinton	1	1	0	0	0	0	0	0	0	0	They've looked at my plans and they've said, OK, if we can do this, and I intend to get it done, we will have 10 million more new jobs, because we will be making investments where we can grow the economy.
Clinton	0	0	0	0	0	0	0	0	0	0	Take clean energy.
Clinton	0	0	0	0	0	0	0	0	0	0	Some country is going to be the clean-energy superpower of the 21st century.
Clinton	6	1	1	1	1	0	0	1	0	1	Donald thinks that climate change is a hoax perpetrated by the Chinese.
Clinton	0	0	0	0	0	0	0	0	0	0	I think it's real.
Trump	5	1	1	0	1	1	1	0	0	0	I did not.

Evaluation. We approach *Task 1* as a ranking task. As in the first edition, we plan to use *Mean Average Precision* (MAP) as the official evaluation measure. Most media rarely check more than 50 claims per debate. Thus, we plan to add $P@k$ for $k \in \{5, 10, 20, 50\}$ as well.

4.2   Task 2: Evidence and Factuality

Task 2 is defined as follows: *Given a claim associated with a set of Web pages P (that constitute the results of Web search in response to using the claim as a search query), identify which of the Web pages (and passages of those Web pages) can be useful in assisting a human who is fact-checking the claim. Finally, judge the claim factuality according to the supporting information in the passages of the Web documents.* This task will be run in Arabic.

The task is divided into several subtasks that target different aspects of the problem:

Subtask A: *Rank the Web pages P based on how useful they are for verifying the target claim.* The systems are required to produce a score for each page, based on which the pages would be ranked. See the definition of "useful" pages below.

Subtask B: *Classify each of the Web pages* as *"very useful for verification", "useful", "not useful", or "unsure."* A page is considered *very useful* for verification if it is *relevant* with respect to the claim (i.e., on-topic and discussing the claim) and it *provides sufficient evidence* to verify the veracity of the claim such that there is no need for another document to be checked for this claim. A page is useful for verification if it is relevant to the claim and provides some valid evidence, but it is not sufficient to determine the claim's veracity on its own. The evidence can be a source, some statistics, a quote, etc. However, a particular piece of evidence is considered not valid if the source cannot be verified (e.g., expressing that "experts say that ..." without mentioning who those experts are), or it is just an opinion of a person/expert instead of objective analysis. Notice that this is different from *stance detection*, as a Web page might agree with a claim, but it might still lack evidence to verify it.

Subtask C: *Find passages* within those Web pages that are *useful* for claim verification. Again, notice that this is different from stance detection.

Subtask D: *Find the claim's factuality* as *"true" or "false."* The claim is considered true if it is accurate and there is nothing significant missing. A claim is false if it is not accurate.

Dataset. *Task 2* is completely new to the lab this year. For the dataset, we will select a set of about 75 claims from multiple sources including an pre-existing set of Arabic claims [3], a survey in which we asked the public to provide examples of claims they have heard of, and some headlines from six Arabic news agencies that we rewrote into claims.

For each claim, we will search (using the claim as a query) a commercial search engine (e.g., Google or Bing) and we will extract the top 50 resulting Web pages. Crowd workers will then be hired to annotate the Web pages for relevance. In-house annotators will then be hired to annotate the relevant pages for the first two subtasks (i.e., those based on usefulness of pages). As for Subtask C, only pages that are labeled as useful for claim verification will be used, and we will split each page into paragraphs (assuming each paragraph is a passage). In-house annotators will then label a paragraph as whether it is useful for verifying the claim or not. Majority voting will be used to determine the final label of a page/passage at the different labelling tasks. The final dataset will consist of 25 training claims and 50 testing claim. As this is a new task, we will work on releasing the training claims with annotations early in the lab schedule in order to support experiments by the participating teams.

Figure 2 is an example for *Task 2*. For the sake of readability, the example is given in English, but this year the task will be offered in Arabic only.

Evaluation. *Task 2* includes both ranking and classification subtasks. For Subtask A, we plan to use ranking measures such as *Mean Average Precision* (MAP) as the official evaluation measure and *Precision at k* ($P@k$). For the classification subtasks B, C, and D, we will use Accuracy, Precision, Recall, and F_1. F_1 will be the official evaluation measure.

Fig. 2. A claim associated with a useful Web page, and a useful passage (in a box).

References

1. Atanasova, P., et al.: Overview of the CLEF-2018 CheckThat! lab on automatic identification and verification of political claims, task 1: check-worthiness. In: CLEF 2018 Working Notes. Working Notes of CLEF 2018 - Conference and Labs of the Evaluation Forum, CEUR Workshop Proceedings, Avignon, France. CEUR-WS.org (2018)
2. Ba, M.L., Berti-Equille, L., Shah, K., Hammady, H.M.: VERA: a platform for veracity estimation over web data. In: Proceedings of the 25th International Conference Companion on World Wide Web, pp. 159–162. International World Wide Web Conferences Steering Committee (2016)
3. Baly, R., Mohtarami, M., Glass, J., Màrquez, L., Moschitti, A., Nakov, P.: Integrating stance detection and fact checking in a unified corpus. In: Proceedings of the 2018 Conference of the North American Chapter of the Association for Computational Linguistics: Human Language Technologies, Volume 2 (Short Papers), vol. 2, pp. 21–27 (2018)
4. Castillo, C., Mendoza, M., Poblete, B.: Information credibility on Twitter. In: Proceedings of the 20th International Conference on World Wide Web, pp. 675–684. ACM (2011)
5. Gencheva, P., Nakov, P., Màrquez, L., Barrón-Cedeño, A., Koychev, I.: A context-aware approach for detecting worth-checking claims in political debates. In: Proceedings of the 2017 International Conference on Recent Advances in Natural Language Processing, RANLP 2017, Varna, Bulgaria (2017)
6. Hardalov, M., Koychev, I., Nakov, P.: In search of credible news. In: Dichev, C., Agre, G. (eds.) AIMSA 2016. LNCS (LNAI), vol. 9883, pp. 172–180. Springer, Cham (2016). https://doi.org/10.1007/978-3-319-44748-3_17
7. Hassan, N., Li, C., Tremayne, M.: Detecting check-worthy factual claims in presidential debates. In: Proceedings of the 24th ACM International on Conference on Information and Knowledge Management, pp. 1835–1838. ACM (2015)
8. Karadzhov, G., Gencheva, P., Nakov, P., Koychev, I.: We built a fake news & clickbait filter: what happened next will blow your mind! In: Proceedings of the 2017 International Conference on Recent Advances in Natural Language Processing, RANLP 2017, Varna, Bulgaria (2017)
9. Karadzhov, G., Nakov, P., Màrquez, L., Barrón-Cedeño, A., Koychev, I.: Fully automated fact checking using external sources. In: Proceedings of the 2017 International Conference on Recent Advances in Natural Language Processing, RANLP 2017, Varna, Bulgaria (2017)

10. Ma, J., et al.: Detecting rumors from microblogs with recurrent neural networks. In: Proceedings of IJCAI (2016)
11. Mukherjee, S., Weikum, G.: Leveraging joint interactions for credibility analysis in news communities. In: Proceedings of the 24th ACM International on Conference on Information and Knowledge Management, pp. 353–362. ACM (2015)
12. Nakov, P., et al.: Overview of the CLEF-2018 CheckThat! lab on automatic identification and verification of political claims. In: Bellot, I., et al. (eds.) CLEF 2018. LNCS, vol. 11018, pp. 372–387. Springer, Cham (2018). https://doi.org/10.1007/978-3-319-98932-7_32
13. Popat, K., Mukherjee, S., Strötgen, J., Weikum, G.: Credibility assessment of textual claims on the web. In: Proceedings of the 25th ACM International on Conference on Information and Knowledge Management, pp. 2173–2178. ACM (2016)
14. Yasser, K., Kutlu, M., Elsayed, T.: Re-ranking web search results for better fact-checking: a preliminary study. In: Proceedings of 27th ACM International Conference on Information and Knowledge Management (CIKM), Turin, Italy, pp. 1783–1786. ACM (2018)
15. Zubiaga, A., Liakata, M., Procter, R., Hoi, G.W.S., Tolmie, P.: Analysing how people orient to and spread rumours in social media by looking at conversational threads. PloS ONE **11**(3), e0150989 (2016)

A Task Set Proposal for Automatic Protest Information Collection Across Multiple Countries

Ali Hürriyetoğlu[✉], Erdem Yörük, Deniz Yüret, Çağrı Yoltar, Burak Gürel, Fırat Duruşan, and Osman Mutlu

Koc University, Rumelifener yolu, Sarıyer, İstanbul, Turkey
{ahurriyetoglu,eryoruk,dyuret,cyoltar,bgurel,fdurusan,omutlu}@ku.edu.tr

Abstract. We propose a coherent set of tasks for protest information collection in the context of generalizable natural language processing. The tasks are news article classification, event sentence detection, and event extraction. Having tools for collecting event information from data produced in multiple countries enables comparative sociology and politics studies. We have annotated news articles in English from a source and a target country in order to be able to measure the performance of the tools developed using data from one country on data from a different country. Our preliminary experiments have shown that the performance of the tools developed using English texts from India drops to a level that are not usable when they are applied on English texts from China. We think our setting addresses the challenge of building generalizable NLP tools that perform well independent of the source of the text and will accelerate progress in line of developing generalizable NLP systems.

Keywords: Natural language processing · Information retrieval · Machine learning · Text classification · Information extraction · Event extraction · Domain adaptation · Transfer learning · Computational social science · Contentious politics · Protest information

1 Introduction

Comparative social studies on social protest requires collecting protest event data from multiple countries. The utility of these collections increases with the number of countries covered, the length of the time span and the weight of the information gathered from local sources. The performance of natural language processing (NLP) tools, those of text classification and information extraction in our setting, has not been satisfactory against the requirements of longer time coverage and working on data from multiple countries [8,17]. In this study, we introduce a set of tasks, supported with the relevant data, for facilitating the creation of protest event databases that are better equipped to handle variations

© Springer Nature Switzerland AG 2019
L. Azzopardi et al. (Eds.): ECIR 2019, LNCS 11438, pp. 316–323, 2019.
https://doi.org/10.1007/978-3-030-15719-7_42

in country settings through both space and time. The setting we propose facilitates testing and improving state-of-the-art methods for text classification and information extraction on English news article texts from India and China. The direction of our work is towards developing generalizable information systems that perform comparatively well on texts from multiple countries.

The need for collecting protest or conflict data has been satisfied by utilizing manual [4,19], semi-automatic [11], and automatic [2,9,10,12,14] methods -each of which presents a different set of challenges that limit the utility of that method. The methods that rely on manual and semi-automated coding, though reliable, require a tremendous amount of effort to replicate on new data as they depend intensely on high quality human effort. On the other hand, text classification and information extraction systems that rely on automated methods yield less reliable results as they tend to perform poorly on texts different from the ones they were developed and validated on [6,13]. The huge amount of news articles that are required to be analyzed and the constant need of repeating the same analyses on new data force us to push limits of automated protest information collection yet again. Furthermore, addressing and remedying performance issues when faced with difficulties presented by variations across datasets requires the tools to be as generalizable as possible.

Much of the difficulty presented by automated methods of data collection on contentious politics events[1] stems from the fact that contentious politics take slightly different forms in different countries and time periods in line with spatial and temporal variation of sociopolitical phenomena. The automated tools run the risk of being biased towards the country and/or time period of the cases that they are trained upon and the need to adapt them to different cases leads developers to either redesign tools from scratch for each individual case or take certain shortcuts which somehow makes variety more manageable. A common such recourse which imposes a level of uniformity to data universe is key term based filtering -a method which relies on an a priori set of keywords related to protest events to filter irrelevant cases out of the training dataset. It is our conviction that this method is arbitrary and possibly cripples the reliability of data collection from the outset by leaving out potentially relevant protest events. Moreover, there is no inbuilt way to determine if or to what extent such unwanted exclusion occurs as the filtering is external to the training-evaluation cycle.

Rather than developing case specific classifiers for every single country or limiting the raw data via key term filters, we strive to develop generalizable information systems that perform comparatively well on multiple country settings and can be applied to any set of random selection of news articles. In order to accommodate the geographical and historical variability of sociopolitical contexts, the chief aspect of our task design takes the tools that are developed on the basis of the data from a certain country and evaluates them on data from a different country. Thus, the evaluation feedback forms a novel basis on which the

---

[1] The term used when referring to these events in collective is "repertoires of contention" [7,15]. We will use "protest events" from here on for the sake of brevity simplicity.

tools are further developed to accommodate even more variation in the future. This rolling training-evaluation cycles is expected to create a virtuous circle of feedback loop which will be more generally applicable with every new country case that is introduced.

This paper describes how we will realize the proposed setting within the lab ProtestNews in the 2019 edition of the Conference and Labs of the Evaluation Forum (CLEF).[2],[3],[4] We introduce the methodology we apply to create the corpus, and the task set we propose, in 2 and 3 respectively. We report our preliminary results in Sect. 4 and conclude our report by pointing to future directions of our work in Sect. 5.

2   Data

We collect online English news articles from a source and a target country, India and China respectively.[5] We first download the freely accessible part of an online news archive and create a random sample of these articles from each source in order to have a representative sample for labelling and annotation for each task.

We apply the same labelling and annotation manuals on data collected from different countries. This approach enables obtaining comparable measures of automatic system performance. Our data preparation process applies state-of-the-art annotation methodology in terms of being based on an annotation manual, sampling the news articles from various sources and periods, and continuously monitoring the annotations to achieve a high inter-annotator agreement.

Annotators that are master students or PhD candidates in social or political sciences work in pairs. In each pair, both annotators annotate the same document, sentence, or token depending on the task.[6] The annotation start by labelling articles in a sample of news articles as containing a protest or not. Sentences of these positively labelled documents are then labelled as containing protest information or not. These sentences should contain either an event trigger or a reference to an event trigger in order to be labelled as positive. Finally, the protest-related sentences are annotated at token level for the information they denote.[7] The supervisor, who is a social scientist and responsible of maintaining the annotation manuals as well, resolves the disagreements between the annotators.

We analyze the annotator agreements as well. To prevent cases where annotators may agree on wrong labelling, we applied the following means of improving

---

[2] http://www.clef-initiative.eu, accessed January 19, 2019.

[3] http://clef2019.clef-initiative.eu, accessed January 19, 2019.

[4] https://emw.ku.edu.tr/clef-protestnews-2019, accessed January 19, 2019.

[5] Using available corpora that are already being allowed to be distributed freely is not an option for our setting due to the requirement of having a representative sample from the source and target countries. Also, the dataset should contain data created in more than one country in order to be useful in our setting.

[6] The overlap ratio is 100%.

[7] We mainly annotate the event trigger, place, time, participant, organizer, and target of the protest.

the corpus. First, we regularly apply a spot check in which the expert double checks a small sample of labels and annotations the annotators agree on the attached label. Second, any erroneous annotation in the positive cases may be captured in the following step where the annotators do a more detailed annotation for the following task. Third, we semi-automatically check the documents labelled as non-protest, by training a classification model on 80% of the all labelled and adjudicated documents or sentences and testing on the remaining 20%. The cases that are predicted as protest by the classifier but labelled as non-protest by the annotators are double checked manually to verify they are indeed non-protest. Finally, in order to eliminate risk of wrong labelling due to lack of knowledge about a country, a domain expert instructs the annotators before they start to do annotation.

We distribute the data in a way that does not violate copyright of the news sources. This involves only sharing information that is needed to reproduce the corpus from the source in cases it is not allowed to distribute the news articles.

3   Tasks

We designed the tasks as depicted in Fig. 1. The analysis should start by predicting whether a random news article mentions a protest. Then, the sentence(s) that contain protest information should be identified. Finally, protest information such as participants, place, and time should be detected in the protest related sentences. This order of tasks provides a controlled setting that enables error analysis and optimization possibility during annotation and tool development efforts.

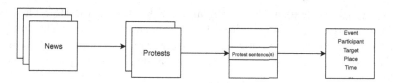

Fig. 1. The lab consists of (a) *Task 1*: News article classification as protest vs. non-protest, (b) *Task 2*: Protest sentence detection, and (c) *Task 3*: Event extraction. Tasks 2 and 3 will be based on news articles labeled for task 1. Participants can choose to participate in one or more of these tasks independent of each other.

The set of tools that will tackle these tasks should be implemented and validated for data originated from a country and tested on data collected from a different country, which are India and China. There will be two level of evaluation, which we refer as Test 1 and Test 2, on data that is not accessible to the lab participants. The first level, which is Test 1, is on test data from the country used for training and developing the methods. The second evaluation, which is Test 2, will be on data from the target country. The primary score for ranking the submissions will be the one on the target country.

We use macro averaged F1 for evaluating the Task 1 and Task 2. The event extraction task, which is Task 3, will be evaluated on F1 score that will be based on the ratio of the match between the prediction and the annotations in the test sets.

Although the annotation effort is continuing to increase amount of news articles for each task, we would like to report the recent approximate number of news articles we have labelled and annotated for each task in Table 1. The training and development columns illustrates the number of documents that will be accessible to the lab participants. These documents are from the source country. The document count for Test 1 and Test 2 columns are from the source and the target countries respectively.

Table 1. Number of annotated news articles for each task

	Training	Development	Test 1	Test 2
Task 1	8,000	1,000	1,000	4,000
Task 2	600	100	100	200
Task 3	300	50	50	20

4   Preliminary Results

We performed various analyses and experiments on the corpus we created in order to further shed light on characteristics of the dataset and the tasks we propose. First, we filtered our corpus with the key terms that were used by Wang et al. [16], Lorenzini et al. [10], and Weidman and Rød [18]. The Table 2 shows the protest coverage of these key terms in our corpus. The low recall demonstrates the difference between the coverage of a random sample and a key term filtered sample.[8] We assume that our random sampling method ensures complete recall.

Table 2. Coverage of the key terms used by recent studies in our corpus

	Precision	Recall	F1-score
Wang et al. [16]	.57	.75	.65
Lorenzini et al. [10]	.42	.88	.57
Weidman and Rød [18]	.60	.58	.59

We have performed automatic classification experiments by training binary machine learning models for task 1 and task 2. For task 1, a support vector

---

[8] The difference between our and these projects' annotation manuals potentially affects the precision and recall as well.

machine (SVM) and a deep neural network (DNN) classifiers were trained using the training data by being optimized on the development data. The SVM model has yielded .85 and .25 F1 score on Test 1 and Test 2 respectively. The pretrained BERT model's [5] performance is .90 and .64 in the same setting. For task 2, three binary sentence classifiers, which are random forest, decision tree, and SVM, were created using the training and development data. The F1 scores of these classifiers are .47, .52, and .56 on Test 1 data. Finally, our experiments for task 3 yielded around .30 lower F1 score than it is reported in publications of these tools on test 1 data [1,3].[9]

5   Conclusion and Future Work

Comparative social science studies deploy concepts, and work on variables that must be applicable across multiple different countries and time periods. As the particular cultural, political and linguistic characteristics of each different geographical and historical context reflect on the news articles, the NLP tools that are utilized to construct news databases used by these studies must have generalized applicability. The preliminary analysis and tool performance results show that the difference in news content and performance differences on data from different countries are significant, which presents a challenge for the text processing systems aiming at such generalizability. The task design we propose in this paper is expected to fulfill such requirements, and will certainly be enriched and moved closer to perfection through contributions in this shared task.

As to the future development path of our line of research, we envision the following improvements to the dataset in line of our broader goal of developing tools for creating a high-quality global protest database with general applicability: (i) the corpus should be extended with English data from additional countries; (ii) data in languages other than English should be included; (iii) instead of labeling only as protest or non-protest, categorization of protest events into types such as demonstration, industrial action, group clash, and armed militancy should be integrated into the task set; and (iv) the problem of distinguishing expressions of events that have not taken place, such as threats and plans of protest events, from events that have taken place must be addressed. Tasks which label planned/threatened events separately from events and non-events promises to tackle this challenge.

Acknowledgments. This work is funded by the European Research Council (ERC) Starting Grant 714868 awarded to Dr. Erdem Yörük for his project Emerging Welfare. (https://emw.ku.edu.tr, accessed January 19) We are grateful to our steering committee members for the CLEF 2019 lab Sophia Ananiadou, Antal van den Bosch, Kemal Oflazer, Arzucan Özgür, Aline Villavicencio, and Hristo Tanev. Finally, we thank to Theresa Gessler and Peter Makarov for their contribution in organizing the CLEF lab by reviewing the annotation manuals and sharing their work with us respectively.

---

[9] https://github.com/emerging-welfare/ie-tools-test-on-India-b1,   accessed   January 19.

References

1. Akdemir, A., Hürriyetoğlu, A., Yörük, E., Gürel, B., Yoltar, C., Yüret, D.: Towards generalizable place name recognition systems: analysis and enhancement of NER systems on English News from India. In: Proceedings of the 12th Workshop on Geographic Information Retrieval, GIR 2018, pp. 8:1–8:10. ACM, New York (2018). https://doi.org/10.1145/3281354.3281363
2. Boschee, E., Natarajan, P., Weischedel, R.: Automatic extraction of events from open source text for predictive forecasting. In: Subrahmanian, V. (ed.) Handbook of Computational Approaches to Counterterrorism, pp. 51–67. Springer, New York (2013). https://doi.org/10.1007/978-1-4614-5311-6_3
3. Büyüköz, B., Hürriyetoğlu, A., Yörük, E., Yüret, D.: Examining existing information extraction tools on manually-annotated protest events in Indian news. In: Proceedings of Computational Linguistics in Netherlands (CLIN), CLIN29 (2019)
4. Chenoweth, E., Lewis, O.A.: Unpacking nonviolent campaigns: introducing the NAVCO 2.0 dataset. J. Peace Res. **50**(3), 415–423 (2013). https://doi.org/10.1177/0022343312471551
5. Devlin, J., Chang, M.W., Lee, K., Toutanova, K.: BERT: pre-training of deep bidirectional transformers for language understanding. arXiv preprint arXiv:1810.04805 (2018)
6. Ettinger, A., Rao, S., Daumé III, H., Bender, E.M.: Towards linguistically generalizable NLP systems: a workshop and shared task. In: Proceedings of the First Workshop on Building Linguistically Generalizable NLP Systems, pp. 1–10. Association for Computational Linguistics (2017). http://aclweb.org/anthology/W17-5401
7. Giugni, M.G.: Was it worth the effort? The outcomes and consequences of social movements. Ann. Rev. Sociol. **24**, 371–393 (1998). http://www.jstor.org/stable/223486
8. Hammond, J., Weidmann, N.B.: Using machine-coded event data for the micro-level study of political violence. Res. Polit. **1**(2) (2014). https://doi.org/10.1177/2053168014539924
9. Leetaru, K., Schrodt, P.A.: GDELT: global data on events, location, and tone, 1979–2012. In: ISA Annual Convention, vol. 2, pp. 1–49. Citeseer (2013)
10. Lorenzini, J., Makarov, P., Kriesi, H., Wueest, B.: Towards a dataset of automatically coded protest events from English-language Newswire documents. In: Paper Presented at the Amsterdam Text Analysis Conference (2016)
11. Nardulli, P.F., Althaus, S.L., Hayes, M.: A progressive supervised-learning approach to generating rich civil strife data. Sociol. Methodol. **45**(1), 148–183 (2015). https://doi.org/10.1177/0081175015581378
12. Schrodt, P.A., Beieler, J., Idris, M.: Three'sa charm? Open event data coding with el: Diablo, Petrarch, and the open event data alliance. In: ISA Annual Convention (2014)
13. Soboroff, I., Ferro, N., Fuhr, N.: Report on GLARE 2018: 1st workshop on generalization in information retrieval: can we predict performance in new domains? SIGIR Forum **52**(2), 132–137 (2018). http://sigir.org/wp-content/uploads/2019/01/p132.pdf
14. Sönmez, Ç., Özgür, A., Yörük, E.: Towards building a political protest database to explain changes in the welfare state. In: Proceedings of the 10th SIGHUM Workshop on Language Technology for Cultural Heritage, Social Sciences, and Humanities, pp. 106–110. Association for Computational Linguistics (2016). https://doi.org/10.18653/v1/W16-2113, http://www.aclweb.org/anthology/W16-2113

15. Tarrow, S.: Power in Movement: Social Movements, Collective Action and Politics. Cambridge Studies in Comparative Politics, Cambridge University Press (1994). https://books.google.com.tr/books?id=hN5nQgAACAAJ
16. Wang, W.: Event detection and extraction from news articles. Ph.D. thesis, Virginia Tech (2018)
17. Wang, W., Kennedy, R., Lazer, D., Ramakrishnan, N.: Growing pains for global monitoring of societal events. Science **353**(6307), 1502–1503 (2016). https://doi.org/10.1126/science.aaf6758. http://science.sciencemag.org/content/353/6307/1502
18. Weidmann, N.B., Rød, E.G.: The Internet and Political Protest in Autocracies, Chap. Coding Protest Events in Autocracies. Oxford University Press, Oxford (2019)
19. Yoruk, E.: The politics of the Turkish welfare system transformation in the neoliberal era: welfare as mobilization and containment. The Johns Hopkins University (2012)

Doctoral Consortium Papers

Exploring Result Presentation
in Conversational IR Using
a Wizard-of-Oz Study

Souvick Ghosh[(✉)]

Rutgers University, New Brunswick, NJ 08901, USA
souvick.ghosh@rutgers.edu
https://comminfo.rutgers.edu/ghosh-souvick

Abstract. Recent researches in conversational IR have explored problems related to context enhancement, question-answering, and query reformulations. However, very few researches have focused on result presentation over audio channels. The linear and transient nature of speech makes it cognitively challenging for the user to process a large amount of information. Presenting the search results (from SERP) is equally challenging as it is not feasible to read out the list of results. In this paper, we propose a study to evaluate the users' preference of modalities when using conversational search systems. The study will help us to understand how results should be presented in a conversational search system. As we observe how users search using audio queries, interact with the intermediary, and process the results presented, we aim to develop an insight on how to present results more efficiently in a conversational search setting. We also plan on exploring the effectiveness and consistency of different media in a conversational search setting. Our observations will inform future designs and help to create a better understanding of such systems.

Keywords: Conversational information retrieval · Spoken search · Information seeking · Result presentation

1 Introduction and Motivation

Information Retrieval systems consist of three components - the user (or the seeker), the knowledge resource, and an intermediary. In traditional information retrieval systems [2]: (1) The user may not know the exact nature of the information problem [3]; (2) The user may fail to find terms that are accurate in describing his information need. By using spoken dialogues, the user can describe his information problem in natural language, which in turn allows for a better understanding of his knowledge gap (or information need). The system can also ask follow-up questions to resolve ambiguities and provide better responses.

The conversation - using natural language dialogues, over multiple turns - between the user and system could be in the form of text (as in the case of

© Springer Nature Switzerland AG 2019
L. Azzopardi et al. (Eds.): ECIR 2019, LNCS 11438, pp. 327–331, 2019.
https://doi.org/10.1007/978-3-030-15719-7_43

chatbots) or audio (as in the case of personal assistants). Such systems provide more human-like interaction [1] to the user who has the freedom to speak to the system (voice requests) instead of typing. In situations like driving, cooking, or exercising [6], where traditional search (through typing) may be difficult or erroneous, the spoken system allows hands-free and eyes-free operation, and so, the user can multitask. Conversational IR systems are also better suited for people with visual or manual impairment [6] or people with limited literacy skills. Although textual interfaces provide autocorrect suggestions, speaking to the system eliminates the need to the correct spelling of complex, difficult or foreign words.

The limitations for audio-only search interfaces can be attributed to the transient and linear nature of speech, which requires information to be transmitted in smaller chunks (short audios or limited results) [7,9] to prevent overloading the users' short-term memory [10]. Thus, audio-only search systems are not suitable in noisy environments (like outdoors) [10] or for presenting complex structures, images, graphs, and videos.

2   Research Questions

The following research question guide the overall direction and objective of the research study:

RQ1: *How does the mode (text, audio) of result (information) presentation influence the users' experiences in a search task?*
RQ2: *Do the users prefer any specific modality over others for result presentation in a conversational search setting?*
RQ3: *Using an audio-only input channel, are the system responses and results presented, as perceived by the user, consistent across all the modalities?*

3   Proposed Methodology

We attempt to answer our research question(s) by conducting an empirical laboratory-based Wizard of Oz experiment. The study will help us to understand how results should be presented in a conversational search system. While the user will be allowed to present his query in only audio form, the results returned by the system could be in the form of text or audio replies. As we observe how users search using audio queries, interact with the intermediary, and process the results presented, we aim to develop an insight on how to present results more efficiently in a conversational search setting.

The study will contain several search sessions, using two entities or participants, the information seeker, and the intermediary. The seeker - who has no access to the internet or any other online or offline information source (apart from a spoken communication channel) and a non-networked computer - will be presented with an information need (through a simulated backstory and a search task). The intermediary – on the other end of the communication channel (audio only), with access to a networked computer but no knowledge of the backstory or the task assigned to the user – will attempt to help the seeker with the search task. The participants will need to understand and create models of each other, envision the limitations, and collaborate to complete the tasks. The seeker and the intermediary will be in different physical spaces and will not be able to see each other or communicate using gestures or otherwise. The wizard represents future versions of conversational agents like Siri, Cortana, or Alexa, where the mode of result presentation is expected to influence the search experience of the user.

3.1   Recruitment and Experimental Protocol

We plan on recruiting $N = 48$ users with similar search skills and experience. While we aim to have a balanced number of male and female participants, the participants will be required to be fluent speakers and listeners of North American English with some familiarity in using internet and search functions. The user and the intermediary will be asked to complete different search tasks in a laboratory setting. The experiment will be a within-subjects design, in which the users will perform different search tasks using the three different systems, a baseline system, and two experimental systems. There will be a total of four tasks, the first being a warm-up task to familiarize the users with the system. The next three tasks will explore the three methods of result presentation: 1. the baseline or control system, where the result is presented as the intermediary retrieves it using a screen sharing mechanism; 2. the text-based system, where the response of the intermediary will be presented as text on the screen; and 3. the spoken system, where the response of the intermediary will be converted back to speech using a text-to-speech synthesizer and presented over an audio-only channel. Three experimental systems will be used to evaluate the three methods of presenting the search results: the baseline system ($S_{baseline}$), the textual system (S_{text}), and the audio system (S_{audio}). The order of the task types and topics and the experimental system will be rotated for different users to prevent any learning effects. For all the three systems, the intermediary will help the seeker in resolving his information problem and will control how the responses are presented back to the participant.

3.2   Tasks and Experimental Procedure

We will create backstories or simulated search situations [4] to place the user in real-life information seeking situation. Such a situation comprises two parts: the backstory and the search task. The backstory provides context about the search task, creates an information need, and situates the user in the simulated task. This promotes a more natural search behavior [5]. The first task will be a warm-up task which will familiarize the participants with the search system, the search process, and the intermediary. The next three tasks will be assigned one after the other and will involve the use of the baseline or experimental systems. To prevent one task or system from influencing the next task-system combination, we plan on using a Graeco-Latin square design [8]. Table 1 shows the experimental design. In this table, we have presented a system-task block for three users, where the task categories are represented as T_{A-C} and the systems are represented using $S_{baseline}$, S_{text}, and S_{audio}. There will be 16 similar blocks of three users for our proposed recruitment of 48 participants. By rotating the experimental systems and the tasks, we aim to prevent any task learning effects. Our tasks, which will aim to simulate naturalistic search behavior among experimental subjects, will be adopted from the literature [11]. The tasks, which have already been developed, range from high to moderate in difficulty and complexity (adapted from [11]) and are expected to initiate a multi-turn conversation between the seeker and the intermediary. Although we have avoided the details of the task in this paper, they range from various topics ranging from vacation planning to policy development.

Table 1. Experimental design using Graeco-Latin square.

User	System/task order		
	1	2	3
1	$S_{baseline}$, T_A	S_{text}, T_B	S_{audio}, T_C
2	S_{audio}, T_B	$S_{baseline}$, T_C	S_{text}, T_A
3	S_{text}, T_C	S_{audio}, T_A	$S_{baseline}$, T_B

Once the participants report for the user study, they will be given a brief overview of the research procedure, the objectives and their role in the study. They will also be required to sign the consent form and fill up the pre-test questionnaire. A warm-up task will be given to familiarize the participants with the functioning of the three systems. The researcher will be available to assist the participants with understanding how each system works and will provide demonstrations if required. Next, the participants will need to complete three search tasks in three consecutive sessions, using different systems, the order of which will be determined using the matrix presented in Table 1. The researcher or an expert searcher will play the role of the wizard. For each search session, the participant will have 15 min to complete the task, after which they will need to answer the post-task questionnaire and move to the next task. At the end of the study, the researchers will conduct a semi-structured interview to assess the experience of the participants. The entire study should not take more than two hours to complete. We aim to collect different types of raw data during our study, like the users' background and demographic information, the details of the search session, and the pre-test and post-task questionnaire, and the exit interview to assess the users' search experience. Subjects will be asked to complete assessments of differentials on task, search experiences, and quality of information after every search session (post-task questionnaire). As the systems will be identical except in how the results are presented, the system-level differentials will allow us to process how the result presentation was perceived by the participants. The quality of information retrieved by the different systems, and the complexity of the search tasks will also be evaluated using Friedman Rank Sum Test or multiway ANOVA.

References

1. Arguello, J., Choi, B., Capra, R.: Factors affecting users' information requests. In: SIGIR 1st International Workshop on Conversational Approaches to Information Retrieval (CAIR 2017), vol. 4 (2017)
2. Begany, G.M., Sa, N., Yuan, X.: Factors affecting user perception of a spoken language vs. textual search interface: a content analysis. Interact. Comput. **28**(2), 170–180 (2015)
3. Belkin, N.J., Oddy, R.N., Brooks, H.M.: Ask for information retrieval: part I. Background and theory. J. Doc. **38**(2), 61–71 (1982)
4. Borlund, P.: Experimental components for the evaluation of interactive information retrieval systems. J. Doc. **56**(1), 71–90 (2000)
5. Borlund, P.: Evaluation of interactive information retrieval systems (2002)
6. Guy, I.: Searching by talking: analysis of voice queries on mobile web search. In: Proceedings of the 39th International ACM SIGIR Conference on Research and Development in Information Retrieval, pp. 35–44. ACM (2016)
7. Lai, J., Yankelovich, N.: Conversational speech interfaces. In: The Human-Computer Interaction Handbook, pp. 698–713. L. Erlbaum Associates Inc. (2002)
8. Tague-Sutcliffe, J.: The pragmatics of information retrieval experimentation, revisited. Inf. Process. Manag. **28**(4), 467–490 (1992)
9. Trippas, J.R., Spina, D., Sanderson, M., Cavedon, L.: Results presentation methods for a spoken conversational search system. In: Proceedings of the First International Workshop on Novel Web Search Interfaces and Systems, pp. 13–15. ACM (2015)
10. Turunen, M., Hakulinen, J., Rajput, N., Nanavati, A.A.: Evaluation of mobile and pervasive speech applications. In: Speech in Mobile and Pervasive Environments, pp. 219–262 (2012)
11. White, R.W., Jose, J.M., Ruthven, I.: An implicit feedback approach for interactive information retrieval. Inf. Process. Manag. **42**(1), 166–190 (2006)

Keyword Search on RDF Datasets

Dennis Dosso$^{(\boxtimes)}$ (ID)

Department of Information Engineering, University of Padua, Padua, Italy
dosso@dei.unipd.it

Abstract. In the last years, the Resource Description Framework (RDF) has gained popularity as the *de-facto* representation format for heterogeneous structured data on the Web. RDF datasets are interrogated via the SPARQL language, which is often not intuitive for a user since it requires the knowledge of the syntax, the underlying structure of the dataset and the IRIs. On the other hand, today users are accustomed to Web-based search facilities that propose simple keyword-based interfaces to interrogate data. Hence, in order to ease the access to the data to users, we aim to develop of an effective and efficient system for keyword search over RDF graphs. Furthermore, we propose a methodology to properly evaluate these systems. Finally, we aim to address the problem of the explainability of the information contained in the answers to non-expert users.

Keywords: RDF graphs · Keyword search · Explainability

1 Motivation

Recently, the continuous growth of knowledge-sharing communities like Wikipedia and the advances in automated information-extraction from Web pages have made it possible to build large-scale knowledge bases [6]. In the meantime the Web of Data emerged as one of the principal means to expose structured data [12]. These datasets are usually built using RDF, a family of W3C specification, where data are represented in the form of a directed graph. RDF is the *de-facto* standard for publishing, accessing and sharing data, because it allows for flexible manipulation, enrichment, discovery, and reuse of data across applications, enterprises, and community boundaries. RDF is used by applications in industry, biology and human science such as Eagle-i [13], Europeana [11], Dbpedia [1], Disgenet [10] and many others, proving its importance and validity [8].

The standard way to interrogate RDF is the SPARQL query language, but it can become very difficult to write complex queries in this language, even for expert users, since it requires the knowledge of the language and of the underlying structure of the dataset.

Our research task is to enable the non-expert user to interrogate real-world RDF databases in a more intuitive and easy way. The keyword search paradigm can become a tool to enable the general user to gain access to these datasets.

© Springer Nature Switzerland AG 2019
L. Azzopardi et al. (Eds.): ECIR 2019, LNCS 11438, pp. 332–336, 2019.
https://doi.org/10.1007/978-3-030-15719-7_44

Amongst the difficulties regarding keyword search systems, as described in [4] and [5], there are the long execution times (more than one hour on average) and the memory required by most of the systems in the literature. These systems often cannot complete their execution even on small databases (1M triples) and thus cannot scale to real-world sizes. Our aim is to develop a keyword search system to enable non-expert users to perform keyword query on real-world RDF datasets. The output of such a system is a list of answer graphs ranked in decreasing order of relevance with respect to the information need.

The system needs to be effective and efficient: (i) to be effective it needs to create answer graphs that correctly cover the information need of the user, ranked from the most relevant to the least relevant; and (ii) to be efficient, it needs to operate in a time in the order of seconds or minutes even on real-world datasets of tens of millions of triples.

Keyword Search has been thoroughly studied in the contest of structured databases such as relational DB and Knowledge Bases. There are good reviews about this topic, such as [2,15] with a particular focus on RDB, and [14] with a focus on graph data.

The rest of the paper is organized as follows: Sect. 2 describes the proposed research and our methodology; Sect. 3 provides an outlook on future challenges.

2   Description of the Proposed Research and Methodology

The description of our research stems from four principal research topics related to keyword search.

Efficiency. The answer time of a system is crucial since users are accustomed to the rapid response of web search engines. A keyword search system needs to be efficient in order to be perceived as useful by the general user. This means that it needs to complete its execution in a reasonable time and with an efficient use of memory. [5] in particular showed that many approaches in the literature cannot complete their task due to time or memory issues.

In our system in particular we execute computations off-line (before the user submits her query) in order to create data structures that will help fasten the execution of the on-line phase. In particular, we create off-line a collection of subgraphs from the dataset. This collection is called "representative collection" since our aim is to build the subgraphs in such a way that they cover one single topic each. Such a collection can be efficiently indexed, searched and used to build the final answers to the query. We were able to achieve high on-line efficiency (in the order of minutes) on a dataset of more than 100M of triples. Our datasets are two orders of magnitude bigger than the ones that can be found in works like [5,7] and [9].

Effectiveness. The output of a keyword search system is a ranked list of answer subgraphs. Effectiveness can be measured on two aspects: the ability to create *accurate* answer graphs, and the ability to rank them accordingly to their *relevance* to the user query. With "accurate" answer graph we mean a graph that contains many relevant triples, i.e. triples that contain useful information for the user. Moreover, it should not contain noisy triples, i.e. triples that are not interesting for the user. Since we are returning a ranking, once the user has seen a relevant triple in one answer graph, the same triple presented in graphs down below in the ranking is considered redundant, and therefore no more useful. This means that the system should also be able to produce answer graphs *different enough* one from the other in order to completely cover the information need without too much redundancy.

In our system, we rely on heuristics to build and then prune graphs. We use the representative collection to look for subgraphs containing all the query words. Then, we create answer subgraphs that contain all the query words and we prune them to remove the noisy triples. We use an adapted version for RDF of the ranking function described in [9] to order the answer graphs using both their topological structure and text content.

In our experiments, we took the IMDB database[1] and parsed it in order to produce an RDF dataset of circa 116M triples. We also took a subset of IMDB of 1M triples, called rIMDB, in order to confront our algorithms to the others that cannot scale. As benchmarks, we implemented the keyword search system SLM described in [7] and adapted the keyword system MRF-KS for RDB in [9] in order for it to work on RDF. Our approach is called TSA+VDP (Topologic Syntactic Aggregator+Virtual Document Pruning).

Evaluation. The Cranfield paradigm is a well-established method to evaluate IR systems [4]. However, in the context of keyword search over structured data (in particular relational DBs), its implementation reported some limits as highlighted in [3]. Our aim is to develop a general evaluation framework for keyword search systems on RDF that follows the Cranfield paradigm and that can face the lacks shown so far in the literature.

In this regard, we developed an evaluation framework based on couples made of a keyword query and its counter-part SPARQL version. The SPARQL version produces the "exact" answer graph, which is used as ground truth. The answer graphs produced by the system are compared to this GT graph in order to understand if they are relevant to the user. Then, their position in the ranking is considered in order to understand the quality of the ranking itself. We designed two different functions: a Signal-To-Noise Ratio function (SNR) to decide if an answer graph is relevant to the user query; and a triple-based Discounted Cumulative Gain (tb-DCG) function to evaluate the quality of a ranking. The higher the tb-DCG, the better.

Table 1 reports the average performances obtained by the two benchmarks SLM and MRF-KS, compared to our TSA+VDP in terms of effectiveness via the tb-DCG and in terms of efficiency via time and memory usage. We used 50 different topics created by hand. This is not the only algorithm we developed, but it is the most promising one in terms of trade-off between efficiency and

---

[1] https://datasets.imdbws.com/.

Table 1. Performances in rIMDB, 1M triples. † indicates the systems in the top performing group with $p < 0.01$. The best system is in bold.

Systems	tb-DCG	Time (sec)	Memory (MB)
SLM	0.0030 ± 0.00	$\mathbf{34.5000 \pm 1.59}^{\dagger}$	3.2733 ± 0.06
MRF-KS	0.4217 ± 0.06	440.6800 ± 189.77	$\mathbf{0.9858 \pm 0.44}$
TSA+VDP	$\mathbf{0.5449 \pm 0.12}^{\dagger}$	$35.8710 \pm 9.53^{\dagger}$	19.8860 ± 3.66

effectiveness. As can be seen, in the rIMDB dataset our algorithms obtain the best performances in terms of effectiveness and efficiency.

Only our algorithm could be deployed on the whole IMDB database, obtaining a tb-DCG of 0.4490 with an average time of 200 s. TSA+VDP is the only system to complete its execution on a dataset of such a dimension. None of the algorithms of the state of the art could be deployed in such instances.

Explainability. Another crucial aspect is the ability of the system to *explain* the answers provided to the user. In other words, since we are assuming that the user is not an expert, it is reasonable that she does not know SPARQL and, possibly, does not know RDF. An output composed of a list of RDF triples can be useless to her, since she will be unable to read it. More useful would be a description of the answers in a more user-friendly format. For example, a text snippet in natural language that describes the content of the answer graphs.

3   Research Issues

In this section, we list some open issues and research directions.

1. We aim to develop a new automatic technique to build explanatory text document to be attached to the answer graphs in order to better explain their meaning to the non-expert user. We will investigate how data provenance and data citation can be deployed and integrated in order to solve this problem.
2. We want to investigate how the creation of fielded documents can affect the ranking function and the possibility to explain them to the user. We will use the text content of the IRIs and the structure of the dataset to better leverage on the presence of specific words and their frequency.
3. Often an RDF dataset does not present human-readable IRIs. These IRIs are made of numbers and acronyms. Thus, many keywords used in the queries are not matched in triples that are relevant to the information need. A possibility in this regard is to implement a *query expansion* technique, with the support of a dictionary, created ad-hoc for every dataset, that will enhance the ability of the keyword query to match with words in the dataset. This will help the algorithms to produce answer graphs and rank them in the best possible way.
4. There are many other databases of big dimensions and wide informative scope (like DBpedia) to be tested. Our aim is to test our system in these other databases and adapt it to the different peculiarities.

References

1. Auer, S., Bizer, C., Kobilarov, G., Lehmann, J., Cyganiak, R., Ives, Z.: DBpedia: a nucleus for a web of open data. In: Aberer, K., et al. (eds.) ASWC/ISWC -2007. LNCS, vol. 4825, pp. 722–735. Springer, Heidelberg (2007). https://doi.org/10.1007/978-3-540-76298-0_52

2. Bast, H., Buchhold, B., Haussmann, H.: Semantic search on text and knowledge bases. Found. Trends Inf. Retr. **10**(2–3), 119–271 (2016)

3. Bergamaschi, S., Ferro, N., Guerra, F., Silvello, G.: Keyword-based search over databases: a roadmap for a reference architecture paired with an evaluation framework. In: Nguyen, N.T., Kowalczyk, R., Rupino da Cunha, P. (eds.) Transactions on Computational Collective Intelligence XXI. LNCS, vol. 9630, pp. 1–20. Springer, Heidelberg (2016). https://doi.org/10.1007/978-3-662-49521-6_1

4. Coffman, J., Weaver, A.C.: A framework for evaluating database keyword search strategies. In: Proceedings of the 19th ACM International Conference on Information and Knowledge Management, pp. 729–738. ACM Press (2010)

5. Coffman, J., Weaver, A.C.: An empirical performance evaluation of relational keyword search systems. IEEE Trans. Knowl. Data Eng. **26**(1), 30–42 (2014)

6. Doan, A., Ramakrishnan, R., Vaithyanathan, S.: Managing information extraction: state of the art and research directions. In: Proceedings of the 2006 ACM SIGMOD International Conference on Management of Data, pp. 799–800. ACM (2006)

7. Elbassuoni, S., Blanco, R.: Keyword search over RDF graphs. In: Proceedings of the 20th ACM Conference on Information and Knowledge Management, CIKM 2011, pp. 237–242. ACM Press, New York (2011)

8. Feigenbaum, L., Herman, I., Hongsermeier, T., Neumann, E., Stephens, S.: The semantic web in action. Sci. Am. **297**(6), 90–97 (2007)

9. Mass, Y., Sagiv, Y.: Virtual documents and answer priors in keyword search over data graphs. In: Proceedings of the Workshops of the EDBT/ICDT 2016 Joint Conference, CEUR Workshop Proceedings, vol. 1558. CEUR-WS.org (2016)

10. Paschke, A., Burger, A., Romano, P., Marshall, M.S., Splendiani., A. (eds.) Proceedings of the 6th International Workshop on Semantic Web Applications and Tools for Life Sciences, Edinburgh, UK, 10 December 2013, CEUR Workshop Proceedings, vol. 1114. CEUR-WS.org (2014)

11. Petras, V., Hill, T., Stiller, J., Gäde, M.: Europeana - a search engine for digitised cultural heritage material. Datenbank-Spektrum **17**(1), 41–46 (2017)

12. Pound, J., Mika, P., Zaragoza, H.: Ad-hoc object retrieval in the web of data. In: Proceedings of the 19th International Conference on World Wide Web, WWW 2010. pp. 771–780. ACM Press, New York (2010)

13. Torniai, C., Bourges-Waldegg, D., Hoffmann, S.: eagle-i: biomedical research resource datasets. Semant. Web **6**(2), 139–146 (2015)

14. Wang, H., Aggarwal, C.C.: A survey of algorithms for keyword search on graph data. In: Aggarwal, C., Wang, H. (eds.) Managing and Mining Graph Data. ADBS, vol. 40, pp. 249–273. Springer, Boston (2010). https://doi.org/10.1007/978-1-4419-6045-0_8

15. Yu, J.X., Qin, L., Chang, L.: Keyword search in relational databases: a survey. IEEE Data Eng. Bull. **33**(1), 67–78 (2010)

Learning User and Item Representations for Recommender Systems

Alfonso Landin[(✉)][iD]

Information Retrieval Lab, Department of Computer Science,
University of A Coruña, A Coruña, Spain
alfonso.landin@udc.es

Abstract. The fields of Recommender Systems (RS) and Information Retrieval (IR) are closely related. A Recommender System can usually be seen as a specialized Information Retrieval system where the information need is implicit in the user profile. This parallelism has been exploited in the past to transfer methods between fields. One popular approach is to put the standard bag-of-words representation of queries and documents in IR at the same level as the user and item representations obtained from the user-item matrix in RS. Furthermore, in the last years, new ways of representing words and documents as densely distributed representations have risen. These embeddings show the ability to capture the syntactic and semantic relationships of words and have been applied both in IR and natural language processing. It is our objective to study ways to adapt those techniques to produce user/item representations, evaluate their quality and find ways to exploit them to make useful recommendations. Moreover, we will study ways to generate those representations leveraging properties particular to collaborative filtering data.

Keywords: Recommender systems · Collaborative filtering · Embedding models

1 Introduction

Information Retrieval (IR) is a discipline that studies the construction of systems that aim to satisfy the information needs of the users [1]. The most common method for expressing those information needs is through a textual query, although it is possible to use additional information when estimating the relevance of the documents, such as the users' context. Given this definition, we can consider a recommender system as an information retrieval system where the information need is implicit and inferred from the users' profile.

Despite the similarities between both paradigms, it has not been until recently that works on adapting successful techniques from one field to the other have been published. Examples include the reformulation of the collaborative filtering (CF) problem under the vector space model [2], the use of Language and Relevance Models [12] for the recommendation task [19,24], or the use of a recommendation technique, SLIM [18], for pseudo-relevance feedback [25].

© Springer Nature Switzerland AG 2019
L. Azzopardi et al. (Eds.): ECIR 2019, LNCS 11438, pp. 337–342, 2019.
https://doi.org/10.1007/978-3-030-15719-7_45

A set of techniques that is in vogue within IR are embedding models [16], that allow obtaining compact representations of terms [15] and documents [13]. These methods have demonstrated to achieve better results with techniques such as language models [9] or the vector space model [17]. For that reason, there is a lot of research effort currently in this direction. We aim in this thesis to undertake the task of generating dense user and item representations from CF data for their use in recommender systems. These representations could be applied to the computation of neighborhoods for memory-based systems or as part of the information used when constructing models for model-based approaches.

2   Background

In this section we will introduce Recommender Systems and word embedding models from the field of Natural Language Processing.

2.1   Recommender Systems

Recommender systems are usually classified into three categories: content-based models, collaborative filtering techniques and hybrid methods [22].

Content-based (CB) recommenders use the information and metadata of the items and/or users to compute the recommendations. As an illustration, for films recommendation, these techniques use the genre, director, plot, etc. to suggest similar movies to the user. In contrast, Collaborative Filtering (CF) techniques exploit past interactions of all the users of the system, using their preferences, whether explicit or implicit. In the previous example, a CF method would recommend to a user a movie that other users with similar preferences have rate positively. At last, hybrid approaches make use of both types of information.

CF methods use the set of users, the set of items and the set of user-item interactions for computing the recommendations. Information is usually presented as a user-item matrix, where each row corresponds to a user and a column to an item [22]. The value of each cell of the matrix is the value associated with the interaction between the user (row) and the item (column). This value can be the rating given by the user, the number of times the user reproduced the item, whether or not the user bought the product, etc. The value for user-item pairs with no interaction it is commonly considered as zero [5,10,18,21]. It is also frequent, especially in the case of implicit feedback, to transform these values to a binarized form, taking all known interactions a value of one, setting the value of unseen interactions to zero [10,18,21].

Representing the interaction is such a way, we can obtain the representations of the users and items by taking the corresponding row and column. This representation is analogous to the bag of words representation used in IR.

2.2   Word Embedding Models

In recent years, several techniques have been developed in the field of Natural Language Processing (NLP) to obtain densely distributed representations of words and documents. Examples include word2vec [14,15], GloVe [20] or doc2vec [13]. While the idea of obtaining this kind of representations is not new [6,8], these models have proven to capture syntactic and semantic relationships between words better, keeping the computational complexity of the training process at manageable levels that allows the techniques to scale with the size of the corpus.

The underlying idea is an old one: "You shall know a word by the company it keeps" [7]. Over this idea, models were proposed to obtain representations trying to keep a couple of properties: words that appear together in several contexts are related, and words that shared common contexts are also related.

3   Proposed Research

Previous works [19] have established an analogy between IR concepts and RS concepts. This parallelism has been proven appropriate to successfully adapt IR models to the recommendation task [4,19,24]. In this analogy users, or more precisely, user profiles, play a dual role, as a query and as documents. In the same way that documents and queries are composed of terms, user profiles are composed of items. Based on this modeling of the problem several research objectives are planned, with the aim of answering various research questions:

Objective 1. Learn user and item representations adapting existing word embedding models [14,15,20]. Two problems must be tackled. First, words in the text present a natural order which conditions the models' search for syntactic and semantic relations, how we will approach the order issue inside the users' profiles? Second, when computing the representations, embedding models implicitly use the term frequency of the words in the texts, how we will translate those frequencies to the graded item preferences? The analogy enables the computation of item embeddings. The computation of user embeddings can be achieved by transposing the user-item matrix. This transposition changes the analogy to one where users are the terms, and item profiles are the documents, and then the designed solutions for item embeddings can be used to compute user embeddings.

Objective 2. Learn user and item embeddings simultaneously. In the previous objective, we stated the embedding learning problem as two independent problems at item and user level. We think that simultaneously tackling both problems can improve the quality of both types of embeddings. Therefore, we plan to adapt models that can learn word and document representations at the same time, such as doc2vec [13], to the task of computing user and item representations.

Objective 3. Assess specific aspects of the quality of the representations. Traditional quality assessment is based on recommendation accuracy. We will use dense representations to compute the similarities between users and/or items and then evaluate if when using those representations we get better figures than with traditional representations. But we also plan to assess whether or not the produced embeddings can capture properties such as item categories.

Objective 4. Exploit the representations with current state-of-the-art models and development of alternative models. The use of neural models is an open research field in the RS area. We can change the input of those models to use the embeddings. We also plan to devise new models that exploit the characteristics of the embeddings more effectively. Preliminary results can be seen in [11].

Objective 5. Formulate new specially tailored models to calculate embeddings that take advantage of the properties of CF data. The models that we plan to leverage in the previous objectives were developed to exploit the properties of natural languages. While there are proven similarities between both worlds, it should be possible to obtain better results if the differences are also taken into account when developing a model from scratch. The knowledge gathered during previous stages should help guide the development of such new solutions.

RQ1. Are the dense representations able to capture the properties of the users and items?

RQ2. Can the embeddings be used to generate effective recommendations?

RQ3. Are the learned representations able to improve the performance of current state-of-the-art models when used as input for them?

RQ4. Is it possible to improve the quality of the embedding if the characteristics of the data are taken into account when constructing models to calculate them?

Recommender systems evaluation is essential to compare the performance between models or to tune parameters. We aim to measure the performance of the systems for the top-N recommendation task, evaluating the ability of the methods to place the relevant items at the top of the list [5]. We will follow the *TestItems* approach described in [2], scoring for each user every item included in the test set. While the restrictions may lead to the underestimation the actual value of the metric, it produces comparable and trustworthy results [2].

The effectiveness of the different approaches will be evaluated with precision based metrics from IR, using metrics that have proven to be robust, such as nDCG and MAP [23]. Attention will be paid to the bias introduced by the different metrics [3]. While accuracy will be a central part of the evaluation, other properties of the system will not be ignored, including novelty and diversity. Metrics such as Mean Self Information (MSI) for novelty or the inverse of the Gini index for diversity will be used as part of the evaluation protocol.

4   Research Issues

This research proposal outlines a path for the development of several solutions to the problem of obtaining quality representations for users and/or items that can be used effectively in the recommendation task. Even though the objectives seem clear several questions are left open: are the proposed adaptations feasible or should adjustments be made? Are there any other approaches that should be taken into consideration? Is the evaluation methodology enough to measure the quality of the representations or should any other steps be taken to ensure the validity of the results and the soundness of the conclusions?

Acknowledgments. This work has received support from accreditation 2016–2019 ED431G/01 (Xunta de Galicia/ERDF) and grant FPU17/03210 (MICIU).

References

1. Baeza-Yates, R., Ribeiro-Neto, B.: Modern Information Retrieval: The Concepts and Technology Behind Search, 2nd edn. Addison-Wesley Publishing Company, Boston (2011)
2. Bellogín, A., Castells, P., Cantador, I.: Precision-oriented evaluation of recommender systems. In: Proceedings of the 5th ACM Conference on Recommender Systems, RecSys 2011, pp. 333–336. ACM (2011). https://doi.org/10.1145/2043932. 2043996
3. Bellogín, A., Castells, P., Cantador, I.: Statistical biases in information retrieval metrics for recommender systems. Inf. Retrieval J. **20**(6), 606–634 (2017). https://doi.org/10.1007/s10791-017-9312-z
4. Bellogín, A., Wang, J., Castells, P.: Text retrieval methods for item ranking in collaborative filtering. In: Clough, P., et al. (eds.) ECIR 2011. LNCS, vol. 6611, pp. 301–306. Springer, Heidelberg (2011). https://doi.org/10.1007/978-3-642-20161-5_30
5. Cremonesi, P., Koren, Y., Turrin, R.: Performance of recommender algorithms on top-n recommendation tasks. In: Proceedings of the Fourth ACM Conference on Recommender Systems, RecSys 2010, pp. 39–46. ACM (2010). https://doi.org/10. 1145/1864708.1864721
6. Deerwester, S., Dumais, S.T., Furnas, G.W., Landauer, T.K., Harshman, R.: Indexing by latent semantic analysis. J. Am. Soc. Inf. Sci. **41**(6), 391–407 (1990). https://doi.org/10.1002/(SICI)1097-4571(199009)41:6⟨391::AID-ASI1⟩3.0.CO;2-9
7. Firth, J.R.: A synopsis of linguistic theory 1930–55. In: Studies in Linguistic Analysis (Special Volume of the Philological Society), vol. 1952–59, pp. 1–32. The Philological Society, Oxford (1957)
8. Furnas, G.W., et al.: Information retrieval using a singular value decomposition model of latent semantic structure. In: Proceedings of the 11th Annual International ACM SIGIR Conference on Research and Development in Information Retrieval, SIGIR 1988, pp. 465–480. ACM (1988). https://doi.org/10.1145/62437. 62487
9. Ganguly, D., Roy, D., Mitra, M., Jones, G.J.: Word embedding based generalized language model for information retrieval. In: Proceedings of the 38th International ACM SIGIR Conference on Research and Development in Information Retrieval, SIGIR 2015, pp. 795–798. ACM (2015). https://doi.org/10.1145/2766462.2767780
10. Hu, Y., Koren, Y., Volinsky, C.: Collaborative filtering for implicit feedback datasets. In: 2008 Eighth IEEE International Conference on Data Mining, pp. 263–272, December 2008. https://doi.org/10.1109/ICDM.2008.22
11. Landin, A., Valcarce, D., Parapar, J., Barreiro, Á.: PRIN: a probabilistic recommender with item priors and neural models. In: Azzopardi, L., et al. (eds.) ECIR 2019. LNCS, vol. 11438, pp. 133–147. Springer, Heidelberg (2019)
12. Lavrenko, V., Croft, W.B.: Relevance based language models. In: Proceedings of the 24th Annual International ACM SIGIR Conference on Research and Development in Information Retrieval, SIGIR 2001, pp. 120–127. ACM (2001). https://doi.org/10.1145/383952.383972

13. Le, Q., Mikolov, T.: Distributed representations of sentences and documents. In: Proceedings of the 31st International Conference on Machine Learning, vol. 32, pp. 1188–1196 (2014)

14. Mikolov, T., Chen, K., Corrado, G., Dean, J.: Efficient estimation of word representations in vector space. CoRR (abs/1301.3), January 2013

15. Mikolov, T., Sutskever, I., Chen, K., Corrado, G.S., Dean, J.: Distributed representations of words and phrases and their compositionality. In: Advances in Neural Information Processing Systems, NIPS 2013, vol. 26, pp. 3111–3119 (2013)

16. Mitra, B., Craswell, N.: An introduction to neural information retrieval. Found. Trends® Inf. Retrieval **13**(1), 1–126 (2018). https://doi.org/10.1561/1500000061

17. Nalisnick, E., Mitra, B., Craswell, N., Caruana, R.: Improving document ranking with dual word embeddings. In: Proceedings of the 25th International Conference Companion on World Wide Web, WWW 2016 Companion, International World Wide Web Conferences Steering Committee, pp. 83–84 (2016). https://doi.org/10.1145/2872518.2889361

18. Ning, X., Karypis, G.: SLIM: sparse linear methods for top-n recommender systems. In: 2011 IEEE 11th International Conference on Data Mining, pp. 497–506, December 2011. https://doi.org/10.1109/ICDM.2011.134

19. Parapar, J., Bellogín, A., Castells, P., Barreiro, A.: Relevance-based language modelling for recommender systems. Inf. Process. Manag. **49**(4), 966–980 (2013). https://doi.org/10.1016/j.ipm.2013.03.001

20. Pennington, J., Socher, R., Manning, C.: GloVe: global vectors for word representation. In: Proceedings of the 2014 Conference on Empirical Methods in Natural Language Processing (EMNLP), pp. 1532–1543. Association for Computational Linguistics (2014). https://doi.org/10.3115/v1/D14-1162

21. Rendle, S., Freudenthaler, C., Gantner, Z., Schmidt-Thieme, L.: BPR: Bayesian personalized ranking from implicit feedback. In: Proceedings of the 25th Conference on Uncertainty in Artificial Intelligence, UAI 2009, pp. 452–461. AUAI Press (2009)

22. Ricci, F., Rokach, L., Shapira, B.: Recommender Systems Handbook, 2nd edn. Springer, Heidelberg (2015)

23. Valcarce, D., Bellogín, A., Parapar, J., Castells, P.: On the robustness and discriminative power of information retrieval metrics for top-n recommendation. In: Proceedings of the 12th ACM Conference on Recommender Systems, RecSys 2018, pp. 260–268. ACM (2018). https://doi.org/10.1145/3240323.3240347

24. Valcarce, D., Parapar, J., Barreiro, Á.: Language models for collaborative filtering neighbourhoods. In: Ferro, N., et al. (eds.) ECIR 2016. LNCS, vol. 9626, pp. 614–625. Springer, Cham (2016). https://doi.org/10.1007/978-3-319-30671-1_45

25. Valcarce, D., Parapar, J., Barreiro, A.: LiMe: linear methods for pseudo-relevance feedback. In: Proceedings of the 33rd Annual ACM Symposium on Applied Computing, SAC 2018, pp. 678–687. ACM (2018). https://doi.org/10.1145/3167132.3167207

Improving the Annotation Efficiency and Effectiveness in the Text Domain

Markus Zlabinger[✉]

Institute of Software Technology and Interactive Systems,
Favoritenstrasse 9-11/188, 1040 Vienna, Austria
markus.zlabinger@tuwien.ac.at

Abstract. Annotated corpora are an important resource to evaluate methods, compare competing methods, or to train supervised learning methods. When creating a new corpora with the help of human annotators, two important goals are pursued by annotation practitioners: Minimizing the required resources (efficiency) and maximizing the resulting annotation quality (effectiveness). Optimizing these two criteria is a challenging problem, especially in certain domains (e.g. medical, legal). In the scope of my PhD thesis, the aim is to create novel annotation methods for an efficient and effective data acquisition. In this paper, methods and preliminary results are described for two ongoing annotation projects: medical information extraction and question-answering.

Keywords: Text annotation · Corpus creation · Data acquisition

1 Introduction and Motivation

Annotated corpora are an important resource in the scientific domain. They are essential to evaluate methods, compare competing methods, or to create supervised learning methods. For some methods, appropriate ground truth is publicly available (e.g. TREC test collection). In cases where no ground truth is available, it can be created from scratch with the help of human annotators.

When creating a new corpora, the annotation practitioners usually pursue two important goals: First, the annotators should be supported as good as possible so that they can complete their annotations quickly (I refer to this as efficiency). And second, the quality of the annotated labels should be as high as possible (I refer to this as effectiveness). Both pursued goals target different aspects of annotation: A more efficient annotation procedure leads to decreased costs, i.e. less resources are required. On the other hand, a more effective annotation procedure leads to annotations of higher quality, which makes the resulting corpora a resource of higher value.

In the scope of my PhD thesis, I will develop annotator support tools, best-practices and guidelines that optimize the annotation process with respect to two dimensions: efficiency and effectiveness. Since the data annotation area is quite broad, the focus will be on text annotation in challenging domains.

L. Azzopardi et al. (Eds.): ECIR 2019, LNCS 11438, pp. 343–347, 2019.
https://doi.org/10.1007/978-3-030-15719-7_46

In this doctoral paper, methodologies and preliminary results for two ongoing annotation projects are presented. The first project is about annotation of named-entities in medical publications. To annotate medical texts with high quality, expert knowledge is usually required to understand the technical terminology [3]. We show that by combining a supporting annotation tool with extensive annotator training runs and guidelines, the task can be performed even by non-experts with sufficient quality.

The second described project is about annotation of questions with the correct answer. For this project, we use questions that were asked to a technical support chat-bot and assign the correct answers based on a static answer catalog. We show that by pre-grouping questions that require the same answer (i.e. redundant questions), the annotation task can be conducted nearly 50% more time-efficient.

The planned contributions of my PhD thesis are the following:

- Novel annotation tools, best-practices and guidelines are created that improve the efficiency and/or effectiveness for domain-specific annotation tasks.
- By evaluating the novel methodologies, several new corpora will be created, which are made publicly available to the research community.

Paper Organization. Related work is discussed in Sect. 2, for both the medical and the question-answer annotation task. Methodologies and preliminary results are reported in Sect. 3. The paper is concluded in Sect. 4.

2   Background and Related Work

Named-Entity Annotation in the Medical Domain

One annotation project that I am working on is about named-entity recognition in the medical domain—more specifically, PICO annotation. The P in PICO stands for Population (e.g. *"women with headache"*), the I for Intervention (e.g. *"aspirin twice per day"*), the C for Comparison (e.g. *"placebo"*) and the O for Outcome (e.g. *"pain reduction"*).

Kim et al. [3] created a PICO corpus consisting of 1,000 abstracts of medical publications. Each sentence in the abstracts was annotated with the appropriate PICO label. The corpus was used by Chabou et al. [2] to create a sentence-level classifier. For sentences that were classified as containing one of the PICO elements, they performed a fine-grained word-level extraction using a rule-based approach. Since no evaluation data existed at that time for word-level PICO extraction, no results were reported in their paper.

Research Gap: There are two research gaps that I intend to fill: First, the creation of a word-level PICO corpus. And second, the creation of annotator support methods that allow laypeople to conduct the task with high quality.

Question-Answer Annotation

Another annotation project that I am working on is about question-answer annotation. Related literature is presented in the following paragraphs.

For the Eight Text Retrieval Conference (8-TREC), one track was about answering 200 short-text, fact-based questions [5]. For each question, participants were asked to compute a ranked list of five possible answers. Those five answers were then manually labeled by human assessors as either *correct* or *incorrect*.

Yang et al. [6] introduced the WikiQA question-answering dataset, which consists of 3,047 Bing questions. For each question, a set of potential answer sentences taken from Wikipedia was manually annotated by crowdsourcing.

One task of the 2015 SemEval workshop [4] was about community question answering (e.g. platforms like Stack Overflow). For the task, answers to questions were manually labeled as being *relevant*, *non-relevant* or *unsure*.

Research Gap: The effectiveness in the related annotation projects was increased by redundantly collecting labels and then, computing a final label via majority voting. There were, however, no methods reported that target the efficiency of the task, which is part of my research.

3   Proposed Methodology and Preliminary Results

In this section, I present the methods and preliminary results for the medical entity annotation task and the question-answering task. For both tasks, the goal is to answer following research question:

- **What annotation methods improve the efficiency or the effectiveness of domain-specific annotation tasks?**

PICO Annotation. The aim for this task is to create a word-level PICO dataset by using appropriate guidelines and a specialized annotation tool. Moreover, by using the tool, laypeople should be able to produce high quality annotations that are similar or only slightly worse compared to the annotations of experts. Following, methods and preliminary results are summarized (see [7,8] for details).

The key component of our effective PICO annotation approach is a specialized annotation tool, which was developed as follows: First, we created an alpha version of the tool that contained only the basic functionality that was needed to create annotations. Second, we hired medical experts and students to produce temporary annotations. Third, the temporary annotations were used to compute the inter-annotator agreement (IAA). Fourth, through analysis of the errors that annotators made, we identified areas for improvement for the tool. Fifth, an improved version was designed and implemented.

The just described steps were repeated three times to obtain incrementally improved versions of the annotation tool. The highest gain in IAA was observed when we pre-labeled the text that was shown to the annotators with semantic labels like *drug* or *disease*. Such labels are strong indicators for the PICO elements—for example, a *drug* might be an indicator for an Intervention (I). The semantic labels were automatically computed using GATE's BioYodie[1]. Another

---

[1] https://gate.ac.uk/applications/bio-yodie.html.

significant gain regarding the IAA was observed when we changed the tool's input method: In the initial version, open text input was allowed—whereas in the final version, annotators could only mark a span of words and then, assign a PICO label to the span (all inputs by clicking).

In total, the IAA increased from 20% to 55%, which is close to the agreement of sentence-level PICO annotation ($\approx$65%) [3]. We observed that the non-experts produced labels that were of similar quality when compared to the experts. Overall, we assume that the increase of IAA resulted from: the tool support, the practical experience that annotators gained from testing the tool and finally, from the written guidelines on how to annotate.

Question-Answer Annotation. The aim for this task is to label questions with the correct answer. As raw data, we were provided with 500 questions asked to a chat-bot and the corresponding answer catalog consisting of 373 answers. The data provider is an Austrian telecommunication company.

By examining the raw data, we noticed that finding the correct answer within the provided catalog is not difficult, however, the look-up is a time-consuming procedure. Furthermore, we observed that a large number of questions is asked redundantly, i.e., two or more users ask the same question. Based on this observation, we created an approach that drastically reduces the number of answer look-ups. The approach consists of following steps: First, a candidate question is shown to the annotator. Second, a list of similar questions with respect to the candidate question are also shown to the annotator. Third, the annotator marks questions in the list that have the same intent as the candidate question (i.e. they require the exact same answer). Finally, only a single answer look-up is performed to find the right label for the candidate plus the marked questions.

The challenging part of the described approach is the retrieval of similar question with respect to a candidate question (i.e. the Second step). That is because questions are user generated and contain colloquial language and spelling mistakes. Moreover, some users submit keyword-based questions (*"extend contract how"*), whereas others submit regular questions (*"How can I extend my contract?"*). We address these problems by using a state-of-the-art short-text similarity method [1] that we adapt for our task.

With the described group-wise annotation approach, we measured a reduction of answer look-ups of 51% without the loss of annotation quality. Note that the described methodology and results are not published yet.

4   Conclusion and Outlook

In this paper, efficient and effective annotation methods for medical named-entity recognition and question-answering of customer support questions were presented. For future research in the scope of my PhD, the plan is to also develop methods for other domains (e.g. legal or patent domain). Another goal will be to find annotation techniques that are not limited to certain domains or specific tasks, i.e. methods that are applicable to a broad palette of annotation problems. For example, research on how annotators should be best schooled for a task.

References

1. Arora, S., Liang, Y., Ma, T.: Simple but tough-to-beat baseline for sentence embeddings. In: International Conference on Learning Representations, p. 16 (2017)
2. Chabou, S., Iglewski, M.: PICO extraction by combining the robustness of machine-learning methods with the rule-based methods. In: 2015 World Congress on Information Technology and Computer Applications (WCITCA), pp. 1–4 (2015)
3. Kim, S.N., Martinez, D., Cavedon, L., Yencken, L.: Automatic classification of sentences to support evidence based medicine. BMC Bioinform. **12**(2), S5 (2011)
4. Nakov, P., Màrquez, L., Magdy, W., Moschitti, A., Glass, J., Randeree, B.: Semeval-2015 task 3: answer selection in community question answering. In: Proceedings of the 9th International Workshop on Semantic Evaluation, SemEval 2015, pp. 269–281 (2015)
5. Voorhees, E.M.: The TREC question answering track. Nat. Lang. Eng. **7**(04) (2001). https://doi.org/10.1017/S1351324901002789
6. Yang, Y., Yih, W.T., Meek, C.: WikiQA: a challenge dataset for open-domain question answering. In: Proceedings of the 2015 Conference on Empirical Methods in Natural Language Processing, pp. 2013–2018 (2015)
7. Zlabinger, M., Andersson, L., Brassey, J., Hanbury, A.: Extracting the population, intervention, comparison and sentiment from randomized controlled trials. Stud. Health Technol. Inform. **247**, 146–150 (2018)
8. Zlabinger, M., Andersson, L., Hanbury, A., Andersson, M., Quasnik, V., Brassey, J.: Medical entity corpus with PICO elements and sentiment analysis. In: Proceedings of the Eleventh International Conference on Language Resources and Evaluation (LREC) (2018)

Logic-Based Models for Social Media Retrieval

Firas Sabbah[✉]

University of Duisburg-Essen, Duisburg, Germany
sabbah@is.inf.uni-due.de

Abstract. The huge amount of information that is available in social media today has resulted in a pressing demand for information retrieval models to support users in different retrieval tasks in this area. Evaluation of products or services based on other consumers' reviews are examples of retrieval tasks where users benefit from the information in social media. However, as information in online reviews is provided from different customers, the reviews differ in their degree of credibility. Furthermore, not every detail of a product is reviewed but needs to be taken care when validating a product. Finally, reviews contain contradictions. In this work, we propose a logic-based pipeline to develop a social media retrieval model in which issues of credibility, omissions and contradictions are considered. For modeling these criteria, our approach depends on a probabilistic four-valued logic combined with an open world assumption which employs the unknown knowledge in retrieval tasks.

Keywords: Logic-based models · User reviews · IR ·
Ranking algorithms · Credibility · Contradictions · Missing knowledge

1 Motivation

Online shopping has become a normal scenario of internet users. Most of the online shops offering products or services have been converted to user-driven platforms where users are able not only to online buy what is offered but also contribute and add their reviews to the items listed by the shops. These reviews are important because they express real user experiences and thus provide valuable information for others to make decisions [1,2].

However, although the reviews are useful they also bring challenges that need to be addressed. First, they suffer from credibility issues, i.e. not every review is trustworthy. Furthermore, the reviews contain some unknown knowledge. Usually, users provide evaluations for products and services as a form of reviews of one or many aspects of the reviewed item. Each of those reviews expresses the sentiment of the user regarding the mentioned aspects. However, a valuable amount of information is still hidden when the users mention some aspects and ignoring others. Finally, there is a huge amount of contradictions between the reviews.

The focus of this research is to investigate methods for improving the ranking of products and services in which the problems of credibility, information omission and contradictions are addressed. For such work, we found that 4-valued

© Springer Nature Switzerland AG 2019
L. Azzopardi et al. (Eds.): ECIR 2019, LNCS 11438, pp. 348–352, 2019.
https://doi.org/10.1007/978-3-030-15719-7_47

logical based models are covering all required aspects of the investigation. Thus we plan to investigate these models to address open challenges in ranking online user reviews.

2   Background and Related Work

The previous works [3–10] have presented different approaches for evaluating and ranking the products. However, four-valued logic based approaches [12–14] were used to deal with contradictions that appear in several parts of a document or more than one document that represent the possible answer of a query in IR. In our proposed logic-based approach, we support the consumer decision by ranking products based on their related reviews, taking into consideration credibility and polarity of each review. Our model is also depending on the four-valued logic concept to represent contradictions among reviews. While the previous approaches are mainly based on feature extraction and machine learning, our logic-based approach however, is more explicit in handling reviews' credibility, contradictions and unknown knowledge.

3   Proposed Solution and Research Questions

To explain the different phases of the proposed pipeline, we start with a real case example. A user is trying to book a hotel online. The required hotel should be clean and near the city center. Obviously, searching for a room in the city center will depend on the hotel description in the website. But to find a clean hotel we need to search in reviews of the previous customers and see how they describe the cleanness of the hotel. Assuming that we have the following two reviews from the available set of reviews about hotel X which is in the city center:

- rev_1 = "quiet, clean and near the main station"
- rev_2 = "not clean and small beds"

As a pipeline process, we first analyze the content of the reviews to know whether the cleanliness aspect has been mentioned in the reviews or not. In the second step, we annotate the reviews with the suitable logical values. In the third phase, we focus on indexing and probabilistic reasoning of the reviews in order to extract a commutative knowledge about the hotel represented by four logical values. In the final step, we use the logical values to generate a ranking score of the hotel regarding the required aspect. Later on, a group of hotels can be ranked based on those scores.

Extraction of Aspects: The process aims to identify the different aspects that mentioned in the reviews. In the experiments so far, we use a keywords-based approach to extract a predefined aspects. However, the process should be extended to extract the implicitly mentioned aspects. In addition to extract the mentioned aspects, this step includes analyzing the sentiment or the polarity of the aspects in order to know whether the users are praising or dispraising the products or services.

Logic-Based Annotation of the Aspects: According to the four-valued logic [11], the reviews are annotated with "true" and "false" values for the positive and the negative sentiments. However, a logical value of "unknown" is assigned for the normal sentiments and for the cases where users do not mention an aspect in the review. In the second step of the annotation process, we define a probability value for each aspect. This value reflects the sentiment's level of the user's evaluation. We use the star-rating score as a basis for defining this probability. In the given example, we assume that rev_1 provides 7 stars out of 10 as a star-rating evaluation, and 5 stars out of 10 for rev_2. According to this assumption, we define explicit probabilities of the cleanliness aspect as the following:

- $P(rev_1$ describe cleanliness positively$) = 0.7$
- $P(rev_2$ describe cleanliness negatively$) = 0.5$

However, the extracted knowledge from the example is still not complete. According to the open word assumption (OWA), the information is defined as "unknown" if it is neither "true" nor "false". The OWA is the apposite concept of what is called the closed world assumption (CWA), where what is not defined as true is defined as false and what is not false is true. From this fact, we define implicit probabilities of the cleanliness aspect as the following:

- $P(rev_1$ describe cleanliness unknowingly$) = 1.0 - 0.7 = 0.3$
- $P(rev_2$ describe cleanliness unknowingly$) = 1.0 - 0.5 = 0.5$

Probabilistic Reasoning Using Hyspirit: After collecting the knowledge about the hotels, we need now to assign a value of credibility for each review. The credibility of a review is a numerical value between 0 and 1 and it indicates the belief level of a review. Indeed, the credibility values are assigned according to the believing style about the reviews. Here, we discuss the disjoint and the independent cases.

In logic, two events are mutually exclusive or disjoint if they cannot both be true at the same time. As an application of this concept on the hotels issue, the two reviews (rev_1 and rev_2) cannot be true at the same time. This means, the credibility values of the reviews in disjoint cases is distributed in a way in which they sum up to less than or equal to 1. In the independent case, the truthiness of a review does not affect the truthiness of other reviews. This case is more general and realistic in which the customer believe in each

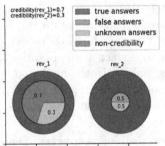

Fig. 1. Disjoint reviews

review with different degree of credibility. Figures 1 and 2 show rev_1 and rev_2 when they regarded as disjoint and independent events respectively. Based on those two cases, we arise here a group of research questions:

- In general, can we have better review-based rankings by modeling credibility, contradiction and missing knowledge using 4-valued logic models? The following questions are more specific to this general question:
- What is the proper method of regarding a review as an independent or disjoint event?
- Is it always possible to regard the reviews as disjoint and independent events regardless the required aspect?
- How could the credibility differs in the disjoint and independent cases, and which factors/attributes should be considered in credibility calculation of each case?

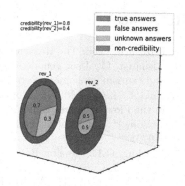

Fig. 2. Independent reviews

The overall knowledge of truth values is constructed using different methods based on the required case. This includes calculating the probabilities of the truth values for all possible answers that lead to those truth values. For example, the "True" truth value of a hotel in the independent case (Fig. 3) is constructed by finding all paths that have at least one true-answer-review and zero false-answer-reviews and zero or more of unknown answers.

However, performing such operations with a consideration of all possible cases requires a reliable probabilistic IR system. Hyspirit is a probabilistic IR interface [13] that perfectly fits to this IR task. The interface produces at the final stage a tuple of 4-valued logic based result. Each one of those values is calculated depending on the participating reviews of the targeted hotel and the targeted aspect.

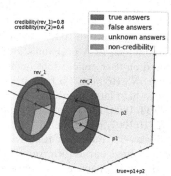

Ranking the Items: In the last step, we use the produced 4 logic values results in order to rank the items. Since we have 4 different values for each item, the ranking procedure is able to rank the items depending on different perspectives regarding those values. Here, we arise another research question:

Fig. 3. True value about the cleanliness aspect

- How can the 4 truth values be employed in order to provide a single score which can be used as a basis of ranking procedure?

References

1. Razorfish. Feed: The Razorfish Consumer Experience Report (2008)
2. Akehurst, G.: User generated content: the use of blogs for tourism organisations and tourism consumers. Serv. Bus. **3**(1), 51 (2009)
3. Katz, E., Lazarsfeld, P.F., Roper, E.: Personal Influence: The Part Played by People in the Flow of Mass Communications. Routledge, Abingdon (2017)
4. Nelson, P.: Information and consumer behavior. J. Polit. Econ. **78**(2), 311–329 (1970)
5. Dowling, G.R., Staelin, R.: A model of perceived risk and intended risk-handling activity. J. Consum. Res. **21**(1), 119–134 (1994)
6. Chen, Y., Xie, J.: Third-party product review and firm marketing strategy. Mark. Sci. **24**(2), 218–240 (2005)
7. Zhang, K., et al.: Mining millions of reviews: a technique to rank products based on importance of reviews. In: Proceedings of the 13th International Conference on Electronic Commerce. ACM (2011)
8. Tian, P., et al.: Research of product ranking technology based on opinion mining. In: Second International Conference on Intelligent Computation Technology and Automation, ICICTA 2009, vol. 4. IEEE (2009)
9. Kong, R., et al.: Customer reviews for individual product feature-based ranking. In: 2011 First International Conference on Instrumentation, Measurement, Computer, Communication and Control. IEEE (2011)
10. Zhang, K., Narayanan, R., Choudhary, A.N.: Voice of the customers: mining online customer reviews for product feature-based ranking. WOSN **10**, 11 (2010)
11. Belnap, N.D.: A useful four-valued logic. In: Dunn, J.M., Epstein, G. (eds.) Modern Uses of Multiple-Valued Logic. EPIS, vol. 2, pp. 5–37. Springer, Dordrecht (1977). https://doi.org/10.1007/978-94-010-1161-7_2
12. Frommholz, I.: A probabilistic framework for information modelling and retrieval based on user annotations on digital objects. Ph.D. thesis (2008)
13. Fuhr, N., Rölleke, T.: HySpirit — a probabilistic inference engine for hypermedia retrieval in large databases. In: Schek, H.-J., Alonso, G., Saltor, F., Ramos, I. (eds.) EDBT 1998. LNCS, vol. 1377, pp. 24–38. Springer, Heidelberg (1998). https://doi.org/10.1007/BFb0100975
14. Rölleke, T., Fuhr, N.: Retrieval of complex objects using a four-valued logic. In: Proceedings of the 19th Annual International ACM SIGIR Conference on Research and Development in Information Retrieval. ACM (1996)

Adapting Models for the Case of Early Risk Prediction on the Internet

Razan Masood[(✉)]

Duisburg-Essen University, Duisburg, Germany
razan.masood@uni-due.de

Abstract. Internet users who suffer from variant types of mental disorders found in Social Media platforms and other online communities an easier way to reveal their minds. The textual content produced by such users proved to be handful in evaluating mental health and detecting related illnesses even in early stages. This paper overviews my recent doctoral work on eRisk 2019 shared task for early detection of Anorexia and early detection of Self-harm. These sub-tasks have many aspects to consider, including textual content, timing, and the similarities to other mental health prediction tasks. Thus, based on these different dimensions, we propose solutions based on Neural-Networks, Multi-task learning, domain adaptation and Markov Models.

Keywords: Mental health · Anorexia · Self-harm · NLP · Machine learning

1 Introduction

The atmosphere of Social Media (SM) made it normal for users to talk about themselves and be the center of interest, which is not the case in the real world. Furthermore, the way the user communicates information and interacts with other users and their language usage turned to reveal much of their personalities and implicit properties. In addition, many users that are likely to have psychological problems like depression, eating disorder and Self-harm issues turn to SM to talk about their problems either to share them with users that have the same problems or to ask for advice. Hence, examining their posts could be part of the treatment cycle. On the other hand, the use of automated natural language processing methods has proven to enhance the prediction and diagnosis of many mental illnesses in early stages [3]. Computational methods like machine learning can contribute in mental health sector in terms of performing early detection and enhancing the access to treatment[1].

In this research we are interested in two mental health issues, namely Anorexia and Self-harm. Anorexia Nervosa (AN) is a type of eating disorder

---

[1] https://www.weforum.org/agenda/2018/03/3-ways-ai-could-could-be-used-in-mental-health/.

© Springer Nature Switzerland AG 2019
L. Azzopardi et al. (Eds.): ECIR 2019, LNCS 11438, pp. 353–358, 2019.
https://doi.org/10.1007/978-3-030-15719-7_48

and is defined as a mental illness where a person is obsessed with body image and weight, by which it affects their diet and yields undesirable physical and psychological symptoms [2]. Anorexia has also the highest mortality rates compared to other disorders [5]. The main issue with Anorexia is the late diagnosis and treatment which leads to difficulties in recovering. Likewise, Self-harm refers to a broad spectrum of abusive behaviours towards oneself like self-cutting and self-poisoning. The harm could be so severe that it ends up with suicide [15]. Some patients of mental illnesses especially of young ages turn to SM communities also to emphasis their illnesses in what is called *Digital Self-harm* [12]. Accordingly, this shows the importance of analyzing the content generated by these types of users to provide a kind of support by the early detection of mental illness signs [11]. Previous work used SM content from different point of views to better understand mental illness and the effect of SM [6,7,22].

2   ERisk 2019 Subtasks and Data

ERisk[2] is a set of challenges related to predicting health problems among users on the Internet using their generated data. The data is extracted from Reddit[3]. This website includes many communities to discuss certain topics and interests in what are called *subreddits*. The subtasks of interest are Early Detection of Signs of Anorexia and Early Detection of Signs of Self-harm. The data consists of users' posts in chronological order [10].

Both tasks are evaluated using early risk detection error (ERDE) measure for early detection systems defined by [9].

3   Related Work

The winning team of eRisk 2018 compares the performance of different models based on linguistic metadata, Bag of Words (BoW), word embedding and Convolutional Neural Network (CNN) [17]. By analyzing different user characteristics on Twitter data, it was found that users with eating disorders have similar writing styles like using emoticons and hash-tags. They also express concerns of death and emotions like sadness [18]. Furthermore, the content of both pro-eating disorder blogs and recovery blogs was analyzed to distinguish the individual writing characteristics. Such findings concluded that pro-ED blogs contain more discussions on food, weight and body shape than any other aspect as well as fewer social references [7,21]. Likewise, online textual content produced by users of Self-harm issues, also have its defining patterns like frequent uni-grams [19].

As mental disorders may accompany and correlate with each other [2], multi-task learning was applied for estimating suicide risks, where they used anxiety estimation, eating disorders, gender detection and many other tasks as auxiliary

---

[2] http://early.irlab.org/.
[3] https://www.reddit.com/.

tasks, and showed an enhancement in the results [1]. Other solutions involved the detection of "Crisis" like suicide and Self-harm joined with prediction explanation. The models they used were mainly based on interpretable neural network and attention mechanisms [8].

4   State of the Art Models in Similar Tasks

4.1   Traditional Machine Learning and Neural Networks

Earlier work on eRisk 2018 applied traditional machine learning like logistic regression ensemble classifier [17]. However, the best performances were achieved using Convolutional Neural Network (CNN) based models with word embeddings which we propose for experiments as well [17]. This model can be updated to include attention mechanism as in [8].

4.2   Multi-task Learning and Domain Adaptation

Multi-task learning is a machine learning approach based on combining multiple related learning tasks to improve the generalization performance of all auxiliary tasks [23]. eRisk sub-tasks are related to Anorexia and Self-harm detection, hence, as mentioned earlier, these disorders are mostly linked to other disorders like depression, personality disorders, substance abuse, Obsessive Compulsive Disorder and Bipolar Disorder. The difference in our case is that eRisk data is only labeled for the Anorexia task. Here we may experiment with a form of multi-task learning which is seen as a mix between multi-task-learning and domain adaptation where both labeled and unlabeled tasks can be joined [13]. Similarly, due to that Self-harm subtask has no training data, we could adapt the model for depression detection using a model trained on eRisk 2018 sub-task.

4.3   Hidden Markov Models

Hidden Markov Model (HMM) is a stochastic model that is generative and probabilistic. HMMs has a sequence of observable variables that is determined by other hidden unknown states [14]. The data provided for eRisk 2019 tasks are sequences of observations for each user that carry multiple features. After a number of observation in a specific sequence, there should be a prediction of the user mental state. Features that might be modeled for this task are Linguistic Inquiry and Word Count (LIWC) features [16], sentiments, stance, and Part-of-Speech tags.

5   Research Questions

To our knowledge, none of previous work have investigated robust models that work across domains of similar illnesses. In addition, none of them have looked at combining different sources of information to better address the issues. This includes combining different aspects of user generated content on SM such as content-based aspects, structural-based aspects and behavioural aspects. In this work we aim to tackle these points:

– Social posts can be regarded from different perspectives. Content of the post, such as a Reddit post or the comments to textual content is one prospect to be taken into account. In addition, the information about the author of the post like their replies' stances towards other posts and their communication network could bring other indicating aspects. These kind of information sources have been already investigated in other tasks that involved SM such as detecting and verifying online rumours [25], whereas mental health studies on SM were more restricted on psychological data analysis with limited computational engagement.
  However, it is difficult to find one method that fits to capture all information sources. For instance, [4] applied Hidden Markov Models(HMMs) on online rumours verification by modelling stance of replies towards social posts related to pre-defined rumours. By this, They showed that modelling stance transitions along with time information is superior in predicting the veracity of such posts.
  Nonetheless, it may be not as effective when another source of information is the learned target. **Thus, our first research question is concerned with the investigation of methods that fit best to a particular source of information.**
– Fitting a model based on a particular source of information of a particular event would more likely result a domain dependent model, i.e. it will perform well on the trained domain but fail to do so for a new one. Hence, Domain adaptation got a great attention lately [20,24]. In our work, we aim to explore general feature extraction that is not only suitable to our domain in hand, but works across similar domains. **Therefore, our second research question is concerned with investigating general features that combine knowledge which applies on many mental illnesses at the same time.**
– Finally, our aim is to construct a model that performs early detection of Anorexia and Self-harm. This implies combining methods that fit best to different information sources and then create a pipeline where these different methods work together on the task. **Hence, our last research question is concerned with the investigation of ways for incorporating such methods to achieve the best performance.**

6   Research Issues

This work is still in its early stages, hence we mainly seek advice that can get us better insights into the problem in hand and extend the suggested solutions. One of the issues is that in the next phase we need to look into particular sides of the task such as how can we extend the mental health information resources in a way that bears the ethical issues in mind and what are the features that make the models applicable? Another issue is what other methods in the literature that can be suggested for application?

References

1. Benton, A., Mitchell, M., Hovy, D.: Multi-task learning for mental health using social media text. arXiv preprint arXiv:1712.03538 (2017)
2. Coopers, P.: The Costs of Eating Disorders Social, Health and Economic Impacts. B-eat, Norwich (2015)
3. Corcoran, C.M., et al.: Prediction of psychosis across protocols and risk cohorts using automated language analysis. World Psychiatry **17**(1), 67–75 (2018)
4. Dungs, S., Aker, A., Fuhr, N., Bontcheva, K.: Can rumour stance alone predict veracity? In: Proceedings of the 27th International Conference on Computational Linguistics, pp. 3360–3370 (2018)
5. Guillaume, S., et al.: Characteristics of suicide attempts in anorexia and bulimia nervosa: a case-control study. PLoS One **6**(8), e23578 (2011)
6. Guillaume, S., et al.: Characteristics of suicide attempts in anorexia and bulimia nervosa: a case-control study. PloS One **6**, e23578 (2011)
7. Juarascio, A.S., Shoaib, A., Timko, C.A.: Pro-eating disorder communities on social networking sites: a content analysis. Eat. Disord. **18**(5), 393–407 (2010)
8. Kshirsagar, R., Morris, R., Bowman, S.: Detecting and explaining crisis. In: Proceedings of the Fourth Workshop on Computational Linguistics and Clinical Psychology - From Linguistic Signal to Clinical Reality, pp. 66–73. Association for Computational Linguistics (2017)
9. Losada, D.E., Crestani, F.: A test collection for research on depression and language use. In: Fuhr, N., et al. (eds.) CLEF 2016. LNCS, vol. 9822, pp. 28–39. Springer, Cham (2016). https://doi.org/10.1007/978-3-319-44564-9_3
10. Losada, D.E., Crestani, F., Parapar, J.: Overview of eRisk: early risk prediction on the internet. In: Bellot, P., et al. (eds.) CLEF 2018. LNCS, vol. 11018, pp. 343–361. Springer, Cham (2018). https://doi.org/10.1007/978-3-319-98932-7_30
11. Marchant, A., et al.: A systematic review of the relationship between internet use, self-harm and suicidal behaviour in young people: the good, the bad and the unknown. PLOS One **12**(8), e0181722 (2017)
12. Patchin, J.W., Hinduja, S.: Digital self-harm among adolescents. J. Adolesc. Health **61**(6), 761–766 (2017)
13. Pentina, A., Lampert, C.H.: Multi-task learning with labeled and unlabeled tasks. arXiv preprint arXiv:1602.06518 (2016)
14. Rabiner, L.R.: A tutorial on hidden Markov models and selected applications in speech recognition. Proc. IEEE **77**(2), 257–286 (1989)
15. Skegg, K.: Self-harm. Lancet **366**(9495), 1471–1483 (2005)
16. Tausczik, Y.R., Pennebaker, J.W.: The psychological meaning of words: LIWC and computerized text analysis methods. J. Lang. Soc. Psychol. **29**(1), 24–54 (2010)
17. Trotzek, M., Koitka, S., Friedrich, C.M.: Word embeddings and linguistic metadata at the CLEF 2018 tasks for early detection of depression and anorexia (2018). http://ceur-ws.org/Vol-2125/
18. Wang, T., Brede, M., Ianni, A., Mentzakis, E.: Detecting and characterizing eating-disorder communities on social media. In: Proceedings of the Tenth ACM International Conference on Web Search and Data Mining, pp. 91–100. ACM (2017)
19. Wang, Y., et al.: Understanding and discovering deliberate self-harm content in social media. In: Proceedings of the 26th International Conference on World Wide Web, pp. 93–102. International World Wide Web Conferences Steering Committee (2017)

20. Weiss, K., Khoshgoftaar, T.M., Wang, D.D.: A survey of transfer learning. J. Big Data **3**, 9 (2016)
21. Wolf, M., Theis, F., Kordy, H.: Language use in eating disorder blogs: psychological implications of social online activity. J. Lang. Soc. Psychol. **32**(2), 212–226 (2013)
22. Zdanow, C., Wright, B.: The representation of self injury and suicide on emo social networking groups. Afr. Sociol. Rev./Rev. Africaine de Sociologie **16**(2), 81–101 (2012)
23. Zhang, Y., Yang, Q.: A survey on multi-task learning. CoRR abs/1707.08114 (2017)
24. Ziser, Y., Reichart, R.: Pivot Based Language Modeling for Improved Neural Domain Adaptation. NAACL Submission, pp. 1–10 (2018)
25. Zubiaga, A., Aker, A., Bontcheva, K., Liakata, M., Procter, R.: Detection and resolution of rumours in social media: a survey. ACM Comput. Surv. (CSUR) **51**(2), 32 (2018)

Integration of Images into the Patent Retrieval Process

Wiebke Thode[(✉)]

University of Hildesheim, Universitaetsplatz 1, 31141 Hildesheim, Germany
thodewi@uni-hildesheim.de
http://www.iwist.org

Abstract. Specialized patent retrieval systems mostly use textual information which is difficult enough because of the specialized characteristics of the text. However, in patents there also are drawings which show the invention. Empirical research has shown that patent experts can use these images to determine relevance very quickly. However, these drawings are binary and sometimes abstract and other times very specific; therefore there has not been an effective way to include the visual information into the information retrieval process. In addition, the number and the quality of drawings differs vastly even inside patent classes. This work focuses on the inclusion of images into the patent retrieval process using a combination of visual and textual information. With this multimodal approach it will hopefully be possible to achieve better results than just by using one modality individually. The goal is to develop a prototypical system using an iterative user-centered process.

Keywords: Patent retrieval · Multimodal information retrieval ·
Visual information seeking

1 Motivation for the Proposed Research

"Patent searchers are often expected to find (almost) all relevant documents in a very limited time frame. The sheer amount of already available patent documents makes ways of supporting the professionals indispensable" ([7], p. 30).

This process is complicated by the characteristics of the patent documents themselves as they involve specific language, e.g. the extensive use of neologisms which often is one of the main reasons why it can be such a challenge to retrieve patent information [3]. Considering the difficulties with using the text of a patent document to facilitate the retrieval of specific information, the question is what elements the experts use in order to understand the content of a document. Moreover, how can these elements be used in order to achieve better performance of the retrieval systems? The basis for information retrieval in this domain mostly is textual information. However, a patent normally includes visual elements (technical drawings, flowcharts, graphs etc.). So far, visual elements are only displayed with the results. Herein lies the motivation for this

© Springer Nature Switzerland AG 2019
L. Azzopardi et al. (Eds.): ECIR 2019, LNCS 11438, pp. 359–363, 2019.
https://doi.org/10.1007/978-3-030-15719-7_49

project: How can the visual elements be included in the patent retrieval process and which effects do they yield?

Empirical work within the patent domain focuses on questioning patent professionals and/or observing their work processes (see e.g. [1,6]). This is necessary since the work processes of patent professionals cannot be compared to "normal" users, as the patent professionals operate in highly specialized fields. Observations (e.g. by [1]) have shown that the patent experts are quicker to adapt new functionalities if they deem them useful in their search strategies and the drawings in the patents are often one of the first things they look at when judging the relevance of a result (interview result). The drawings are described as "paramount to the understanding of the invention and yet almost completely unsearchable" ([9], p. 1). The main motivation for this research is therefore to facilitate the integration of images into the search process of patent experts. This is done by eliciting requirements through expert interviews and the prototypical development of a patent information retrieval system. The initial interviews have indicated that the experts do use the drawings to judge relevance. It is therefore interesting, how the drawings are used in searches in the patent domain. The use of CBIR methods have shown to have some limitations when they are used on patent images (see e.g. [11]), therefore the usage of multimodal methods (the combination of text and images) is pursued here.

2   Background and Related Work

Patents are "multi-page, multi-modal, multi-language, semi-structured, and metadata rich documents" however it is necessary to construct complex queries in order to retrieve the information within the patents ([10], p. 1). The "state-of-the-art performance in automatic PR [...] [is] still around average. These observation motivates the need for interactive search tools which provide cognitive assistance to patent professionals with minimal efforts." ([10], p. 1). Research shows that the images are used by the experts when analyzing the patents. So why is it not yet included in commercial systems? The main reason lies in the characteristics of the images themselves. By law (except for some specific design patents) the images have to be "bi-level images (black and white), colorless and mostly textureless" ([3], p. 2). This means the methods have to be adapted to the characteristics of the drawings. This mostly leaves the exploitation of shape information and spatial information. Common CBIR algorithms cannot be used since they mostly rely on color and texture information [3]. The work on exploiting contextual and spatial information is done by [4] who use contextual local primitives, a method which is "based on the detection of the junction and end points, classification of the local primitives to local primitive words and establishment of the geodesic connections of the local primitives" ([4], p. 9111). Experiments have shown this method outperforming other approaches e.g. the SIFT-based methods. Other than that, "Deep learning, and particularly deep convolutional neural networks (CNN), have become an extremely powerful tool in computer vision" ([5], p. 237). So far deep learning methods were rarely used on patent images, with one exception being [8] who use Convolutional Neural Networks on patent images achieving an accuracy of 51% on the training data (3000 patents) and an accuracy of 30% on the test data (800 patents), indicating that more training is needed [8]. However, [8] simply used all patents they could crawl without first choosing a specific IPC to work with. Their corpus includes patents registered in Russia from 1994 on [8]. It is possible that the results could be improved if there would be a focus on a specific class of patents.

The approach of this project attempts to leverage the potential benefits of including textual information specifically referring to the images. One such multimodal approach has been attempted by [2] who use deep neural networks to pursue a multimodal approach. The experiments here have not been conducted in the patent domain.

3   Description of Proposed Research

The multimodal approach combines the visual with the textual information in order to compensate for the deficiencies of using one of them individually. The multimodal approach of [2] has shown to outperform other methods in a total of six retrieval tasks with "statistically significant differences" ([2], p. 7). However, the question remains how to efficiently combine visual and textual information in the patent domain. Following the user centered approach, the expert are interviewed to elicit points on how this can be achieved.

This is done in order to answer the following research questions:

RQ1: Which requirements do the users have of a system which includes patent drawings? Which modality (image/text only or a combination) do the users prefer in practice?

RQ2: How do expert users include drawings into their information seeking process and how can this be implemented into a prototypical system?

RQ3: Can the combination of different modalities be beneficial in the patent domain?

In order to answer RQ1 and RQ2, interviews were conducted with experts. As these were preliminary interviews, the selection criteria were very broad. The only criterion was that the expert works within the patent domain professionally. There was no differentiation between speciality, however all of them agreed that the drawings are more important in utility patents which cover products, processes and machines. Going forward, this will be the focus. A total of four expert interviews were conducted so far with more interviews planned iteratively after the development of initial functionalities. So far, all the experts can see a benefit when including images into their retrieval process, as they often use them during their search. An important result of these interviews is that the experts know about the limitations when trying to retrieve patent drawings. Therefore, there is a clear preference for the use of visual information in combination with textual information which has a direct connection to the images.

This text might have stronger substantive significance during the retrieval process and the experts expect it to help them finding other patent documents. It might provide them with new keywords or new concepts to retrieve the patents that cannot be found by using citations or their initial keywords. Another result is that the experts cannot yet imagine how the drawings can be integrated into the retrieval process. It is therefore necessary to evaluate multiple options in order to elicit the best features.

RQ3 will be answered by iteratively developing functionalities based on the interviews and evaluate them further with user involvement by conducting more interviews. How these functionalities can be implemented is currently worked on. After concluding the work on RQ3, an evaluation of the prototype will be conducted with expert users using interviews or a focus group.

4   Research Methodology

The thesis is located in Information Science where an empirical research approach is pushed. The Information Science research paradigm involved users at any part of the research process. First the users' requirements are elicited using expert interviews with a semi-structured interview guide. The interviews are analyzed using a structured content analysis which results in a list of requirements and ideas which can be incorporated into the prototypical system. The retrieval functionality is supposed to be implemented using a multimodal approach, which is a clear result from the interviews. This prototype has to be evaluated by actual users which is why focused interviews or a focus group will be conducted. The methodology therefore uses a user-centered approach when developing a prototypical system which incorporates multimodality in patent retrieval. This approach has the distinct advantage that the user is more involved in the development process so that it is possible to elicit the requirements the specific users have for the system. This has already been done. One limitation of the first interviews is that the term "patent expert" is very loosely defined and the expertise is very heterogeneous. To qualify this point, the focus of this work is on utility patents, which still is a very broad field. This means all future interviews will be conducted with experts who work with this type of patent. The iterative nature of the research design ensures the user involvement in the whole process as the evaluation results will be used to improve the prototypical system. The prototypical system has to be evaluated iteratively by the user.

5   Issues for Discussion at the Doctoral Consortium

During the consortium, I want to be able to get another point of view on my work as well as being able to discuss different ideas and approaches. When I will present at the consortium I will have developed first features and have some prototypical functionalities. This consortium will give me the opportunity to present them to Information Retrieval experts and discuss different aspect, e.g. possible improvements of the methodology or first results. One specific question is about the query modality: How will the users effectively query visual information and how can this be implemented in the patent domain.

References

1. Azzopardi, L., Joho, H., Vanderbauwhede, W.: A survey on patent users search behavior, search functionality and system requirements (2010). https://www.semanticscholar.org/paper/A-Survey-on-Patent-Users-Search-Behavior%2C-Search-Azzopardi-Joho/a9f927787f0236fd04ebe706ff1b1a37ce502ff7?tab=abstract
2. Balaneshin-kordan, S., Kotov, A.: Deep neural architecture for multi-modal retrieval based on joint embedding space for text and images. In: Proceedings of the Eleventh ACM International Conference on Web Search and Data Mining, WSDM 2018, pp. 28–36. ACM, New York (2018). https://doi.org/10.1145/3159652.3159735
3. Bhatti, N., Hanbury, A.: Image search in patents: a review. Int. J. Doc. Anal. Recogn. (IJDAR) 16(4), 309–329 (2013). https://doi.org/10.1007/s10032-012-0197-5
4. Bhatti, N., Hanbury, A., Stottinger, J.: Contextual local primitives for binary patent image retrieval. Multimedia Tools Appl. 77(7), 9111–9151 (2018). https://doi.org/10.1007/s11042-017-4808-5
5. Gordo, A., Almazán, J., Revaud, J., Larlus, D.: End-to-end learning of deepvisual representations for image retrieval. Int. J. Comput. Vis. 124(2), 237–254 (2017). https://doi.org/10.1007/s11263-017-1016-8
6. Hansen, P.: Task-based information seeking and retrieval in the patent domain: processes and relationships. Ph.d. thesis. University of Tampere, Tampere (2011)
7. Jürgens, J.J., Hansen, P., Womser-Hacker, C.: Going beyond CLEF-IP: the 'reality' for patent searchers? In: Catarci, T., Forner, P., Hiemstra, D., Peñas, A., Santucci, G. (eds.) CLEF 2012. LNCS, vol. 7488, pp. 30–35. Springer, Heidelberg (2012). https://doi.org/10.1007/978-3-642-33247-0_4
8. Kravets, A., Lebedev, N., Legenchenko, M.: Patents images retrieval and convolutional neural network training dataset quality improvement. In: Proceedings of the IV International Research Conference Information Technologies in Science, Management, Social Sphere and Medicine (ITSMSSM 2017). Atlantis Press, Paris, France, 12 May 2017–12 August 2017. https://doi.org/10.2991/itsmssm-17.2017.59
9. Lupu, M., Schuster, R., Mörzinger, R., Piroi, F., Schleser, T., Hanbury, A.: Patent images - a glass-encased tool. In: Lindstaedt, S., Granitzer, M. (eds.) Proceedings of the 12th International Conference on Knowledge Management and Knowledge Technologies - i-KNOW 2012, p. 1. ACM Press, New York, New York, USA (2012). https://doi.org/10.1145/2362456.2362477
10. Shalaby, W., Zadrozny, W.: Patent retrieval: a literature review. http://arxiv.org/pdf/1701.00324v1
11. Vrochidis, S., Papadopoulos, S., Moumtzidou, A., Sidiropoulos, P., Pianta, E., Kompatsiaris, I.: Towards content-based patent image retrieval: a framework perspective. World Patent Inf. 32(2), 94–106 (2010). https://doi.org/10.1016/j.wpi.2009.05.010. http://www.sciencedirect.com/science/article/pii/S0172219009000489

Dialogue-Based Information Retrieval

Abhishek Kaushik$^{(\boxtimes)}$

ADAPT Centre, School of Computing, Dublin City University, Dublin 9, Ireland
`abhishek.kaushik2@mail.dcu.ie`

Abstract. Conversational search presents opportunities to support users in their search activities to improve the effectiveness and efficiency of search while reducing cognitive load. Since conversation is a natural means of human information inquiry, framing the information retrieval process within a dialogue is expected to make the search process more natural for the user, in terms of query entry, interaction to locate relevant content, and engaging with the system output. My PhD Research project seeks to make progress toward realizing the vision of conversational search systems.

Keywords: Conversational search · User search behaviour · Dialogue-based search

1 Introduction

In the operation of a current information retrieval (IR) system, a user enters a text query describing their information need. In response to this, the system returns a list of potentially relevant items ranked in order of their estimated likelihood of relevance [2]. The user then selects one or more relevant items with which to satisfy their information need. However, often the user is not able to fully describe what they want to know, their query may be ambiguous or fail to match well with the content of relevant items. If any of these situations arose while seeking information from a human intermediary, the natural response would be to enter into a dialogue to resolve the problem [2]. The objective of this research is to advance IR systems by developing and implementing a framework to support the use of dialogue-based approaches in IR.

2 Motivation

Current Search Systems: In order to satisfy their information need using a current search system, a user may need to perform multiple passes using queries modified based on information gained in previous searches. This strategy has various limitations:

Supported by Science Foundation Ireland as part of the ADAPT Centre at DCU (Grant No. 13/RC/2106) www.adaptcentre.ie.
A. Kaushik—Under the Supervision of Gareth J. F. Jones.

L. Azzopardi et al. (Eds.): ECIR 2019, LNCS 11438, pp. 364–368, 2019.
https://doi.org/10.1007/978-3-030-15719-7_50

1. The user must seek to completely describe their information need in their query. Insufficient knowledge of the subject of their information need or the available search targets can make this process frustrating and inefficient.
2. There is a high cognitive load in forming such queries.
3. The IR system is designed to satisfy the information need in a single pass based on the query entered.
4. The user generally inspect multiple returned items to find relevant information.

The Potential for Conversational Search: Conversational search seeks to support more natural interaction between users and content information archives via dialogue-based engagement. Existing studies have been conducted to explore the potential for conversational search using humans as an intermediate agent between user and the search engine. For example, the work by Trippas *et al.* at RMIT [5,6] which focuses on speech only based search for users seeking information on general topics such as those that a user might pose to a web search engine [5,6].

By contrast, my research starts from the analysis of current standard IR systems, and seeks to understand the scope for conversational interventions in the search process and how these might be facilitated. Since the incorporation of conversational methods into IR processes is currently poorly explored, my research assumes the use of text-based engagement with IR systems leaving alternative engagement using speech interfaces for future studies.

Conversational and Dialogue Systems: Outside the area of IR, there is currently much interest in dialogue and conversational systems for engagement with information systems. These applications generally focus on fixed tasks in very narrow domains. In search, while the scope of the user tasks can generally be well defined, the topics and material over which the system must operate are diverse and unstructured. This poses unique challenges not faced by standard conversational systems which perform clearly defined tasks of limited scope.

In conversational search, we can think in terms of an agent taking the role of a human intermediary supporting the user's search activities. Such a conversational search agent should exhibit autonomic behaviour with the ability of self-adaptation according to the searcher's information seeking activities and potentially relevant content identified. It is not clear what the most appropriate technology to drive such an agent will be, but given the potential to define the activities and responsibilities, and the absence of large training data sets, I propose to explore rule-based approaches, at least as a mechanism to bootstrap the abilities of the agent. The study conducted by Stein *et al.* [4] explained dialogue strategies during information seeking activities in a collaborative environment.

Cognitive Models of the Search Process: In order to develop sound conversational search methods, it is important to examine the relevant cognitive issues in the search process. Early contributions in understanding and modeling information needs and activities of searchers were made in Belkin's work on the development of the Anomalous States of Knowledge (ASK) model [1]. This highlights the difficulties of users in specifying the details of their information needs for engagement with IR systems. Various extensions to the ASK models taking alternative or complementary approaches to model information seeking have been developed since then, e.g. Kuhlthau's Information Search Process (ISP) [3] and Vakkari's [7] learning model of search. I propose to explore the integration of these models in the rule base of an agent, in order to model and direct support of the user's search activities [2].

3   Research Questions

In order to undertake my PhD research, I have identified the following research questions which I am seeking to address.

RQ1: How can conversational interventions be used to support user search activities? What opportunities are there for conversational support in current search engines? I carried out and reported a study examining the behaviour of users when using a standard web search engine, designed to enable me to identify opportunities to support their search activities using a conversational agent [2].

RQ2: What are the requirements of a conversational search interface? User cognitive load: The whole search process places cognitive load onto the user. We observed users can take a lot of time to read a single long document to satisfy their information need [2]. A search agent could potentially reduce this cognitive load by making suggestions to the user, e.g. of extracted significant information which assist in either understanding or resolving their information need.

Challenges and limitations: We need to study engagement with potential conversational interfaces in user-based studies and consider how our system can maintain a record of previous conversations.

RQ3: How should my work on conversational search be implemented and evaluated? Consideration of the requirements of my conversational search system will form the basis of the design and implementation of technologies and prototypes to enable them. In addition, methods for evaluation will need to be explored.

4   Methodology

The overarching approach to my PhD project is propose and evaluate conversational search methods into established search processes. To achieve this I am working on the following elements:

1. Examining the actions undertaken by searchers using current systems, relating these to cognitive models of the search process, and proposing how the functionality of a conversation search agent might be used to enable more effective and satisfying search experiences.

2. Proposing strategies to enable conversational strategies within the search process.
3. Building prototypes of the proposed conversational search components, and their incorporation within a standard search engine framework using an open source toolkit.

4.1 Current Status

Completed Work

1. Initial review of existing relevant literature.
2. Study of user search behaviours using current search tool [2].

Current Work

1. Development of methods for automatic content analysis for use in conversational engagement, e.g. highlighting significant and diverse elements of retrieved content.
2. Proposal of dialogue strategies and agent models for use in conversational search.

These are currently under active investigation with some parts already implemented.

Future Work

1. Development of evaluation strategy and test data.
2. Implementation of models of prototype conversational search components and system.
3. Evaluation of proposed models.

5 Research Issues for Discussion at the Doctoral Consortium

Conversational search is a rapidly emerging research area which is currently attracting considerable interest from the information retrieval research community.

The scope of potential research in conversational search is very broad. As this time, I need to work towards determining the exact form of the main focus in my PhD research. As part of deciding the topics that I will concentrate on, I am interested in identifying potential risks in my plans, and in developing potential response plans for these.

Regardless of the exact topics investigated, evaluation is an important component of experimental IR research. The emerging nature of research in conversational search means that there are no standard evaluation methods or datasets available for work in this area. I am thus very interested to get input on this aspect of my work.

References

1. Belkin, N.: Anomalous states of knowledge as a basis for information retrieval. Can. J. Inf. Sci. **5**, 133–143 (1980)
2. Kaushik, A., Jones, G.J.F.: Exploring current user web search behaviours in analysis tasks to be supported in conversational search. In: Second International Workshop on Conversational Approaches to Information Retrieval (CAIR 2018), July 2018, Ann Arbor, Michigan, USA (2018)
3. Kuhlthau, C.C.: Inside the search process: information seeking from the user's perspective. J. Am. Soc. Inf. Sci. **42**(5), 361–371 (1991)
4. Stein, A., Maier, E.: Structuring collaborative information-seeking dialogues. Knowl.-Based Syst. **8**(2–3), 82–93 (1995)
5. Trippas, J.R., Spina, D., Cavedon, L., Joho, H., Sanderson, M.: Informing the design of spoken conversational search: perspective paper. In: Proceedings of the 2018 Conference on Human Information Interaction and Retrieval, CHIIR 2018, pp. 32–41. ACM, New York (2018). https://doi.org/10.1145/3176349.3176387
6. Trippas, J.R., Spina, D., Cavedon, L., Sanderson, M.: How do people interact in conversational speech-only search tasks: a preliminary analysis. In: Proceedings of the 2017 Conference on Conference Human Information Interaction and Retrieval, CHIIR 2017, pp. 325–328. ACM, New York (2017). https://doi.org/10.1145/3020165.3022144
7. Vakkari, P.: Searching as learning: a systematization based on literature. J. Inf. Sci. **42**(1), 7–18 (2016)

Dynamic Diversification for Interactive Complex Search

Ameer Albahem[✉]

RMIT University, Melbourne, Australia
ameer.albahem@rmit.edu.au

Abstract. Many real-world searches are examples of complex information needs such as exploratory, comparative or survey oriented searches. In these search scenarios, users engage interactively with search systems to tackle their information needs. On one hand, user interactions can be leveraged to induce search intents and reformulate queries. On the other hand, the nature of these scenarios introduces constraints in the search process. For instance, systems are expected to satisfy the information needs earlier in the interaction. This research investigates a dynamic diversification approach that observes user interactions and dynamically changes its behaviour in response. In this research, we investigated how dynamic diversification methods should be evaluated. In that regard, we studied and analysed a wide range of offline metrics that model topical relevance novelty and user effort. In addition, this research investigates how to exploit user interactions to develop dynamic diversification methods. In particular, we study the impact of the different dimensions of user relevance feedback, the internal components of relevance feedback algorithms and diversification methods on the overall performance of dynamic diversification methods. Lastly, we intend to measure user satisfaction with these methods using a controlled user study.

Keywords: Evaluation · Dynamic search · Relevance feedback · Diversification

1 Motivation

Many search tasks are complex, exploratory in nature and run over multiple iterations. Users in such tasks provide signals of search intents and relevance over the entire search session. In multi-page search [15], user clicks of documents and dwell times are implicit signals of relevance. In professional search such as patent prior art, users are willing to rate relevance of search results if search systems can learn and improve [17]. In conversational search, users might instruct search systems to focus or discard certain content. These tasks are examples of *dynamic search* where systems can learn from user interaction and dynamically modify their outputs.

In these search scenarios, factors such as topical relevance, novelty and the amount of user effort impact the overall user satisfaction with search results

© Springer Nature Switzerland AG 2019
L. Azzopardi et al. (Eds.): ECIR 2019, LNCS 11438, pp. 369–374, 2019.
https://doi.org/10.1007/978-3-030-15719-7_51

[8]. Managing relevance and novelty have been studied extensively within search result diversification where the aim is to tackle query ambiguity (e.g. apple the fruit vs Apple the company) and underspecified information needs [14]. The essence of search result diversification is how to rank documents to balance topical relevance and novelty (covering various intents or subtopics), so the ranking can satisfy users with different query intents in a single-iteration search session. In an interactive scenario or medium, diversification methods could leverage user feedback and session history to deliver better results. For instance, user explicit feedback can be used to penalize unseen documents that are similar to the seen non-relevant documents.

In this research, we hypothesize that a diversification approach that changes its output dynamically based on search session history (user feedback) will improve user satisfaction. We call this approach dynamic diversification.

2   Research Questions

Several effectiveness metrics have been proposed to model topical relevance, novelty and the amount of user effort in their formulations. As these metrics address multiple (and sometimes competing) aspects of performance, it is unclear whether they behave as intended by rewarding systems that retrieve relevant and novel documents while minimizing user effort. As a result, the first research question we investigate is:

RQ 1: How can interactive complex search be evaluated? There are a lot of metrics that aim to evaluate some or all of the dimensions of dynamic search quality, such as topical relevance, novelty and user effort. In answering this question, we investigate which of these metrics model which dimensions as well as the intuitiveness and reliability of these metrics in capturing the dimensions.

Depending on the search task, studies have utilized different approaches to tackle user information needs by leveraging various types and granularity-levels of user feedback. Sloan and Wang [15] instantiated a Markov Hidden model in an exploratory multi-page search where they used simulated user clicks to update and rank documents in later pages. In the TREC Dynamic Domain track [17], many runs used various combinations of relevance feedback algorithms, feedback representations and diversification methods to diversify search results over multiple iterations. However, it is not clear from these studies what contributes to the performance of these methods. For instance, Moraes et al. [10] and Joganah et al. [9] used passage- and subtopic-level simultaneously in their runs[1], but it is not clear whether the observed improvements were due to the exploitation

---

[1] Topic-level feedback is when a user rates the relevance of an information unit to the general topic of the user information need. In a subtopic-level feedback, the relevance judgments are based on the subtopics of the query. Document-level feedback is a user rates the overall relevance of documents (whole content) to the topic or a subtopic of the information need. A passage-level is when relevance judgments are provided based on passages (text chunks) from the documents.

of the subtopic-level relevance information, passage-oriented representation, utilization of diversification methods or simply using multiple queries in their runs. Furthermore, an overlap exists between the internal components of diversification frameworks and relevance feedback algorithms. For instance, the Explicit Query Aspect Diversification [13] (xQuAD) framework relies on its sub-query representations to diversify search results. Similarly, researches have found using multiple queries improves performance in interactive document retrieval [4]. In this research, we systemically study the impact of the components on dynamic diversification methods. Our second research question is:

RQ 2: How do the user feedback dimensions, relevance feedback algorithms and diversification methods impact the performance of dynamic diversification?

The outcomes of **RQ 1** informed choosing metrics to evaluate methods in **RQ 2**. Nevertheless, the evaluation is offline using metrics with underlined assumptions about user behavior. To test our hypothesis, we intend to study the quality of dynamic diversification methods from a user perspective. In particular, we aim to answer the following research question:

RQ 3: How do users rate using dynamic diversification to tackle complex information needs?

3   Research Methodology and Proposed Experiments

3.1   Results So Far

To answer **RQ 1**, we performed a meta-analysis study of dynamic search metrics. This analysis consists two experiments. In the first experiment, we performed an axiomatic-based analysis, which was published as a long paper in ADCS 2018 [2]. In the second experiment, published in ECIR 2019 [1], we adapted two meta-analysis frameworks: the Intuitiveness Test [12] and Metric Unanimity [3] to study dynamic search metrics. We also studied how well these two approaches agree with each other.

3.2   Research Methodology and Experiments

Anatomy of Dynamic Diversification Component Performances. To answer **RQ 2**, we plan to instantiate the point-grid performance analysis framework of Ferro and Silvello [7] on dynamic diversification. In particular, we will first build a grid of points (GoP) containing all the combinations of dynamic diversification component configurations. Then we will perform General Linear Mixed Model (GLMM) and analysis of variance (ANOVA). We have identified four main components in dynamic diversification: initial ranking, diversification methods, relevance feedback methods and feedback dimensions. In particular, we used the following configurations:

- Initial ranking: query language model, Sequential Dependence model, BM25 and Vector Space Model.
- Relevance feedback algorithms: Relevance Model [18], the Rocchio method [11], Mixture model [4].
- Diversification: Maximum Marginal Relevance [5], xQuAD [13], PM2 [6].
- Feedback dimensions: which dimensions of the user feedback a dynamic system utilizes to update its dynamic ranking. This can be decomposed into multiple dimensions: feedback level (topic-level vs subtopic-level relevance), feedback type (positive vs negative), feedback unit (document vs passage).

We will instantiate the framework in the TREC Dynamic Domain collections, TREC Web Diversity, and TREC Session collection.

User-Based Evaluation of Dynamic Diversification. As described previously, dynamic diversification can be instantiated in multiple search scenarios, such as multi-page search, specialized domain search and conversational search. To answer **RQ 3** and evaluate dynamic diversification from the user perspective, we will instantiate dynamic diversification in the third scenario. In particular, we aim to answer the following sub questions:

1. Does dynamic diversification lead to a better user satisfaction as compared to non-dynamic methods?
2. How well do dynamic search metrics correlate with user satisfaction?

One of the challenges in studying user satisfaction in conversational search is that user satisfaction is affected by factors related to presentation modality (speech vs text), conversational aspect (dialogue management) and system components (e.g. Automatic Speech Recognition quality and Natural Language Understanding).

One approach to tackle this is the experimental setup used in Trippas et al. [16] where two persons take the roles of the search process actors: the user and the search system. The first person acts an information receiver (the search user), whereas the other person acts as an information retriever (the mediator between the search system and the user).

Another approach is to evaluate dynamic diversification in an interactive search interface using different configurations. In particular, given two systems (a static and a dynamic systems), we could measure user satisfaction under various variables, such as presentation modality and communication method. The first focuses on how search results are presented to users. There are two options: text and speech. The second variable focuses on how user gives inputs and receives outputs. Options are buttons and voice commands.

We will recruit users to perform search tasks using two systems: a dynamic diversification based system and a system employing the best static method from the previous experiments (**RQ 2**). At the end of each session, users will be asked to rate the overall search system performance. After collecting data, we measure correlations using Pearson's r and Spearman's p. We evaluate a metric by how well it predicts user-rated performance. As for comparing the dynamic diversification system to the static system, the average user rating of the system performance will be used to judge whether dynamic diversification could lead to a better user satisfaction.

Acknowledgement. This research was partially supported by Australian Research Council (projects LP130100563 and LP150100252), and Real Thing Entertainment Pty Ltd.

References

1. Albahem, A., Spina, D., Scholer, F., Cavedon, L.: Meta-evaluation of dynamic search: How do metrics capture topical relevance, diversity and user effort? In: Proceedings of ECIR (2019, to appear)
2. Albahem, A., Spina, D., Scholer, F., Moffat, A., Cavedon, L.: Desirable properties for diversity and truncated effectiveness metrics. In: Proceedings of Australasian Document Computing Symposium, pp. 9:1–9:7 (2018)
3. Amigó, E., Spina, D., Carrillo-de Albornoz, J.: An axiomatic analysis of diversity evaluation metrics: introducing the rank-biased utility metric. In: Proceedings of SIGIR, pp. 625–634 (2018)
4. Brondwine, E., Shtok, A., Kurland, O.: Utilizing focused relevance feedback. In: Proceedings of SIGIR, pp. 1061–1064 (2016)
5. Carbonell, J., Goldstein, J.: The use of MMR, diversity-based reranking for reordering documents and producing summaries. In: Proceedings of SIGIR, pp. 335–336 (1998)
6. Dang, V., Croft, W.B.: Diversity by proportionality: an election-based approach to search result diversification. In: Proceedings of SIGIR, pp. 65–74 (2012)
7. Ferro, N., Silvello, G.: Toward an anatomy of IR system component performances. J. Assoc. Inf. Sci. Technol. **69**(2), 187–200 (2018)
8. Jiang, J., He, D., Allan, J.: Comparing in situ and multidimensional relevance judgments. In: Proceedings of SIGIR, pp. 405–414 (2017)
9. Joganah, R., Khoury, R., Lamontagne, L.: Laval university and lakehead university at TREC dynamic domain 2015: combination of techniques for subtopics coverage. In: Proceedings of TREC (2015)
10. Moraes, F., Santos, R.L.T., Ziviani, N.: UFMG at the TREC 2016 dynamic domain track. In: Proceedings of TREC (2016)
11. Rocchio, J.: Relevance Feedback in Information Retrieval. Prentice Hall, Upper Saddle River (1971)
12. Sakai, T.: Evaluation with informational and navigational intents. In: Proceedings of WWW, pp. 499–508 (2012)
13. Santos, R.L., Macdonald, C., Ounis, I.: Exploiting query reformulations for web search result diversification. In: Proceedings of WWW, pp. 881–890 (2010)
14. Santos, R.L., Macdonald, C., Ounis, I.: Search result diversification. Found. Trends Inf. Retrieval **9**(1), 1–90 (2015)
15. Sloan, M., Wang, J.: Dynamic information retrieval: Theoretical framework and application. In: Proceedings of ICTIR, pp. 61–70 (2015)

16. Trippas, J.R., Spina, D., Cavedon, L., Sanderson, M.: How do people interact in conversational speech-only search tasks: a preliminary analysis. In: Proceedings of CHIIR, pp. 325–328 (2017)
17. Yang, H., Frank, J., Soboroff, I.: TREC 2015 dynamic domain track overview. In: Proceedings of TREC (2015)
18. Zhai, C., Lafferty, J.: Model-based feedback in the language modeling approach to information retrieval. In: Proceedings of CIKM, pp. 403–410 (2001)

The Influence of Backstories on Queries with Varying Levels of Intent in Task-Based Specialised Information Retrieval

Manuel Steiner[✉]

RMIT University, Melbourne, Australia
manuel.steiner@rmit.edu.au

Abstract. Various ways of determining ambiguity of search queries exist for general search. Relevancy of result documents for searches however is not determined by the query alone. The current user intent is what drives a search and determines if results are useful. Intent ambiguity describes queries that might have multiple intents. Conventional disambiguation methods might not work in specialised search where a goal is usually similar or the same (e.g. finding job offerings in job search). Research described in this document investigates how to determine single and possible multi-intent queries for job search and how contextual information, especially backstories, affect the job search process. Results will lead to a better understanding of how important backstories and the handling of intent ambiguity are in specialised information retrieval. The importance of test collections with built-in ambiguity to better test performance will also be indicated. The proposed research is conducted with data from a major Australian job search platform.

Keywords: Information need · Backstory · Intent ambiguity

1 Motivation

Ambiguity in information retrieval processes is a common problem which retrieval systems need to handle to provide suitable results for search queries. In case of word-sense ambiguity, the query itself can have multiple meanings [13]. In order to detect and handle such queries, thesauri in various forms can be used to either concentrate on one specific meaning or incorporate documents matching different meanings into the search results. An example of an online thesaurus service is WordNet [15]. The online encyclopedia Wikipedia employs disambiguation pages, on which the user can select a specific concept in case the search term has multiple meanings [18]. An other form of ambiguity is intent ambiguity. This means that different users have different intents in mind when they conduct a search with the same search query [12]. It is more difficult for IR systems to deal with this type of ambiguity because the user's intent is not

© Springer Nature Switzerland AG 2019
L. Azzopardi et al. (Eds.): ECIR 2019, LNCS 11438, pp. 375–379, 2019.
https://doi.org/10.1007/978-3-030-15719-7_52

known. This intent is expressed in search queries which might not convey the complete information need. In order to handle queries with multiple possible intents more effectively, contextual information is required.

The proposed research looks at how intent ambiguity can be determined in specialised, task-based IR. In such domains, conventional disambiguation measures might be ineffective. Furthermore, this research investigates how backstories influence the search process. A backstory is a narrative of a searcher's current situation that determines the intent [5]. This intent leads to a formulated search query that is processed by an IR system. Backstories are not only potentially useful for query disambiguation but are the standard way in general IR test collections to convey a search scenario or intention. Backstories from general test collections however might not be useful in specialised IR since the task is usually different (e.g. specifically finding a job in the case of job search). Thus, new backstory collections are necessary for test collections and research. Job search is chosen as instantiation of specialised IR due to the availability of production search log data from a major Australian job search platform.

2   Related Work

We will present existing literature related to ambiguous queries as well as work associated with backstories.

Ambiguity in Information Retrieval. For ambiguous queries, different result documents are relevant depending on the information need [12]. Next to word-sense ambiguity databases, such as WordNet [15], multiple approaches exist to mine ambiguous queries from search engine logs. The use of overall click entropy is described in Mei and Church [14]. An extension is the user averaged click entropy by Wang and Agichtein [19]. Query reformulations by multiple users might also indicate ambiguity [17]. Phan et al. [16] examined if longer search queries imply a narrower information need. They found that broad needs correspond to shorter queries but narrower needs are not necessarily expressed in longer queries. Hafernik and Jansen [11] manually classified queries into the categories broad and narrow based on occurrences of certain features. The findings concur with these of Phan et al. [16]. Sanderson [18] explained that test collections represent topics unambiguously. He showed that ambiguous queries are common in modern search and argued that test collections with built-in ambiguity are needed. IR systems might employ result diversification techniques to deliver results relevant to different meanings or intents of ambiguous queries [1,7–10].

The Importance of Backstories. To evaluate the relationship between search query length and information need specificity, Phan et al. [16] had experiment participants think of backstories and suitable search queries. Bailey et al. [4] manually created backstories for TREC topics and crowd-sourced multiple queries for each backstory to create a test collection with query variability. Backstories also constitute an integral part in an alternative evaluation method developed by Borlund and Ingwersen [5]. In comparison to TREC environments, where topic narratives provide clear instructions about what is considered relevant, Borlund and Ingwersen [5] let users formulate their own search queries and determine relevant results. The queries are formulated based on what they call *simulated work task situations* (a description of a user's current situation that requires them to use an IR system). It is used to describe the environment to the users and make them understand the problem that needs to be solved. Users will generate different search queries and deem different results relevant based on their own interpretation of the information need. This better resembles real-life IR system usage. In follow-up research, Borlund and Ingwersen [6] demonstrated that real information needs can be substituted with simulated ones.

3 Proposed Research

In this section, we present the proposed research questions and methodology.

3.1 Research Questions

The following main research questions are proposed during the candidature. All research questions are investigated in the area of specialised IR, more specifically job search. A comparison to other search areas would be a valuable addition.

RQ1 – In Which Ways Do Different Types of Queries Affect Backstories Associated with Searches? Does a varying degree of diversity in the background contexts of searchers exist, depending on if a search query has a single, specific or multiple intents? If so, it is important to evaluate search engines with suitable test collections that account for these situations. Relevant documents should change with the user intent. It would also mean that search result diversification is a useful method to cater for multiple different intents.

RQ2 – How Do Backstories Affect Relevance Judgements? Result relevance would presumably change for ambiguous queries when considering contextual information. If backstories are significantly important for relevance judgements, the suggestion of more diverse test collections would be strengthened [18].

RQ3 – How Do Backstories Affect Search Query Formulations? Do diverse query result sets lead to more specific queries with specific backstories in mind, compared to ambiguous queries without knowledge about the search situation? Do diverse backstories for a given search result lead to more diverse queries as opposed to similar ones that convey closely related scenarios? Do diverse backstories encourage the formulation of ambiguous or specific queries? Are query filters used if they are present in the backstory?

3.2   Methodology

After determining a suitable method to distinguish between single and possible multiple intent job search queries, crowd-sourcing experiments are proposed to gather backstories and explore how they affect different search aspects.

Research Question 1. If current methods to distinguish between ambiguous and unambiguous queries for general purpose search [17,19] are unsuitable for specialised IR systems, a suitable method needs to be developed.

A methodology to crowd-source backstories will be established so they can be generated at a large scale for a relatively low price. Writing a narrative for a search intent entails higher imaginative loads compared to possibly simpler tasks that can successfully be crowd-sourced, such as judging relevance [2,3]. Crowd-sourced backstories are then compared via different methods to determine if they vary in similarity based on the intent class of the query for which they were generated. A significant difference would inform the usefulness of search result diversification based on queries.

The crowd-sourcing experiment is extended with query filters to determine if they have any effect on formulated backstories, i.e. if the settings are reflected in the backstories and which settings are more important.

Research Question 2. To determine if evaluation processes for specialised IR systems would benefit from ambiguity in test collections, relevance assessments will be crowd-sourced for query-result pairs of different intent levels. It is evaluated if judgements differ significantly based on the query intent level.

Relevance judgements made with backstories in mind compared to ones with just the search query present are compared to evaluate if backstories are important for relevance judgements as well as query disambiguation or if judgements are primarily made based on the query string alone. Furthermore, it is evaluated if backstories aid in assessor agreement when judging relevance.

Research Question 3. Search queries are crowd-sourced for multiple results where the query leading to the results was ambiguous. As optional information, a backstory is presented. Resulting queries are then analysed in terms of diversity to evaluate the disambiguation factor of backstories.

Furthermore, search queries are crowd-sourced for given pairs of backstories and search results. Closely related, specific backstories will be presented as well as a diverse range of different backstories. It is evaluated if specific backstories lead to similarly formulated search queries and different backstories lead to a wide variety of search queries.

Queries as well as additional query filters are gathered to evaluate if filter information present in backstories is actually used as part of the search or if backstories are merely transformed into a keyword query. This would inform the importance of different query parameters.

Acknowledgements. This research is partially supported by the Australian Research Council Project LP150100252 and SEEK Ltd.

References

1. Agrawal, R., Gollapudi, S., Halverson, A., Ieong, S.: Diversifying search results. In: WSDM 2009, pp. 5–14 (2009)
2. Alonso, O., Mizzaro, S.: Using crowdsourcing for TREC relevance assessment. Inf. Process. Manag. **48**(6), 1053–1066 (2012)
3. Alonso, O., Rose, D.E., Stewart, B.: Crowdsourcing for relevance evaluation. SIGIR Forum **42**(2), 9–15 (2008)
4. Bailey, P., Moffat, A., Scholer, F., Thomas, P.: UQV100: a test collection with query variability. In: SIGIR 2016, pp. 725–728 (2016)
5. Borlund, P., Ingwersen, P.: The development of a method for the evaluation of interactive information retrieval systems. J. Doc. **53**(3), 225–250 (1997)
6. Borlund, P., Ingwersen, P.: The application of work tasks in connection with the evaluation of interactive information retrieval systems: empirical results. In: MIRA 1999 (1999)
7. Carbonell, J.G., Goldstein, J.: The use of MMR, diversity-based reranking for reordering documents and producing summaries. In: SIGIR 1998, pp. 335–336 (1998)
8. Chapelle, O., Ji, S., Liao, C., Velipasaoglu, E., Lai, L., Wu, S.: Intent-based diversification of web search results: metrics and algorithms. Inf. Retrieval **14**(6), 572–592 (2011)
9. Drosou, M., Pitoura, E.: Search result diversification. SIGMOD Rec. **39**(1), 41–47 (2010)
10. Gollapudi, S., Sharma, A.: An axiomatic approach for result diversification. In: WWW 2009, pp. 381–390 (2009)
11. Hafernik, C.T., Jansen, B.J.: Understanding the specificity of web search queries. In: CHI 2013, pp. 1827–1832 (2013)
12. Jones, K.S., Robertson, S.E., Sanderson, M.: Ambiguous requests: implications for retrieval tests, systems and theories. SIGIR Forum **41**(2), 8–17 (2007)
13. Krovetz, R., Croft, W.B.: Lexical ambiguity and information retrieval. ACM Trans. Inf. Syst. **10**(2), 115–141 (1992)
14. Mei, Q., Church, K.W.: Entropy of search logs: how hard is search? With personalization? With backoff? In: WSDM 2008, pp. 45–54 (2008)
15. Miller, G.A.: WordNet: a lexical database for English. Commun. ACM **38**(11), 39–41 (1995)
16. Phan, N., Bailey, P., Wilkinson, R.: Understanding the relationship of information need specificity to search query length. In: SIGIR 2007, pp. 709–710 (2007)
17. Radlinski, F., Dumais, S.T.: Improving personalized web search using result diversification. In: SIGIR 2006, pp. 691–692 (2006)
18. Sanderson, M.: Ambiguous queries: test collections need more sense. In: SIGIR 2008, pp. 499–506 (2008)
19. Wang, Y., Agichtein, E.: Query ambiguity revisited: clickthrough measures for distinguishing informational and ambiguous queries. In: NAACL 2010, pp. 361–364 (2010)

Workshops

Proposal for the 1st Interdisciplinary Workshop on Algorithm Selection and Meta-Learning in Information Retrieval (AMIR)

Joeran Beel[1]([⊠]) and Lars Kotthoff[2]

[1] School of Computer Science and Statistics, Artificial Intelligence Discipline, ADAPT Centre, Trinity College Dublin, Dublin, Ireland
`beelj@tcd.ie`
[2] Department of Computer Science, Meta Algorithmics, Learning and Large-Scale Empirical Testing Lab, University of Wyoming, Laramie, USA
`larsko@uwyo.edu`

Abstract. The algorithm selection problem describes the challenge of identifying the best algorithm for a given problem space. In many domains, particularly artificial intelligence, the algorithm selection problem is well studied, and various approaches and tools exist to tackle it in practice. Especially through meta-learning impressive performance improvements have been achieved. The information retrieval (IR) community, however, has paid little attention to the algorithm selection problem, although the problem is highly relevant in information retrieval. This workshop will bring together researchers from the fields of algorithm selection and meta-learning as well as information retrieval. We aim to raise the awareness in the IR community of the algorithm selection problem; identify the potential for automatic algorithm selection in information retrieval; and explore possible solutions for this context. In particular, we will explore to what extent existing solutions to the algorithm selection problem from other domains can be applied in information retrieval, and also how techniques from IR can be used for automated algorithm selection and meta-learning.

1 Motivation

There is a plethora of algorithms for information retrieval applications, such as search engines and recommender systems. There are about 100 approaches to recommend research papers alone [1]. The question that researchers and practitioners alike are faced with is which one of these approaches to choose for their particular problem. This is a difficult choice even for experts, compounded by ongoing research that develops ever more approaches.

This publication has emanated from research conducted with the financial support of Science Foundation Ireland (SFI) under Grant Number 13/RC/2106.

© Springer Nature Switzerland AG 2019
L. Azzopardi et al. (Eds.): ECIR 2019, LNCS 11438, pp. 383–388, 2019.
https://doi.org/10.1007/978-3-030-15719-7_53

The challenge of identifying the best algorithm for a given application is not new. The so-called "algorithm selection problem" was first mentioned in the 1970s [2] and has attracted significant attention in various disciplines since then. Particularly in artificial intelligence, impressive performance achievements have been enabled by algorithm selection systems. A prominent example is the SATzilla system [3].

More generally, algorithm selection is an example of meta-learning, where the experience gained from solving problems informs how to solve future problems. Meta-learning and automating modelling processes has gained significant traction in the machine learning community, in particular with so-called AutoML approaches that automate the entire machine learning and data mining process from ingesting the data to making predictions. An example of such a system is Auto-WEKA [4]. There have also been multiple competitions [5, 6] and workshops, symposia and tutorials [7–11], including a Dagstuhl seminar [8]. The OpenML platform was developed to facilitate the exchange of data and machine learning models to enable research into meta-learning [12].

Despite the significance of the algorithm selection problem and notable advances in solving it in many domains, the information retrieval community has paid little attention to it. There are a few papers that investigate the algorithm selection problem in the context of information retrieval, for example in the field of recommender systems [13–21]. However, the number of researchers interested in this topic is limited, and results so far have been not as impressive as in other domains.

There is potential for applying IR techniques in meta-learning as well. The algorithm selection problem can be seen as a traditional information retrieval task, i.e. the task of identifying the most relevant item (an algorithm) from a large corpus (thousands of potential algorithms and parameters) for a given information need (e.g. classifying photos or making recommendations). We see great potential for the information retrieval community contributing to solving the algorithm selection problem.

2   Objectives, Outcomes and Vision for the Workshop

We propose the 1st Interdisciplinary Workshop on Algorithm Selection and Meta-Learning in Information Retrieval (AMIR)[1]. We aim at achieving the following goals:

- Raise awareness in the information retrieval community of the algorithm selection problem.
- Identify the potential for automated algorithm selection and meta learning in IR applications.
- Familiarize the IR community with algorithm selection and meta-learning tools and research that has been published in related disciplines such as machine learning.
- Find solutions to address and solve the algorithm selection problem in IR.

The expected outcome is a workshop proceedings book, which we aim to publish at http://www.ceur-ws.org/. Our vision is to establish a regular workshop at ECIR or

---

[1] http://amir-workshop.org/.

related venues (e.g. SIGIR, UMAP, RecSys) and eventually – in the long run – solve the algorithm selection problem in information retrieval. We hope to stimulate collaborations between researchers in IR and meta-learning through presentations and discussions at the workshop, which will ultimately lead to joint publications and research proposals.

3   Topical Outline

We will explore (a) how existing solutions for algorithm selection and meta-learning can be applied to identify the best algorithm for a given information retrieval problem and (b) how information retrieval techniques may be applied to solve the algorithm selection problem in IR and other domains.

More precisely, topics relevant for the workshop are

- Automated Machine Learning
- Algorithm Selection
- Algorithm Configuration
- Meta-Learning
- Hyper-Parameter Optimization
- Evaluation Methods and Metrics
- Benchmarking
- Meta Heuristics
- Learning to Learn
- Recommender Systems for Algorithms
- Algorithm Selection as User Modeling Task
- Search Engines for Algorithms
- Neural Network Search

4   Planned Format and Structure

We envision a half-day workshop with the following submission types.

- Research Papers, Position Papers, Case Studies (6 or 12 pages, LNCS format) with 5–15 min presentations at the Workshop
- Posters, Demonstrations, Nectar[2], Datasets (4 pages, LNCS format) with a 1–2 min teaser presentation, and a poster session.

The tentative schedule is as follows.

8:30	Poster setup	10:00	Coffee Break and Posters
9:00	Welcome and	10:30	Paper Presentations
9:10	Keynote Talk	12:00	Lunch, Outlook, Discussion
9:45	Poster Pitches		

---

[2] A Nectar track allows well known researchers to present a summary of their recent work relating to the workshop topic. Compare http://www.ecmlpkdd2018.org/nectar-track/.

Depending on the number of submissions, part of the paper session may be replaced by a panel discussion with experts from IR and machine learning to discuss the algorithm selection problem in the context of information retrieval in depth.

5 Expected Audience and Attendees

Workshops on the algorithm selection problem in the field of machine learning attracted significant attention. For instance, the *Meta-Learning and Algorithm Selection Workshop* at ECMLPKDD in 2015 resulted in 15 publications [9]. The *Workshop on Meta-Learning* (MetaLearn 2017) at NIPPS resulted in 29 publications, including posters [10]. Given that the algorithm selection problem is less known in the IR community, we expect around 5 publications (short and full paper) plus a few posters.

We are confident to receive a significant number of manuscripts and expect a high number of attendees as the algorithm selection problem is relevant for everyone in information retrieval, particularly for everyone who wants to deploy a real-world system. It is an easy-to-understand problem that every researcher has faced himself/herself; and has attracted a lot of attention in other communities already.

6 Organizers

6.1 Joeran Beel

Joeran Beel[3] is Assistant Professor in Intelligent Systems at the School of Computer Science and Statistics at Trinity College Dublin. He is also affiliated with the ADAPT Centre, an interdisciplinary research center that cooperates with industry partners including Google, Deutsche Bank, and Huawei. Joeran is further a Visiting Professor at the National Institute of Informatics in Tokyo. His research focuses on information retrieval, recommender systems, user modeling and machine learning. He has developed novel algorithms in these fields and conducted research on the question of how to evaluate information retrieval systems. Joeran is serving as general co-chair of the 26th Irish Conference on Artificial Intelligence and Cognitive Science and served on program committees for major information retrieval venues including SIGIR, ECIR, UMAP, RecSys, and ACM TOIS.

6.2 Lars Kotthoff

Lars Kotthoff[4] is Assistant Professor at the University of Wyoming. He leads the Meta-Algorithmics, Learning and Large-scale Empirical Testing (MALLET) lab and has acquired more than $400 K in external funding to date. Lars is also the PI for the Artificially Intelligent Manufacturing center (AIM) at the University of Wyoming. He co-organized multiple workshops on meta-learning and automatic machine learning

---

[3] https://www.scss.tcd.ie/joeran.beel/.

[4] http://www.cs.uwyo.edu/~larsko/.

(e.g. [9]) and the Algorithm Selection Competition Series [5]. He was workshop and masterclass chair at the CPAIOR 2014 conference and organized the ACP summer school on constraint programming in 2018. His research combines artificial intelligence and machine learning to build robust systems with state-of-the-art performance. Lars' more than 60 publications have garnered ~ 800 citations and his research has been supported by funding agencies and industry in various countries.

References

1. Beel, J., Gipp, B., Langer, S., Breitinger, C.: Research paper recommender systems: a literature survey. Int. J. Digit. Libr. 305–338 (2016)
2. Rice, J.R.: The algorithm selection problem (1975)
3. Xu, L., Hutter, F., Hoos, H.H., Leyton-Brown, K.: SATzilla: portfolio-based algorithm selection for SAT. J. Artif. Intell. Res. **32**, 565–606 (2008)
4. Kotthoff, L., Thornton, C., Hoos, H.H., Hutter, F., Leyton-Brown, K.: Auto-WEKA 2.0: automatic model selection and hyperparameter optimization in WEKA. J. Mach. Learn. Res. **18**, 826–830 (2017)
5. Lindauer, M., van Rijn, J.N., Kotthoff, L.: The algorithm selection competition series 2015-17. arXiv preprint arXiv:1805.01214 (2018)
6. Tu, W.-W.: The 3rd AutoML challenge: AutoML for lifelong machine learning. In: NIPS 2018 Challenge (2018)
7. Brazdil, P.: Metalearning and algorithm selection. In: 21st European Conference on Artificial Intelligence (ECAI) (2014)
8. Hoos, H.H., Neumann, F., Trautmann, H.: Automated algorithm selection and configuration. Report from Dagstuhl Seminar 16412, vol. 6 (2016)
9. Vanschoren, J., Brazdil, P., Giraud-Carrier, C., Kotthoff, L.: Meta-learning and algorithm selection workshop at ECMLPKDD. In: CEUR Workshop Proceedings (2015)
10. Calandra, R., Hutter, F., Larochelle, H., Levine, S.: Workshop on meta-learning (MetaLearn 2017) @NIPS (2017). http://metalearning.ml
11. Miikkulainen, R., Le, Q., Stanley, K., Fernando, C.: Metalearning symposium @NIPS (2017). http://metalearning-symposium.ml
12. Vanschoren, J., Van Rijn, J.N., Bischl, B., Torgo, L.: OpenML: networked science in machine learning. ACM SIGKDD Explor. Newsl. **15**, 49–60 (2014)
13. Ahsan, M., Ngo-Ye, L.: A Conceptual model of recommender system for algorithm selection. In: AMCIS 2005 Proceedings, p. 122 (2005)
14. Collins, A., Tkaczyk, D., Beel, J.: A novel approach to recommendation algorithm selection using meta-learning. In: Proceedings of the 26th Irish Conference on Artificial Intelligence and Cognitive Science (AICS), CEUR-WS, pp. 210–219 (2018)
15. Cunha, T., Soares, C., de Carvalho, A.C.: Metalearning and recommender systems: a literature review and empirical study on the algorithm selection problem for collaborative filtering. Inf. Sci. **423**, 128–144 (2018)
16. Cunha, T., Soares, C., de Carvalho, A.C.: CF4CF: recommending collaborative filtering algorithms using collaborative filtering. arXiv preprint arXiv:1803.02250 (2018)
17. Cunha, T., Soares, C., de Carvalho, A.C.P.L.F.: Selecting collaborative filtering algorithms using metalearning. In: Frasconi, P., Landwehr, N., Manco, G., Vreeken, J. (eds.) ECML PKDD 2016. LNCS (LNAI), vol. 9852, pp. 393–409. Springer, Cham (2016). https://doi.org/10.1007/978-3-319-46227-1_25

18. Matuszyk, P., Spiliopoulou, M.: Predicting the performance of collaborative filtering algorithms. In: Proceedings of the 4th International Conference on Web Intelligence, Mining and Semantics (WIMS 2014), p. 38. ACM (2014)
19. Mısır, M., Sebag, M.: ALORS: an algorithm recommender system. Artif. Intell. **244**, 291–314 (2017)
20. Romero, C., Olmo, J.L., Ventura, S.: A meta-learning approach for recommending a subset of white-box classification algorithms for Moodle datasets. In: Educational Data Mining (2013)
21. Vartak, M., Thiagarajan, A., Miranda, C., Bratman, J., Larochelle, H.: A meta-learning perspective on cold-start recommendations for items. In: Advances in Neural Information Processing Systems, pp. 6907–6917 (2017)

The 2nd International Workshop on Narrative Extraction from Text: Text2Story 2019

Alípio Mário Jorge[1,2] , Ricardo Campos[1,3(✉)] , Adam Jatowt[4] ,
and Sumit Bhatia[5]

[1] INESC TEC, Porto, Portugal
[2] University of Porto, Porto, Portugal
amjorge@fc.up.pt
[3] Polytechnic Institute of Tomar - Smart Cities Research Center,
Tomar, Portugal
ricardo.campos@ipt.pt
[4] Kyoto University, Kyoto, Japan
adam@dl.kuis.kyoto-u.ac.jp
[5] IBM Research AI, Delhi, India
sumitbhatia@in.ibm.com

Abstract. Building upon the success of the first edition, we organize the second edition of the *Text2Story Workshop on Narrative Extraction from Texts* in conjunction with the 41st European Conference on Information Retrieval (ECIR 2019) on April 14, 2019. Our objective is to further consolidate the efforts of the community and reflect upon the progress made since the last edition. Although the understanding of natural language has improved over the last couple of years – with research works emerging on the grounds of information extraction and text mining – the problem of constructing consistent narrative structures is yet to be solved. It is expected that the state-of-the-art has been advancing in pursuit of methods that automatically identify, interpret and relate the different elements of narratives which are often spread among different sources. In the second edition of the workshop, we foster the discussion of recent advances in the link between Information Retrieval (IR) and formal narrative representations from text.

Keywords: Information extraction · Narrative extraction

1 Background and Motivation

The continuous growth of social networks such as Facebook and Twitter, together with an ever-increasing presence of traditional news media outlets on the Web [11, 12] has changed the way information is being generated and consumed. Rather than relying on a few sources of information about an event or a news item topic (e.g. Trump and Russia), readers now have easy access to content via multiple sources (news websites, Facebook posts, etc.) produced by journalists and social media influencers. Further, active reader participation on social media and comments section of news articles are very common, with discussions lasting over days, weeks or possibly, months. Such a stream of continuously evolving information makes it extremely unwieldy and

L. Azzopardi et al. (Eds.): ECIR 2019, LNCS 11438, pp. 389–393, 2019.
https://doi.org/10.1007/978-3-030-15719-7_54

time-consuming for an interested reader to track, read, and process different sources of information in order to keep abreast of all the developments and different aspects of the topic of interest. One way to overcome these problems is to grasp a vast set of interconnected news articles (or similar unstructured collections), extract the key narrative/story elements, and represent them in intermediate data structures (structured data) that convey the key-points of the story in a way that can be easily and quickly consumed by the readers. Games are very attractive and can be particularly compelling means to capture an audience [4], but other simpler forms of narrative representation such as keyword clouds [1], timeline summarizations [13, 15], visual storytelling [10], video narratives [16] or semi-automatic generation of slide shows [3] are also challenging, useful and widely applicable [6]. These alternate narrative representations can help the readers to quickly understand who the main actors of a story are, their interplay and their trajectories in time and space, motivations, main events, causal relations of events and outcomes without having the need to read the original sources (a.k.a. distant reading [5]). Although information extraction and natural language processing have made significant progress towards automatic interpretation of texts, the problem of fully identifying and relating the different elements of a narrative present in a document (set) still present significant unsolved challenges [7].

Building upon the success of the first edition of the Text2Story workshop [8], we organize its second edition at the 41st European Conference on Information Retrieval (ECIR 2019) on April 14, 2019. The workshop is devoted to narrative extraction from texts as a way to raise awareness about the problem of creating a text-to-narrative-structure and its related tasks. This is a very rich line of research that has been conducted over the last few years by many research groups [2, 9, 14]. The main goal of the workshop is to bring together scientists conducting relevant research in the field of identifying and producing narratives/stories from textual sources, such as journalistic texts, scientific articles or even fragmented sources. Scientists will present their latest breakthroughs with an emphasis on the application of their findings on relevant research from a wide range of areas including information extraction, information retrieval, natural language processing, text mining, artificial intelligence, machine learning and augmented writing (automatic production of media content).

Research works submitted to the workshop foster the scientific advance on all aspects of storyline generation from texts including, but not limited to, narrative and content generation, formal representation, and visualization of narratives. This includes the following topics:

- Event Identification
- Narrative Representation Language
- Sentiment and Opinion Detection
- Argumentation Mining
- Narrative Summarization
- Multi-modal Summarization
- Storyline Visualization
- Temporal Aspects of Storylines
- Story Evolution and Shift Detection
- Causal Relation Extraction and Arrangement

- Evaluation Methodologies for Narrative Extraction
- Big data applied to Narrative Extraction
- Resources and Dataset showcase
- Personalization and Recommendation
- User Profiling and User Behavior Modeling
- Credibility
- Models for detection and removal of bias in generated stories
- Ethical and fair narrative generation
- Fact Checking
- Bots Influence
- Bias in Text Documents
- Automatic Timeline Generation

2 Program Committee and Support Chairs

The program committee members consist of researchers from industry and academia. The following members form the program committee:

- Nicola Ferro (University of Padova)
- Miguel Martinez-Alvarez (Signal)
- João Magalhães (New University of Lisbon)
- Federico Nanni (University of Mannheim)
- Dhruv Gupta (Max Planck Institute for Informatics)
- Yihong Zhang (Kyoto University)
- Nuno Moniz (LIAAD/INESC TEC)
- Bruno Martins (IST and INESC-ID - IST, University of Lisbon)
- Mark Finlayson (Florida International University)
- Marc Spaniol (Université de Caen Normandie)
- Nina Tahmasebi (University of Gothenburg)
- Florian Boudin (Université de Nantes)
- Henrique Lopes Cardoso (University of Porto)
- Udo Kruschwitz (University of Essex)
- Mengdie Zhuang (The University of Sheffield)
- Daniel Loureiro (University of Porto)
- Daniel Gomes (FCT/Arquivo.pt)
- Pablo Gamallo (University of Santiago de Compostela)
- Paulo Quaresma (Universidade de Evora)
- Sérgio Nunes (University of Porto)
- Álvaro Figueira (University of Porto)
- Gaël Dias (Normandie University)
- Gerasimos Lampouras (The University of Sheffield)
- Conceição Rocha (CPES INESC TEC)
- João Paulo Cordeiro (LIAAD INESC TEC; University of Beira Interior)
- Arian Pasquali (LIAAD INESC TEC)
- Vítor Mangaravite (UFMG; LIAAD INESC TEC)

Proceedings Chair

- Conceição Rocha (CPES INESC TEC)
- João Paulo Cordeiro (LIAAD INESC TEC; University of Beira Interior)

Web Chair

- Arian Pasquali (LIAAD INESC TEC)

Dissemination Chair

- Vítor Mangaravite (UFMG; LIAAD INESC TEC)

Acknowledgments. First two authors of this paper are financed by the ERDF – European Regional Development Fund through the Operational Programme for Competitiveness and Internationalisation – COMPETE 2020 Programme within project «POCI-01-0145-FEDER-006961», and by National Funds through the FCT – Fundação para a Ciência e a Tecnologia (Portuguese Foundation for Science and Technology) as part of project UID/EEA/50014/2013.

References

1. Campos, R., Mangaravite, V., Pasquali, A., Jorge, A.M., Nunes, C., Jatowt, A.: A text feature based automatic keyword extraction method for single documents. In: Pasi, G., Piwowarski, B., Azzopardi, L., Hanbury, A. (eds.) ECIR 2018. LNCS, vol. 10772, pp. 684–691. Springer, Cham (2018). https://doi.org/10.1007/978-3-319-76941-7_63
2. Caselli, T., et al.: Proceedings of the Events and Stories in the News Workshop@ACL 2017, 4 August, Vancouver, Canada, pp. 1–121. Association for Computational Linguistics (2017)
3. Hullman, J., Drucker, S., Henry, N., Lee, R., Fisher, D., Adar, E.: A deeper understanding of sequence in narrative visualization. IEEE Trans. Vis. Comput. Graph. **19**(12), 2406–2415 (2013)
4. Kessing, J., Tutenel, T., Bidarra, R.: Designing semantic game worlds. In: PCG 2012 – Workshop on Procedural Content Generation for Games (2012)
5. Moretti, F.: Graphs, Maps, Trees: Abstract Models for a Literary History. Verso, London (2005)
6. Liu, S., Wu, Y., Wei, E., Liu, M., Liu, Y.: StoryFlow: tracking the evolution of stories. IEEE Trans. Vis. Comput. Graph. **19**(12), 2436–2445 (2013)
7. John, M., Lohmann, S., Koch, S., Wörner, M., Ertl, T.: Visual analytics for narrative text: visualizing characters and their relationships as extracted from novels. In: Proceedings of the 7th International Conference on Information Visualization Theory and Applications (IVAPP 2016). SciTePress (2016)
8. Jorge, A., Campos, R., Jatowt, A., Nunes, S.: First international workshop on narrative extraction from text (Text2Story 2018). In: Pasi, G., et al. (eds.) ECIR 2018. LNCS, vol. 10772, pp. 833–834. Springer, Cham (2018)
9. Jorge, A., Campos, R., Jatowt, A., Nunes, S.: Proceedings of the First Workshop on Narrative Extraction from Texts (Text2Story 2018@ECIR 2018), Grenoble, France, 26 March, pp. 1–51 (2018). CEUR Workshop Proceedings 2077
10. Lukin, S., Hobbs, R., Voss, C.: A pipeline for creative visual storytelling. In: Proceedings of the First Workshop on Storytelling (StoryNLP 2018@NAACL 2018), New Orleans, USA, 5 June, pp. 20–32 (2018)

11. Martinez-Alvarez, M., et al.: First international workshop on recent trends in news information retrieval (NewsIR 2016). In: Ferro, N., et al. (eds.) ECIR 2016. LNCS, vol. 9626, pp. 878–882. Springer, Cham (2016)

12. Martinez-Alvarez, M., Albakour, D., Corney, D., Gonzalo, J., Poblete, B., Vlachos, A.: Second international workshop on recent trends in news information retrieval (NewsIR 2018). In: Pasi, G., et al. (eds.) ECIR 2018. LNCS, vol. 10772, p. 231. Springer, Cham (2018)

13. McCreadie, R., Santos, R., Macdonald, C., Ounis, I.: Explicit diversification of event aspects for temporal summarization. ACM Trans. Inf. Syst. **36**(3), 25:1–25:31 (2018)

14. Mitchell, M., Huang, T-H., Ferraro, F., Misra, I.: Proceedings of the First Workshop on Storytelling (StoryNLP 2018@NAACL 2018), New Orleans, USA, 5 June. Association for Computational Linguistics (2018)

15. Pasquali, A., Mangaravite, V., Campos, R., Jorge, A., Jatowt, A.: Interactive system for automatically generating temporal narratives. In: Azzopardi, L., et al. (eds.) Advances in Information Retrieval – 41st European Conference on Information Retrieval (ECIR 2019), 14–18 April, Cologne, Germany. Lecture Notes in Computer Science, vol. 11438, pp. 251–255. Springer, Heidelberg (2019)

16. Zsombori, V., et al.: Automatic generation of video narratives from shared UGC. In: Proceedings of the 22nd ACM Conference on Hypertext and Hypermedia, HT 2011, Eindhoven, The Netherlands, 06–09 June, pp. 325–334 (2011)

Bibliometric-Enhanced Information Retrieval: 8th International BIR Workshop

Guillaume Cabanac[1]([✉])(iD), Ingo Frommholz[2](iD), and Philipp Mayr[3](iD)

[1] Computer Science Department, IRIT UMR 5505,
University of Toulouse, Toulouse, France
guillaume.cabanac@univ-tlse3.fr
[2] Institute for Research in Applicable Computing,
University of Bedfordshire, Luton, UK
ifrommholz@acm.org
[3] GESIS – Leibniz-Institute for the Social Sciences, Cologne, Germany
philipp.mayr@gesis.org

Abstract. The Bibliometric-enhanced Information Retrieval workshop series (BIR) at ECIR tackles issues related to academic search, at the crossroads between Information Retrieval and Bibliometrics. BIR is a hot topic investigated by both academia (e.g., ArnetMiner, CiteSeer^X, Doc-Ear) and the industry (e.g., Google Scholar, Microsoft Academic Search, Semantic Scholar). An 8th iteration of the one-day BIR workshop was held at ECIR 2019.

Keywords: Academic search · Information retrieval ·
Digital Libraries · Bibliometrics · Scientometrics

1 Motivation and Relevance to ECIR

Searching for scientific information is a long-lived information need. In the early 1960s, Salton was already striving to enhance information retrieval by including clues inferred from bibliographic citations [21]. The development of citation indexes pioneered by Garfield [6] proved determinant for such a research endeavour at the crossroads between the nascent fields of Information Retrieval (IR) and Bibliometrics[1]. The pioneers who established these fields in Information Science—such as Salton and Garfield—were followed by scientists who specialised in one of these [26], leading to the two loosely connected fields we know of today.

The purpose of the BIR workshop series founded in 2014 is to tighten up the link between IR and Bibliometrics. We strive to get the 'retrievalists' and 'citationists' [26] active in both academia and the industry together, who are

---

[1] Bibliometrics refers to the statistical analysis of the academic literature [20] and plays a key role in scientometrics: the quantitative study of science and innovation [9].

© Springer Nature Switzerland AG 2019
L. Azzopardi et al. (Eds.): ECIR 2019, LNCS 11438, pp. 394–399, 2019.
https://doi.org/10.1007/978-3-030-15719-7_55

developing search engines and recommender systems such as ArnetMiner [24], CiteSeer^X [27], DocEar [1], Google Scholar [25], Microsoft Academic Search [23], and Semantic Scholar [2], just to name a few.

Bibliometric-enhanced IR systems must deal with the multifaceted nature of scientific information by searching for or recommending academic papers, patents [7], venues (i.e., conferences or journals), authors, experts (e.g., peer reviewers), references (to be cited to support an argument), and datasets. The underlying models harness relevance signals from keywords provided by authors, topics extracted from the full-texts, coauthorship networks, citation networks, and various classifications schemes of science.

Bibliometric-enhanced IR is a hot topic whose recent developments made the news—see for instance the Initiative for Open Citations [22] and the Google Dataset Search [5] launched on September 4, 2018. We believe that BIR@ECIR is a much needed scientific event for the 'retrievalists' and 'citationists' to meet and join forces pushing the knowledge boundaries of IR applied to literature search and recommendation.

2 Past Related Activities

The BIR workshop series was launched at ECIR in 2014 [18] and it was held at ECIR each year since then [12–14,17]. As our workshop has been lying at the crossroads between IR and NLP, we also ran it as a joint workshop called BIRNDL (for Bibliometric-enhanced IR and NLP for Digital Libraries) at the JCDL [3] and SIGIR [10,11] conferences. All workshops had a large number of participants, demonstrating the relevance of the workshop's topics. The BIR and BIRNDL workshop series gave the community the opportunity to discuss latest developments and shared tasks such as the CL-SciSumm [8], which was introduced at the BIRNDL joint workshop.

The authors of the most promising workshop papers were offered the opportunity to submit an extended version for a Special Issue for the *Scientometrics* journal [4,19] and of the *International Journal on Digital Libraries* [16].

The target audience of our workshop are researchers and practitioners, junior and senior, from Scientometrics as well as Information Retrieval. These could be IR researchers interested in potential new application areas for their work as well as researchers and practitioners working with, for instance, bibliometric data and interested in how IR methods can make use of such data.

3 Objectives and Topics for BIR@ECIR 2019

We called for original research at the crossroads of IR and Bibliometrics. The accepted papers report on new approaches using bibliometric clues to enhance the search or recommendation of scientific information or significant improvements of existing techniques. Thorough quantitative studies of the various corpora to be indexed (papers, patents, networks or else) were also welcome.

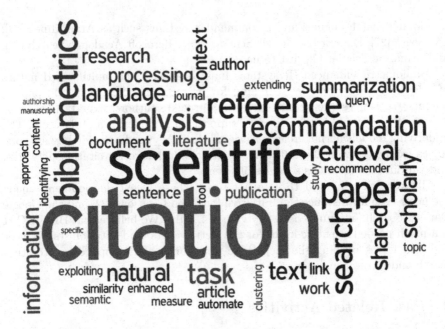

Fig. 1. Main topics of the BIR and BIRNDL workshop series (2014–2018) as extracted from the titles of the papers published in the proceedings, see https://dblp.org/search? q=BIR.ECIR

The topics of the workshop are in line with those of the past BIR and BIRNDL workshops (Fig. 1): a mixture of IR and Bibliometric concepts and techniques. More specifically, the call for papers featured current research issues regarding three aspects of the search/recommendation process:

1. User needs and behaviour regarding scientific information, such as:
   – Finding relevant papers/authors for a literature review;
   – Measuring the degree of plagiarism in a paper;
   – Identifying expert reviewers for a given submission;
   – Flagging predatory conferences and journals.
2. The characteristics of scientific information:
   – Measuring the reliability of bibliographic libraries;
   – Spotting research trends and research fronts.
3. Academic search/recommendation systems:
   – Modelling the multifaceted nature of scientific information;
   – Building test collections for reproducible BIR.

4   Peer Review Process and Organization

The 8th BIR edition ran as a one-day workshop, as it was the case for the previous editions. Keynote talks by leading scientists working at the crossroads between IR and Scientometrics kicked off the day.

Two types of papers were presented: long papers (15-min talks) and short papers (5-min talks). Two interactive sessions closed the morning and evening sessions with posters and demos. These sessions allowed us to discuss the latest developments in the field and opportunities. The interactive sessions were announced with the workshop program. We invited anyone attending to demonstrate their prototypes during flash presentations (5 min). These interactive sessions served as ice-breakers, sparking interesting discussions that usually continued during lunch and the cocktail party. The sessions were also an opportunity for our speakers to further discuss their work.

We ran the workshop with peer review supported by EasyChair[2]. Each submission was assigned to 2 to 3 reviewers, preferably at least one expert in IR and one expert in Bibliometrics. The stronger submissions were accepted as long papers while weaker ones were accepted as short papers, posters, or demos. All authors were instructed to revise their submission according to the reviewers' reports. All accepted papers are planned to be in the workshop proceedings hosted at ceur-ws.org, an established open access repository with no author-processing charges.

As a follow-up of the workshop, the co-chairs will write a report summing up the main themes and discussions to *SIGIR Forum* [15, for instance] and BCS Informer[3], as a way to advertise our research topics as widely as possible among the IR community. All authors are encouraged to submit an extended version of their papers to the Special Issue of the *Scientometrics* journal that will be announced in Spring 2019.

References

1. Beel, J., Langer, S., Gipp, B., Nürnberger, A.: The architecture and datasets of Docear's research paper recommender system. D-Lib Mag. **20**(11/12) (2014). https://doi.org/10.1045/november14-beel
2. Bohannon, J.: A computer program just ranked the most influential brain scientists of the modern era. Science (2016). https://doi.org/10.1126/science.aal0371
3. Cabanac, G., et al. (eds.): BIRNDL 2016: Proceedings of the Joint Workshop on Bibliometric-Enhanced Information Retrieval and Natural Language Processing for Digital Libraries Co-located with the Joint Conference on Digital Libraries, vol. 1610. CEUR-WS, Aachen (2016)
4. Cabanac, G., Frommholz, I., Mayr, P.: Bibliometric-enhanced information retrieval: preface. Scientometrics **116**, 1225–1227 (2018). https://doi.org/10.1007/s11192-018-2861-0
5. Castelvecchi, D.: Google unveils search engine for open data. Nature (2018). https://doi.org/10.1038/d41586-018-06201-x. (News & Comment)
6. Garfield, E.: Citation indexes for science: a new dimension in documentation through association of ideas. Science **122**(3159), 108–111 (1955). https://doi.org/10.1126/science.122.3159.108
7. Garfield, E.: Patent citation indexing and the notions of novelty, similarity, and relevance. J. Chem. Doc. **6**(2), 63–65 (1966). https://doi.org/10.1021/c160021a001

---

[2] https://easychair.org.
[3] http://irsg.bcs.org/informer/.

8. Jaidka, K., Chandrasekaran, M.K., Rustagi, S., Kan, M.Y.: Insights from CL-SciSumm 2016: the faceted scientific document summarization shared task. Int. J. Digit. Libr. **19**(2–3), 163–171 (2018). https://doi.org/10.1007/s00799-017-0221-y

9. Leydesdorff, L., Milojević, S.: Scientometrics. In: Wright, J.D. (ed.) International Encyclopedia of the Social & Behavioral Sciences, vol. 21, 2nd edn, pp. 322–327. Elsevier, Amsterdam (2015). https://doi.org/10.1016/b978-0-08-097086-8.85030-8

10. Mayr, P., Chandrasekaran, M.K., Jaidka, K. (eds.): BIRNDL 2017: Proceedings of the 2nd Joint Workshop on Bibliometric-Enhanced Information Retrieval and Natural Language Processing for Digital Libraries Co-located with the Joint Conference on Digital Libraries, vol. 1888. CEUR-WS, Aachen (2017)

11. Mayr, P., Chandrasekaran, M.K., Jaidka, K. (eds.): BIRNDL 2017: Proceedings of the 3rd Joint Workshop on Bibliometric-Enhanced Information Retrieval and Natural Language Processing for Digital Libraries Co-located with the Joint Conference on Digital Libraries, vol. 2132. CEUR-WS, Aachen (2018)

12. Mayr, P., Frommholz, I., Cabanac, G. (eds.): BIR 2016 Proceedings of the 3rd Workshop on Bibliometric-Enhanced Information Retrieval Co-located with the 38th European Conference on Information Retrieval, vol. 1567. CEUR-WS, Aachen (2016)

13. Mayr, P., Frommholz, I., Cabanac, G. (eds.): BIR 2017 Proceedings of the 5th Workshop on Bibliometric-Enhanced Information Retrieval Co-located with the 39th European Conference on Information Retrieval, vol. 1823. CEUR-WS, Aachen (2017)

14. Mayr, P., Frommholz, I., Cabanac, G. (eds.): BIR 2018 Proceedings of the 7th Workshop on Bibliometric-Enhanced Information Retrieval Co-located with the 40th European Conference on Information Retrieval, vol. 2080. CEUR-WS (2018)

15. Mayr, P., Frommholz, I., Cabanac, G.: Report on the 7th International Workshop on Bibliometric-Enhanced Information Retrieval (BIR 2018). SIGIR Forum **52**(1), 135–139 (2018). https://doi.org/10.1145/3274784.3274798

16. Mayr, P., et al.: Special issue on bibliometric-enhanced information retrieval and natural language processing for digital libraries. Int. J. Digit. Libr. **19**(2–3), 107–111 (2018). https://doi.org/10.1007/s00799-017-0230-x

17. Mayr, P., Frommholz, I., Mutschke, P. (eds.): BIR 2015 Proceedings of the 2nd Workshop on Bibliometric-Enhanced Information Retrieval Co-located with the 37th European Conference on Information Retrieval, vol. 1344. CEUR-WS, Aachen (2015)

18. Mayr, P., Schaer, P., Scharnhorst, A., Larsen, B., Mutschke, P. (eds.): BIR 2016 Proceedings of the 1st Workshop on Bibliometric-Enhanced Information Retrieval Co-located with the 36th European Conference on Information Retrieval, vol. 1143. CEUR-WS, Aachen (2014)

19. Mayr, P., Scharnhorst, A.: Scientometrics and information retrieval: weak-links revitalized. Scientometrics **102**(3), 2193–2199 (2015). https://doi.org/10.1007/s11192-014-1484-3

20. Pritchard, A.: Statistical bibliography or bibliometrics? J. Doc. **25**(4), 348–349 (1969). https://doi.org/10.1108/eb026482. (Documentation notes)

21. Salton, G.: Associative document retrieval techniques using bibliographic information. J. ACM **10**(4), 440–457 (1963). https://doi.org/10.1145/321186.321188

22. Shotton, D.: Funders should mandate open citations. Nature **553**(7687), 129 (2018). https://doi.org/10.1038/d41586-018-00104-7

23. Sinha, A., et al.: An overview of Microsoft Academic Service (MAS) and applications. In: Gangemi, A., Leonardi, S., Panconesi, A. (eds.) WWW 2015: Proceedings of the 24th International Conference on World Wide Web, pp. 243–246. ACM, New York (2015). https://doi.org/10.1145/2740908.2742839

24. Tang, J., Zhang, J., Yao, L., Li, J., Zhang, L., Su, Z.: ArnetMiner: extraction and mining of academic social networks. In: KDD 2008: Proceeding of the 14th ACM SIGKDD International Conference on Knowledge Discovery and Data Mining, pp. 990–998. ACM, New York (2008). https://doi.org/10.1145/1401890.1402008

25. Van Noorden, R.: Google scholar pioneer on search engine's future. Nature (2014). https://doi.org/10.1038/nature.2014.16269

26. White, H.D., McCain, K.W.: Visualizing a discipline: an author co-citation analysis of information science, 1972–1995. J. Am. Soc. Inf. Sci. **49**(4), 327–355 (1998). https://doi.org/b57vc7

27. Williams, K., Wu, J., Choudhury, S.R., Khabsa, M., Giles, C.L.: Scholarly big data information extraction and integration in the CiteSeer^X digital library. In: ICDE 2014: Proceedings of the 30th IEEE International Conference on Data Engineering Workshops, pp. 68–73. IEEE (2014). https://doi.org/10.1109/icdew.2014.6818305

Third Workshop on Social Media for Personalization and Search (SoMePeAS 2019)

Ludovico Boratto[1]([⊠])[iD] and Giovanni Stilo[2][iD]

[1] Data Science and Big Data Analytics, EURECAT,
Carrer de Bilbao 72 (Edifici A), 08005 Barcelona, Spain
ludovico.boratto@acm.org
[2] Dipartimento di Ingegneria e Scienze dell'Informazione e Matematica,
Università degli Studi dell'Aquila, L'Aquila, Italy
giovanni.stilo@univaq.it

Abstract. Social media platforms have become powerful tools to collect the preferences of the users and get to know them more. Indeed, in order to build profiles about what they like or dislike, a system does not only have to rely on explicitly given preferences (e.g., ratings) or on implicitly collected data (e.g., from the browsing sessions). In the middle, there lie opinions and preferences expressed through likes, textual comments, and posted content. Being able to exploit social media to mine user behavior and extract additional information leads to improvements in the accuracy of personalization and search technologies, and to better targeted services to the users. In this workshop, we aim to collect novel ideas in this field and to provide a common ground for researchers working in this area.

Keywords: Social media · Personalization · Search

1 Introduction

In order to improve the web experience of the users, classic personalization technologies (e.g., recommender systems) and search engines usually rely on static schemes. Indeed, users are allowed to express ratings in a fixed range of values for a given catalogue of products, or to express a query that usually returns the same set of webpages/products for all the users.

With the advent of social media, users have been allowed to create new content and to express opinions and preferences through likes and textual comments. Moreover, the social network itself can provide information on who influences whom. Being able to mine usage and collaboration patterns in social media and to analyze the content generated by the users opens new frontiers in the generation of personalization services and in the improvement of search engines. Moreover, recent technological advances, such as deep learning, are able to provide a context to the analyzed data. Even though social media platforms offer

© Springer Nature Switzerland AG 2019
L. Azzopardi et al. (Eds.): ECIR 2019, LNCS 11438, pp. 401–402, 2019.
https://doi.org/10.1007/978-3-030-15719-7

an abundance of data, recent regulations like the GDPR, limit the way in which personal data is collected, treated, and stored. Hence, users' consent and privacy has become a prominent and timely issue.

The aim of this workshop is to discuss novel ideas related to employing social media for personalization and search purposes, focused (but not limited) to the following list:

- Recommender systems
- Search and tagging
- Query expansion
- User modeling and profiling
- Advertising and ad targeting
- Content classification, categorization, and clustering
- Using social network features/community detection algorithms for personalization and search purposes
- Privacy-aware algorithms

2 Workshop Structure

The workshop has two main objectives. The first is to solicit contributions from researchers active in these fields, gaining shared insights on existing approaches, recent advances, and open issues. The second objective is the consolidation of the community of researchers that works on these topics and that can foster discussion, ideas, and sparks for current challenges and future research in this area.

To meet these two objectives, the format of the workshop will be divided into two parts (namely morning and afternoon). In the first part, an academic keynote speaker will give a talk on how social media can be exploited in the literature for personalization and search purposes. Then, we will proceed with the presentation of accepted papers. In the second part, a second keynote speaker will open up on the opportunities that the exploitation of social media can create for the industry. Then, we will continue with the presentation of the papers and we will try to summarize the different problems and positions emerged during the day.

Acknowledgements. We thank the ECIR 2019 organizing committee for giving us the opportunity to host this workshop in conjunction with ECIR 2019 in Cologne, Germany.

Tutorials

Conducting Laboratory Experiments Properly with Statistical Tools: An Easy Hands-On Tutorial

Tetsuya Sakai[✉]

Waseda University, Tokyo, Japan
tetsuyasakai@acm.org

The IR community relies heavily on experimentation and therefore it is of utmost importance that we researchers design and conduct experiments properly and report on the results so that our efforts will add up. One concern regarding the experimental practices in IR is the misuse and misinterpretation of statistical significance tests (e.g. [21, 25]). To name a few examples: if a researcher evaluated four search engines and is interested in the difference between every system pair, conducting a standard t-test $4 * 3/2 = 6$ times is not the correct approach (misuse); the p-value does *not* represent the probability that your research hypothesis is correct (misinterpretation). Moreover, in the IR community, it appears that researchers seldom learn from prior art when designing the sample size of an experiment. This can lead to heavily underpowered experiments, which means that there is a high chance that the researchers will miss the differences actually present, despite having spent their effort and resources for these experiments.

The advent of R as a statistical tool has pros and cons. Pros: anyone can download R for free and conduct statistical significance tests very easily. Cons: anyone can conduct statistical significance tests *without understanding the underlying principles, assumptions, and their limitations.* If the IR community chooses to continue to use statistical significance testing (although there are alternatives: see, for example, [4, 24]), then it must be understood and used properly.

This tutorial will cover the following topics at least: paired and two-sample t-tests; confidence intervals; ANOVA; familywise error rate; Tukey's HSD test; simultaneous confidence intervals; paired randomisation test; randomised Tukey HSD test; limitations of statistical significance tests; relationships among the significance level, statistical power, effect sizes and sample sizes; topic set size design; power analysis; appropriate ways to report on experimental results in a paper.

The tutorial slide deck will at least touch upon the following publications: [1–36].

© Springer Nature Switzerland AG 2019
L. Azzopardi et al. (Eds.): ECIR 2019, LNCS 11438, pp. 405–407, 2019.
https://doi.org/10.1007/978-3-030-15719-7

References

1. Allan, J., Carterette, B., Aslam, J.A., Pavlu, V., Dachev, B., Kanoulas, E.: Million query track 2007 overview. In: Proceedings of TREC 2007 (2008)
2. Bakan, D.: The test of significance in psychological research. Psychol. Bull. **66**(6), 423–437 (1966)
3. Carterette, B.: Multiple testing in statistical analysis of systems-based information retrieval experiments. ACM TOIS **30**(1) (2012)
4. Carterette, B.: Bayesian inference for information retrieval evaluation. In: Proceedings of ACM ICTIR 2015, pp. 31–40 (2015)
5. Cohen, J.: Statistical Power Analysis for the Bahavioral Sciences, 2nd edn. Psychology Press (1988)
6. Cohen, J.: The earth is round ($p < .05$). Am. Psychol. **49**(12), 997–1003 (1994)
7. Crawley, M.J.: Statistics: An Introduction Using R, 2nd edn. Wiley (2015)
8. Deming, W.E.: On probability as a basic for action. Am. Stat. **29**(4), 146–152 (1975)
9. Good, P.: Permutation, Parametric, and Bootstrap Tests of Hypothesis, 3rd edn. Springer, New York (2005). https://doi.org/10.1007/b138696
10. Harlow, L.L., A.Mulaik, S., Steiger, J.H.: What If There Were No Significance Tests? (Classic Edition). Routledge (2016)
11. Kelley, K., Preacher, K.J.: On effect size. Psychol. Methods **17**(2), 137–152 (2012)
12. Loftus, G.R.: On the tyranny of hypothesis testing in the social sciences. Contemp. Psychol. **36**(2), 102–105 (1991)
13. Nagata, Y.: How to Design the Sample Size. Asakura Shoten (2003). (in Japanese)
14. Nagata, Y., Yoshida, M.: Introduction to Multiple Comparison Procedures. Scientist Press (1997). (in Japanese)
15. Olejnik, S., Algina, J.: Generalized eta and omega squared statistics: measures of effect size for some common research designs. Psychol. Res. **8**(4), 434–447 (2003)
16. Robertson, S.E.: The methodology of information retrieval experiment. In: Sparck Jones, K. (ed.) Information Retrieval Experiment. Butterworths (1981). Chapter 1
17. Robertson, S.E., Kanoulas, E.: On per-topic variance in IR evaluation. In: Proceedings of ACM SIGIR 2012, pp. 891–900 (2012)
18. Rothman, K.J.: Writing for epidemiology. Epidemiology **9**(3), 333–337 (1998)
19. Sakai, T.: Evaluating evaluation metrics based on the bootstrap. In: Proceedings of ACM SIGIR 2006, pp. 525–532 (2006)
20. Sakai, T.: Metrics, statistics, tests. In: Ferro, N. (ed.) PROMISE 2013. LNCS, vol. 8173, pp. 116–163. Springer, Heidelberg (2014). https://doi.org/10.1007/978-3-642-54798-0_6
21. Sakai, T.: Statistical significance, power, and sample sizes: a systematic review of SIGIR and TOIS. In: Proceedings of ACM SIGIR 2016, pp. 5–14 (2016)
22. Sakai, T.: Topic set size design. Inf. Retr. **19**(3), 256–283 (2016)
23. Sakai, T.: Two-sample t-tests for IR evaluation: student or welch? In: Proceedings of ACM SIGIR 2016, pp. 1045–1048 (2016)
24. Sakai, T.: The probability that your hypothesis is correct, credible intervals, and effect sizes for IR evaluation. In: Proceedings of ACM SIGIR 2017, pp. 25–34 (2017)
25. Sakai, T.: Laboratory Experiments in Information Retrieval: Sample Sizes, Effect Sizes, and Statistical Power. Springer, Singapore (2018). https://doi.org/10.1007/978-981-13-1199-4
26. Salton, G., Lesk, M.E.: Computer evaluation of indexing and text processing. J. ACM **15**(1), 8–36 (1968)

27. Savoy, J.: Statistical inference in retrieval effectiveness evaluation. Inf. Process. Manag. **33**(4), 495–512 (1997)
28. Smucker, M.D., Allan, J., Carterette, B.: A comparison of statistical significance tests for information retrieval evaluation. In: Proceedings of ACM CIKM 2007, pp. 623–632 (2007)
29. Sparck Jones, K., van Rijsbergen, C.J.: Report on the need for and provision of an 'ideal' information retrieval test collection. Technical report, Computer Laboratory, University of Cambridge, British Library Research and Development Report No.5266 (1975)
30. Sparck Jones, K. (ed.): Information Retrieval Experiment. Butterworths (1981)
31. Sparck Jones, K., Willet, P. (eds.): Readings in Information Retrieval. Morgan Kaufmann (1997)
32. Toyoda, H.: Introduction to Statistical Power Analysis: A Tutorial with R. Tokyo Tosyo (2009). (in Japanese)
33. Van Rijsbergen, C.J.: Information Retrieval. Butterworths (1979). Chapter 7
34. Voorhees, E.M.: Topic set size redux. In: Proceedings of ACM SIGIR 2009, pp. 806–807 (2009)
35. Webber, W., Moffat, A., Zobel, J.: Statistica power in retrieval experimentation. In: Proceedings of ACM CIKM 2008, pp. 571–580 (2008)
36. Ziliak, S.T., McCloskey, D.N.: The Cult of Statistical Significance: How the Standard Error Costs Us Jobs, Justice, and Lives. The University of Michigan Press (2008)

Text Categorization with Style

Jacques Savoy[(✉)]

University of Neuchâtel, 2000 Neuchâtel, Switzerland
Jacques.Savoy@unine.ch
http://members.unine.ch/Jacques.Savoy/

Abstract. Text categorization [1] problems (e.g., automatic indexing, filtering) are recurrent in IR. This tutorial focuses on the text style to provide answers to different questions such as authorship attribution [2, 3] (e.g., who is the secret hand behind Ferrante's novels? [4]), author verification [5, 6] (did St. Paul write this letter? [7]), author profiling [8] (determine the author gender, age range, some psychological traits, etc.), or author linking [9] (cluster these n texts into k groups, one per distinct author).

To solve these questions, the best stylistic markers must be determined according to the underlying problem (e.g., using almost all words [10], the most frequent terms (MFT) [11], only functional words [12, 13], n-grams of words, POS tags or sequence of them [9], n-grams of letters [2], layout [14], etc.).

Based on such feature sets, multivariate analysis can be applied such as principal component analysis (PCA) [15], hierarchical clustering [10], or discriminant analysis [16]. Second, different distance-based strategies have been proposed as, for example, Burrows' Delta [12] (using the top m MFT with $m = 40$ to 1,000), Kullback-Leibler divergence [17] (e.g., using a predefined set of 363 English words), or Labbé's method [10] (based on almost the entire vocabulary and opting for a variant of the Tanimoto distance [18]).

As a third paradigm, different machine learning approaches have been suggested [3, 16] such as decision trees, neural networks, k-NN, random forests, or support vector machines (SVM), the latter being a popular approach in various PAN-CLEF campaigns. The k-NN approach tends to produce better effectiveness than both the naïve Bayes or decision tree [14]. Even if deep learning models have been proposed [19], their effectiveness were lower than expected. Moreover, [16] shows that the Delta [12] scheme could surpass performance levels achieved by the SVM method.

Using the **R** stylo package [20], some hand-on examples (e.g., about the US history [21], political domain [22, 23], or literature [4]) will illustrate the presentation.

Keywords: Text categorization · Authorship attribution · Stylometry

© Springer Nature Switzerland AG 2019
L. Azzopardi et al. (Eds.): ECIR 2019, LNCS 11438, pp. 408–409, 2019.
https://doi.org/10.1007/978-3-030-15719-7

References

1. Sebastiani, F.: Machine learning in automatic text categorization. ACM Comput. Surv. **14**(1), 1–27 (2002)
2. Juola, P.: Authorship attribution. Found. Trends Inf. Retr. **1**(3), 1–104 (2006)
3. Stamatatos, E.: A survey of modern authorship attribution methods. J. Am. Soc. Inf. Sci. Technol. **60**(3), 538–556 (2009)
4. Savoy, J.: Is Starnone really the author behind Ferrante? Digit. Scholarsh. Humanit. **33**(4), 902–918 (2018)
5. Koppel, M., Schler, J., Bonchek-Dokow, E.: Measuring differentiability: unmasking pseudonymous authors. J. Mach. Learn. Res. **8**(6), 1261–1276 (2007)
6. Koppel, M., Seidman, S.: Detecting pseudoepigraphic texts using novel similarity measures. Digit. Scholarsh. Humanit. **33**(1), 72–81 (2018)
7. Savoy, J.: Authorship of Pauline epistles revisited. J. Am. Soc. Inf. Sci. Technol, (2019, to appear)
8. Argamon, S., Koppel, M., Pennebaker, J.W., Schler, J.: Automatically profiling the author of an anonymous text. Commun. ACM **52**(2), 119–123 (2009)
9. Kocher M., Savoy J.: Evaluation of text representation schemes and distance measures for authorship linking? Digit. Scholarsh. Humanit. (2019, to appear)
10. Labbé, D.: Experiments on authorship attribution by intertextual distance in English. J. Quant. Linguist. **14**(1), 33–80 (2007)
11. Savoy, J.: Comparative evaluation of term selection functions for authorship attribution. Digit. Scholarsh. Humanit. **30**(2), 246–261 (2015)
12. Burrows, J.F.: Delta: a measure of stylistic difference and a guide to likely authorship. Lit. Linguist. Comput. **17**(3), 267–287 (2002)
13. Pennebaker, J.W.: The Secret Life of Pronouns. What Our Words Say About Us. Bloomsbury Press, New York (2011)
14. Zheng, R., Li, J., Chen, H., Huang, Z.: A framework for authorship identification of online messages: writing-style features and classification techniques. J. Am. Soc. Inf. Sci. Technol. **57**(3), 378–393 (2006)
15. Craig, H., Kinney, A.F. (eds.): Shakespeare, Computers, and the Mystery of Authorship. Cambridge University Press, Cambridge (2009)
16. Jockers, M.L., Witten, D.M.: A comparative study of machine learning methods for authorship attribution. Lit. Linguist. Comput. **25**(2), 215–223 (2010)
17. Zhao, Y., Zobel, J.: Entropy-based authorship search in large document collections. In: Amati, G., Carpineto, C., Romano, G. (eds.) ECIR 2007. LNCS, vol. 4425, pp. 381–392. Springer, Heidelberg (2007). https://doi.org/10.1007/978-3-540-71496-5_35
18. Kocher, M., Savoy, J.: Distance measures in author profiling. Inf. Process. Manag. **53**(5), 1103–1119 (2017)
19. Kocher, M., Savoy, J.: Distributed language representation for authorship attribution. Digit. Scholarsh. Humanit. **33**(2), 425–441 (2018)
20. Eder, M., Rybicki, J., Kestemont, M.: Stylometry with R: a package for computational text analysis. R J. **8**(1), 107–121 (2016)
21. Savoy, J.: The federalist papers revisited: a collaborative attribution scheme. In: Proceedings ASIST 2013, Montreal, November 2013
22. Savoy, J.: Analysis of the style and the rhetoric of the 2016 US presidential primaries. Digit. Scholarsh. Humanit. **33**(1), 143–159 (2018)
23. Savoy, J.: Text clustering: an application with the State of the Union addresses. J. Am. Soc. Inf. Sci. Technol. **66**(8), 1645–1654 (2015)

Concept to Code: Neural Networks for Sequence Learning

Omprakash Sonie[1](✉), Muthusamy Chelliah[1], Surender Kumar[1], and Bidyut Kr. Patra[2]

[1] Flipkart Pvt Ltd., Bangalore, India
{omprakash.s,muthusamy.c,surender.k}@flipkart.com
[2] National Institute of Technology Rourkela, Rourkela, India
patrabk@nitrkl.ac.in

Abstract. Deep Learning has shown significant performance information retrieval (IR) domain. In this tutorial, we provide conceptual understanding of state-of-the-art deep learning techniques such as Embedding
methods and Recurrent Neural Networks (RNNs). The RNNs are effective in modelling sequential data (e.g. clicks, add to cart, purchase data)
that is generated by users in a session and across sessions. We provide a
hands-on case study for sequence and session aware recommender system
and their evaluation methods. This tutorial also covers various models
based on hierarchical representation and RNNs: attention and attribute
aware, memory based models, multi-layer LSTMs, and combining RNNs
with CNNs.

Keywords: Recommender system · Embedding ·
Deep learning concepts · Coding deep learning

1 OUTLINE

1 INTRODUCTION and EMBEDDING (30 min) [1] We go deeper in explaining embedding for learning representation of product, users and other data.
2 RECURRENT NEURAL NETWORKS (30 min): We cover LSTMs and how
they are used to capture temporal dependencies of users & items [2, 3].
3 SEQUENCE AND SESSION LEARNING (60 min): We talk about the
models which tackle sparse real-world datasets, pre-train a session representation and learn user & item features which co-evolve over time [4–6].
4 SEQUENCE MODELS (60 min): We cover the models which learn hierarchical representation & transaction, model whole session, and learn dynamic
representation & global sequential features [7–9].
5 ATTENTION AND ATTRIBUTE MODELS (75 min): We cover the models
which capture users' general interests and current interests, learn expressive
portions of sequences and capture inter-session dependencies, and use historical records to enhance model expressiveness [10–12].

© Springer Nature Switzerland AG 2019
L. Azzopardi et al. (Eds.): ECIR 2019, LNCS 11438, pp. 410–412, 2019.
https://doi.org/10.1007/978-3-030-15719-7

6 ADVANCED MODELS (75 min): We discuss techniques which model users'
  general preferences & sequential patterns, attribute-level user preference and
  extract local dependency & discover long-term patterns [13–15].
7 CASE STUDY AND CONCLUSION (30 min): We walk through Jupyter
  notebooks for recommender system on e-commerce dataset [16].

References

1. Mikolov, T., Sutskever, I., Chen, K., Corrado, G.S., Dean, J.: Distributed representations of words and phrases and their compositionality. In: Advances in Neural
   Information Processing Systems, pp. 3111–3119. Lake Tahoe (2013)
2. Hochreiter, S., Schmidhuber, J.: Long short-term memory. Neural Comput. **9**(8),
   1735–1780 (1997)
3. Wu, C.Y., Ahmed, A., Beutel, A., Smola, A.J., Jing, H.: Recurrent recommender
   networks. In: Proceedings of the Tenth ACM International Conference on Web
   Search and Data Mining, pp. 495–503. ACM, UK (2017)
4. He, R., McAuley, J.: Fusing similarity models with Markov chains for sparse sequential recommendation. In: IEEE 16th International Conference on Data Mining,
   pp. 191–200. IEEE, Barcelona (2016)
5. Wu, C., Yan, M.: Session-aware information embedding for e-commerce product
   recommendation. In: Proceedings of the ACM on Conference on Information and
   Knowledge Management, pp. 2379–2382. ACM, Singapore (2017)
6. Dai, H., Wang, Y., Trivedi, R., Song, L.: Deep coevolutionary network: embedding
   user and item features for recommendation. In: Recsys Workshop on Deep Learning
   for Recommendation Systems, DLRS 2016 (2016)
7. Wang, P., Guo, J., Lan, Y., Xu, J., Wan, S., Cheng, X.: Learning hierarchical representation model for next basket recommendation. In: Proceedings of the 38th
   International ACM SIGIR Conference on Research and Development in Information Retrieval, pp. 403–412. ACM, Chile (2015)
8. Hidasi, B., Quadrana, M., Karatzoglou, A., Tikk, D.: Parallel recurrent neural network architectures for feature-rich session-based recommendations. In: Proceedings
   of the 10th ACM Conference on Recommender Systems. ACM, Boston (2016)
9. Yu, F., Liu, Q., Wu, S., Wang, L., Tan, T.: A dynamic recurrent model for next
   basket recommendation. In: Proceedings of the 39th International ACM SIGIR
   Conference on Research and Development in Information Retrieval. ACM, Pisa
   (2016)
10. Liu, Q., Zeng, Y., Mokhosi, R., Zhang, H.: STAMP: short-term attention/memory
    priority model for session-based recommendation. In: Proceedings of the 24th ACM
    SIGKDD International Conference on Knowledge Discovery and Data Mining,
    pp. 1831–1839. ACM, London (2018)
11. Loyola, P., Liu, C., Hirate, Y.: Modeling user session and intent with an attention-
    based encoder-decoder architecture. In: Proceedings of the Eleventh ACM Confer-
    ence on Recommender Systems, pp. 147–151. ACM, Como (2017)
12. Chen, X., et al.: Sequential recommendation with user memory networks. In: Pro-
    ceedings of the Eleventh ACM International Conference on Web Search and Data
    Mining. ACM, CA (2018)
13. Tang, J., Wang, K.: Personalized top-n sequential recommendation via convolu-
    tional sequence embedding. In: Proceedings of the Eleventh ACM International
    Conference on Web Search and Data Mining, pp. 565–573. ACM, CA (2018)

14. Huang, J., Zhao, W.X., et. al.: Improving sequential recommendation with knowledge-enhanced memory networks. In: The 41st International ACM SIGIR Conference on Research and Development in IR. ACM, Michigan (2018)
15. Lai, G., Chang, W.C., et. al.: Modeling long-and short-term temporal patterns with deep neural networks. In: The 41st International ACM SIGIR Conference on Research and Development in Information Retrieval. ACM, Michigan (2018)
16. Sonie, O., Sarkar, S., Kumar, S.: Concept to code: learning distributed representation of heterogeneous sources for recommendation. In: Proceedings of the 12th ACM Conference on Recommender Systems, pp. 531–532. ACM, Vancouver (2018)

A Tutorial on Basics, Applications and Current Developments in PLS Path Modeling for the IIR Community

Markus Kattenbeck$^{(\boxtimes)}$ and David Elsweiler

Information Science, University of Regensburg, 93040 Regensburg, Germany
{markus.kattenbeck,david.elsweiler}@ur.de

Keywords: PLS path modeling · IIR · Structural equation modeling

Despite the strong empirical tradition in our field, recent articles have highlighted the misuse of statistical methods and the consequeces they have (see e.g. [10]). It has been suggested that our community extends the toolkit of statistical methods to further progress in the field. We view Structural Equation Models (SEMs) as an example of such a statistical technique that has become popular in neighbouring fields, such as RecSys (see [7]) and would be particularly valuable for user evaluations.

This full-day tutorial, which is introductory in nature, introduces participants to SEMs and their estimation using Partial Least Squares Path Modeling (PLS Path Modeling). Structural Equation Models are a statistical technique to simultaneously assess the relationships between unobserved factors (i.e. latent variables) and the way these factors are measured based on observable variables. In our opinion, the researchers who would benefit most are those focused on studying human-aspects and the (hands-on) examples in the tutorial reflects this. We view SEMs as a tool which offers researchers in our community great utility due to its focus on prediciton and has the potential to contribute to our field in at least three ways: first, by providing a statistically solid option to use ad-hoc surveys; second through offering a possibility to foster advancing theories from a solid basis in empirical data; finally, by providing the possibility to combine log data with survey-based data in a single model.

In general, our tutorial complements the efforts to improve the empirical practices within the IR community (see e.g. the tutorials on controlled user studies by Diane Kelly [3–6], Tetsuya Sakai's tutorials on statistical and power analysis [9, 11, 12], Pia Borlund's ACM CHIIR 2016 keynote on the proper design of IR experiments [1] or Heather O'Brien's perspectives paper on proper ways of survey design [8].

The authors have given a half-day tutorial on the same topic at ACM CHIIR 2018. Therefore, this text is a heavily shortened version of [2].

© Springer Nature Switzerland AG 2019
L. Azzopardi et al. (Eds.): ECIR 2019, LNCS 11438, pp. 413–415, 2019.
https://doi.org/10.1007/978-3-030-15719-7

References

1. Borlund, P.: Interactive information retrieval: an evaluation perspective. In: Proceedings of the 2016 ACM on Conference on Human Information Interaction and Retrieval, CHIIR 2016, p. 151. ACM, New York (2016). https://doi.org/10.1145/2854946.2870648, http://doi.acm.org/10.1145/2854946.2870648

2. Kattenbeck, M., Elsweiler, D.: Estimating models combining latent and measured variables: a tutorial on basics, applications and current developments in structural equation models and their estimation using PLS path modeling. In: Proceedings of the 2018 Conference on Human Information Interaction & Retrieval, CHIIR 2018, pp. 375–377. ACM, New York (2018). https://doi.org/10.1145/3176349.3176899, http://doi.acm.org/10.1145/3176349.3176899

3. Kelly, D., Crescenzi, A.: From design to analysis: conducting controlled laboratory experiments with users. In: Proceedings of the 39th International ACM SIGIR Conference on Research and Development in Information Retrieval, SIGIR 2016, pp. 1207–1210. ACM, New York (2016). https://doi.org/10.1145/2911451.2914809, http://doi.acm.org/10.1145/2911451.2914809

4. Kelly, D., Crescenzi, A.: From design to analysis: conducting controlled laboratory experiments with users. In: Proceedings of the 40th International ACM SIGIR Conference on Research and Development in Information Retrieval, SIGIR 2017, pp. 1411–1414. ACM, New York (2017). https://doi.org/10.1145/3077136.3082063, http://doi.acm.org/10.1145/3077136.3082063

5. Kelly, D., Radlinski, F., Teevan, J.: Choices and constraints: research goals and approaches in information retrieval (part 1). In: Proceedings of the 37th International ACM SIGIR Conference on Research & #38; Development in Information Retrieval, SIGIR 2014, p. 1283. ACM, New York (2014). https://doi.org/10.1145/2600428.2602289, http://doi.acm.org/10.1145/2600428.2602289

6. Kelly, D., Radlinski, F., Teevan, J.: Choices and constraints: research goals and approaches in information retrieval (part 2). In: Proceedings of the 37th International ACM SIGIR Conference on Research & #38; Development in Information Retrieval, SIGIR 2014, p. 1284. ACM, New York (2014). https://doi.org/10.1145/2600428.2602290, http://doi.acm.org/10.1145/2600428.2602290

7. Knijnenburg, B.P., Willemsen, M.C., Gantner, Z., Soncu, H., Newell, C.: Explaining the user experience of recommender systems. User Model. User-Adap. Inter. **22**(4), 441–504 (2012)

8. O'Brien, H.L., McCay-Peet, L.: Asking "good" questions: questionnaire design and analysis in interactive information retrieval research. In: Proceedings of the 2017 Conference on Conference Human Information Interaction and Retrieval, CHIIR 2017, pp. 27–36. ACM, New York (2017). https://doi.org/10.1145/3020165.3020167, http://doi.acm.org/10.1145/3020165.3020167

9. Sakai, T.: Topic set size design and power analysis in practice. In: Proceedings of the 2016 ACM International Conference on the Theory of Information Retrieval, ICTIR 2016, pp. 9–10. ACM, New York (2016). https://doi.org/10.1145/2970398.2970443, http://doi.acm.org/10.1145/2970398.2970443

10. Sakai, T.: Two sample t-tests for IR evaluation: student or welch? In: Proceedings of the 39th International ACM SIGIR Conference on Research and Development in Information Retrieval, SIGIR 2016, pp. 1045–1048. ACM, New York (2016). https://doi.org/10.1145/2911451.2914684, http://doi.acm.org/10.1145/2911451.2914684

11. Sakai, T.: The probability that your hypothesis is correct, credible intervals, and effect sizes for IR evaluation. In: Proceedings of the 40th International ACM SIGIR Conference on Research and Development in Information Retrieval, SIGIR 2017, pp. 25–34. ACM, New York (2017). https://doi.org/10.1145/3077136.3080766, http://doi.acm.org/10.1145/3077136.3080766
12. Sakai, T.: Conducting laboratory experiments properly with statistical tools: an easy hands-on tutorial. In: The 41st International ACM SIGIR Conference on Research & #38; Development in Information Retrieval, SIGIR 2018, pp. 1369–1370. ACM, New York (2018). https://doi.org/10.1145/3209978.3210182, http://doi.acm.org/10.1145/3209978.3210182

Author Index

Printed in the United States
By Bookmasters